BIOLOGY

Foundation for Class VIII

BIOLOGY

Foundation for Class VIII

Abhimanyu Kumar Jha
Ph.D. (Biotechnology)
Professor and Head, Department of Biotechnology
School of Engineering and Technology (SET), Sharda University
Greater Noida, Uttar Pradesh

Abhinav Datta
M.Tech. (Biotechnology)
Teaching Faculty - Biology, Shiv Nadar School
Noida, Uttar Pradesh

Mahek Sharan
M.Tech. (Biotechnology)
Research Scholar in Zoology, Banaras Hindu University
Varanasi, Uttar Pradesh

Universities Press

BIOLOGY FOUNDATION FOR CLASS VIII

UNIVERSITIES PRESS (INDIA) PRIVATE LIMITED

Registered Office
3-6-747/1/A & 3-6-754/1, Himayatnagar, Hyderabad 500 029, Telangana, India
info@universitiespress.com; www.universitiespress.com

Distributed by
Orient Blackswan Private Limited

Registered Office
3-6-752 Himayatnagar, Hyderabad 500 029, Telangana, India

Other Offices
Bengaluru / Chennai / Guwahati / Hyderabad / Kolkata
Mumbai / New Delhi / Noida / Patna / Visakhapatnam

First published by Universities Press (India) Private Ltd 2024

Cover and book design

0501958

ISBN: 978-93-93330-55-0

Typeset in Times New Roman PS Std 10 *by*
SRS Publishing Services, Puducherry

Printed in India by
B.B. Press, Noida 201 301

Published by
Universities Press (India) Private Limited
3-6-747/1/A & 3-6-754/1, Himayatnagar, Hyderabad 500 029, Telangana, India

About the Authors

Abhimanyu Kumar Jha is a molecular biologist with vast experience in teaching NEET UG (previously AIPMT), NEET PG and medical students from around the world with a very high success ratio. He is currently working as Professor and Head of the Department of Biotechnology at Sharda University, Greater Noida. Previously he worked as Dean, Faculty of Life Sciences, and Head of the Department of Biotechnology at the Institute of Applied Medicines and Research, Ghaziabad. He also worked as Head of the Department of Biotechnology at IMS Engineering College, Ghaziabad. He obtained his Ph.D. in Biotechnology from Panjab University, Chandigarh. He has published more than 125 papers in national and international journals and contributed 28 chapters in books published by Elsevier, Springer and Apple Academic Press. He has authored four books and worked on an edited book. He was awarded the honorary degree of Doctor of Science (Honorary D.Sc.) with specialisation in Biological Science by the American University, Washington D.C., USA. He has won many awards such as Distinguished Faculty, Best Indian Research and Young Scientist. His current area of research is on the epidemiology and epigenetics of oral squamous cell carcinoma and liver cancer.

Abhinav Datta completed his M.Tech. in Biotechnology from Dr A.P.J. Abdul Kalam Technical University (formerly Uttar Pradesh Technical University). He is currently a member of the Biology Faculty at Shiv Nadar School, Noida. He is the proud recipient of awards such as Best Edutainment: Making Teaching and Learning Fun from JagranJosh Education Awards in 2022, Global Teacher Award 2020–21 from AKS Education Awards and Best Teacher Award 2016–17 in Science and Technology from The Khaitan School.

Mahek Sharan is currently pursuing her doctoral degree from Banaras Hindu University, Varanasi. She graduated with an M.Tech. in Biotechnology from Chaudhary Charan Singh University, Meerut, and was awarded the Chancellor's and Vice-Chancellor's gold medals. She has more than 10 publications in reputable national and international journals and has contributed to books published by various outlets, including Apple Academic Press.

Preface

Welcome to the fascinating world of Biology! In this book, we delve into the intricacies of life, exploring the diverse organisms that inhabit our planet and the remarkable processes that sustain them. Biology is a field of study that deepens our understanding of the living world and inspires awe and curiosity about the wonders of nature.

This book aims to provide students of Class VIII with a comprehensive and engaging overview of the fundamental principles of biology. It has been designed to foster a sense of wonder and nurture scientific thinking, and also encourage students to develop a lifelong passion for exploring the living world around them. This unique book covers the topics and recent trends of questions for competitive examinations like NEET, NTSE and Olympiads as well as the syllabi of both the CBSE and ICSE boards.

With a strong foundation in these core principles, students will gain the necessary knowledge to understand a topic from its basic level which plays a key role in competitive examinations. Some highlights of the book are as follows:

- The book's greatest asset is simplifying complicated concepts into fundamental ideas and then expanding on them.
- Throughout this book, we have integrated real-world examples, captivating illustrations and thought-provoking exercises to enhance the learning experience.
- These elements aim to stimulate critical thinking, encourage scientific inquiry and foster an appreciation for the interconnectedness of life.
- The book covers all the topics specified in the syllabi in great detail.
- The authors are subject matter experts and have brought their vast experience in teaching these topics to make the subject interesting and easy to learn.
- Each chapter contains questions of different categories (MCQs, Fill in the Blanks, True or False, Match the Following, Very Short Answer, Short Answer, Long Answer, Assertion–Reason, One-word Answer, Odd One Out, Define the Term, Groups of Terms, Image-Based, Case Study-Based and NCERT) with a competitive section with previous years' questions.
- The App accompanying the book contains additional chapter-wise questions for competition preparation.

In addition to providing a solid foundation in Biology, this book also emphasizes the importance of ethical considerations, responsible practices and the significance of sustainability in our interactions with the environment. We hope to instil in students a sense of stewardship for the natural world and empower them to make informed decisions for the betterment of both themselves and the planet.

We would like to express our heartfelt gratitude to the educators, researchers and scientists whose contributions have shaped the field of biology and made this book possible. Their tireless efforts have broadened our knowledge and paved the way for countless discoveries.

Every effort has been taken to make the book error free. Constructive criticism and valuable suggestions from the readers are most welcome for further improvement.

Lastly, we extend our best wishes to the students embarking on this exciting journey through the world of Biology. May your exploration of the living world fill you with wonder, inspire curiosity and ignite a lifelong passion for the pursuit of knowledge. Remember, the beauty and complexity of life await your discovery!

Happy knowledgeable reading!

Abhimanyu Kumar Jha
Abhinav Datta
Mahek Sharan

Acknowledgements

We owe a lot of individuals our gratitude for supporting us throughout our lives. We would like to express our gratitude and acknowledgement to those who have contributed most to the success of this book in order to keep this short.

First and foremost, we would like to express our gratitude to the student audience that we have been serving over the past few years and to Universities Press, our publisher.

Without the ongoing guidance and foresight of the team at Universities Press, including Thomas Mathew Rajesh, Kallol Das and Madhavi Sethupathi, who edited this book with exceptional talent, compassion and diligence, this book would not have been possible.

We would like to express our heartfelt gratitude to the management of Sharda University, Greater Noida, especially Sh. Pradeep Kumar Gupta, Chancellor; Sh. Yatendra Kumar Gupta, Pro-Chancellor; and Prof. (Dr.) Sibaram Khara, Vice Chancellor, for providing constant motivation and support. We are especially indebted to Prof. (Dr) Parma Nand, Dean Academics, Sharda University, and Prof. (Dr.) Shankar Ramamoorthy, Dean, School of Engineering and Technology, Sharda University, who have been a constant support and source of strength.

Moreover, we would like to take this opportunity to thank our family members, especially Sh. Arjun Kumar Jha and Smt. Subhadra Jha (parents of Abhimanyu Kumar Jha), Mrs. Meenakshi Jha (wife of Abhimanyu Kumar Jha), Master Aayushman Jha (son of Abhimanyu Kumar Jha); Mr. Yashveer Sharan Srivastava and Mrs. Rajani Srivastava (parents of Mahek Sharan) and Mrs. Ghazal Srivastava and Mr. Sunny Baghla (sister and brother-in-law of Mahek Sharan), for their endurance, love, wishes and support, which helped us to complete our work. You inspire us every day with your incredible thoughtfulness.

We would also like to acknowledge the blessings and support of our in-laws. We also want to thank and appreciate all our family members who offered their help willingly.

Additionally, heartfelt gratitude has to be conveyed to Abhinav Datta's principal, Ms. Anju Soni, for her constant guidance and motivation.

We want to convey our admiration and gratitude to the scientific community for everything they have done and are now doing to provide us with the best scientific advancements.

Finally, we want to express our gratitude to the Supreme Power, who has blessed us abundantly.

Abhimanyu Kumar Jha
Abhinav Datta
Mahek Sharan

Contents

To
Lord Shiva, Mata Parvati and their family
and
My parents, Sh. Arjun Kumar Jha and Smt. Subhadra Jha

—Abhimanyu Kumar Jha

To
My guiding light of hope and my loved ones

—Mahek Sharan

CHAPTER 1

Cells: Structure and Functions

PREREQUISITES

You must have knowledge of:

- Basic biology

LEARNING OBJECTIVES

At the end of this chapter, you will be able to:

- Describe a call
- Explain cell theory
- List the types of cells
- Describe the structure and functions of a cell

INTRODUCTION

The world is made up of living (biotic) and non-living (abiotic) things. Living things carry out some basic functions, such as respiration, digestion, excretion, circulation and metabolism. These functions are performed by different organs of the body.

The smallest unit of life is called a cell. **Cells** are the basic functional, structural and biological units of all organisms. They are the building blocks of life. The study of cells is known as **cytology**. Cells are the smallest units of life that can perform essential functions; they have the ability to survive independently.

In the human body, every organ such as the brain, muscle, skin and even the bone is composed of cells. Even plant parts such as the root, flower, leaves and wood are composed of cells. Every cell has its own lifespan and is replaced on its death. Cells can only be seen under a microscope. The number of cells present in animals and plants varies from one species to another.

Fun Fact!

Did you know that humans contain approximately 40 trillion cells! The brain itself has 80 billion cells!

The cells that are present in every living organism are much larger than atoms but still very small and are not visible to the unaided eye. The average dimension of a plant and animal cell is between 1 and 100 micrometres.

Mycoplasma is the group of bacteria that are the smallest known cells. The diameter of these cells is around 0.2 micrometres. They look like small, single-celled spheres under an electron microscope with a total mass of 10^{-14}. The mass of a human cell is around 400,000 times heavier than that of a single *Mycoplasma* bacterium. The average size of a human cell is about 20 micrometres.

Fun Fact!

Did you know that about 10,000 human cells would be required to cover the head of a pin?

DISCOVERY OF THE CELL

In 1665, Robert Hooke (1635–1703) observed cork slices under a simple magnifying device. The outer part of the bark of the tree is known as cork. Robert Hooke (Fig. 1.1) sliced the cork into thin pieces and then observed the slice under a microscope. He drew the structure that he observed and described it in his book, *Micrographia*. This book contains around 60 detailed observations of many objects viewed under a simple magnifying glass or coarse compound microscope.

Fig. 1.1 Robert Hooke
(Memorial portrait for the Royal Academy of Engineering, London; Rita Greer; 2011; https://commons.wikimedia.org/wiki/File:Robert_Hooke_1635-1703_Engineer.jpg. Licensed under: FAL)

Hooke observed compartments or partitioned boxes in the slice of cork. These boxes looked like a honeycomb. He noticed that the boxes were separated by a partition or wall. Hooke also saw many tiny pores that he termed 'cell'. This term is derived from the Latin '*cella*' meaning 'a small room', a place where monks lived, and '*cellulae*' or 'honeycomb', which has six-sided structure. Hooke only observed the structure; he did not know the functions served by these cells.

The walls that were observed by Hooke were actually the plant's cell wall. Later, it was realised that what Hooke had seen were actually dead cells. This was deduced from the fact that he did not mention any organelles or a nucleus, which would have been seen in a live cell. Hooke also observed mould, which he described as being bluish in colour.

About 150 years after Hooke's discovery, another scientist, Antonie van Leeuwenhoek, viewed cells under a microscope. He was trying to perfect the design of the microscope with better and improved lenses that had a magnification of about 300 times. Under this microscope, Leeuwenhoek observed motile objects – a characteristic of living organisms. Leeuwenhoek named these living organisms, which were bacteria and protozoa, '*animalcules*'.

Fun Fact!

Antonie van Leeuwenhoek also observed the sperm cells of humans and animals. Later, he viewed the fertilisation process.

Most of the time, the cells that were observed were taken from whole dead tissue or a whole organism as the cells of live animals and plants were very fragile. The cells of living organisms were observed only after better and improved microscopes were invented.

INVENTION OF THE MICROSCOPE

Antonie van Leeuwenhoek (1632–1723) was a land surveyor who developed an interest in observing tissues and organisms under the microscope. This Dutch scientist (Fig. 1.2) is believed to have constructed approximately 400 microscopes in his lifetime! All these microscopes consisted of a single convex lens and were known as simple microscopes. What was impressive about them was that they had considerable magnification power (around 200 times).

Fig. 1.2 Antonie van Leeuwenhoek
(Portrait of Anthonie van Leeuwenhoek; Rijksmuseum; 1680; https://commons.wikimedia.org/wiki/File:Anthonie_van_Leeuwenhoek_(1632-1723)._Natuurkundige_te_Delft_Rijksmuseum_SK-A-957.jpeg. Licensed under: CC-PD-Mark)

In the microscope built by Leeuwenhoek, the object was mounted on the opposite side of the lens and the eyepiece was placed near the lens (Fig. 1.3).

Hooke's microscope used two lenses for greater magnification (Fig. 1.4). This is known as the compound microscope. In his microscope, the object was placed on

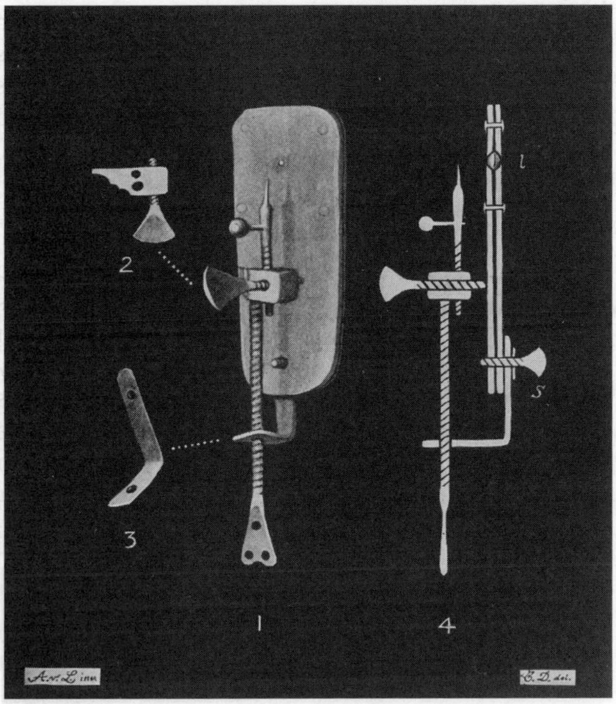

Fig. 1.3 Microscope invented by Antonie van Leeuwenhoek
(Antony van Leeuwenhoek and his "Little animals"...; Interactive Archive Book Images; 2015; https://commons.wikimedia.org/wiki/File:Antony_van_Leeuwenhoek_and_his_%22Little_animals%22;_being_some_account_of_the_father_of_protozoology_and_bacteriology_and_his_multifarious_discoveries_in_these_disciplines;_(1932)_(19123930303).jpg. Licensed under: Flickr's The Commons)

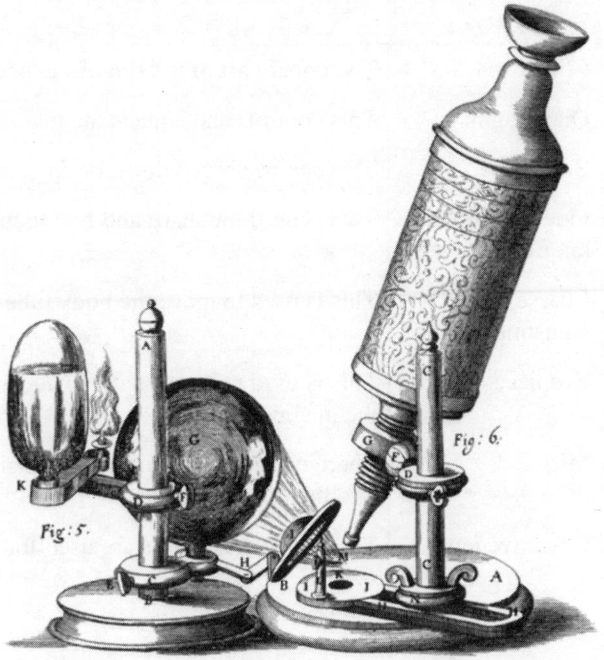

Fig. 1.4 Microscope invented by Robert Hooke
(microscope used by Robert Hooke to observe a "cell"; Royal Society of London; 2017; https://commons.wikimedia.org/wiki/File:Robert_Hooke%27s_microscope.png. Licensed under: CC-PD-Mark)

a stage under the lens and an external light source such as a lamp reflected by a concave mirror was used to observe the object.

Compound microscope: Current compound microscopes are nothing but improved designs of Hooke's first compound microscope. The characteristics of a compound microscope are as follows:

- The magnification ranges from 40X to 1000X
- Two-dimensional images are observed
- Two or more convex lenses are used
- Different configurations are available like monocular, binocular and multi-head
- Only one objective is used at a time

The different parts of a compound microscope perform different functions (Table 1.1).

Electron microscope: The popularity of the compound microscope led to a demand for a more detailed study of cells and its organelles; this gave rise to the invention of the electron microscope. The magnification power of an electron microscope is about 200,000 times that of a normal microscope. The basic difference between a compound microscope and an electron microscope is that the former uses light rays whereas that latter uses beam of

Table 1.1 Parts and functions of a compound microscope

Parts	Functions
Foot or base	This is a U-shaped structure that supports the weight of the microscope
Pillar	This supports the stage
Arm	The entire microscope is handled using this curved structure
Stage	The specimen to be examined is placed on this rectangular and flat plate

(*Continued*)

Parts	Functions
Clips	The upper part of the stage is connected to two clips to hold the slide in position
Diaphragm	This controls and adjusts the intensity of light that passes into the microscope
Body tube	This can be shifted down or up using the adjustment knobs
Fine adjustment knob	This is used for sharp and fine focusing of the object
Coarse adjustment knob	This is used to move the body tube down or up in order to obtain exact focus
Eyepiece	This is used to observe the image and has markings such as 5X, 10X and 15X depending on the magnification
Mirror	There is a concave mirror on one side and a plane mirror on the other side to reflect light rays into the microscope
Objective lens	Three objective lenses are used for magnification of the object: • 10X • 45X • Oil immersion 100X

electrons which are bent by magnets. These microscopes have diverse applications in biology and are used to view the ultrastructure of cells, biopsy samples, metals, crystals, microbes and macromolecules. They are also used on a large scale in sample analysis for failure and quality control.

CELL THEORY

In 1838, a German botanist, Matthias Schleiden, hypothesised that plants are made up of cells and every cell performs a different life process. After a year, Theodor Schwann made similar discoveries, but in animals. This German zoologist declared that, "*all animals and plants are composed of cells, which serve as the units of structure and function*". This hypothesis is popularly known as cell theory. Schleiden and Schwann's theory was proposed in 1839, and Rudolf Virchow modified it in 1858 by adding that cells arise from pre-existing cells.

The important points of cell theory are as follows:

- The cell is the smallest unit of function for all living things
- The cell is the smallest unit of structure for all living things
- All cells arise from pre-existing cells

Let us examine this in detail.

Cell is the functional unit of life: The cells in a monkey or guava perform different activities or functions. For example, in a monkey, all the main functions from digestion of food by enzymes to metabolism and absorption to excretion are carried out by cells. Similarly, in a guava plant, all the life processes from photosynthesis to absorption of water by the roots from the soil are performed by cells.

Cell is the structural unit of life: If you take any organ or part of a monkey or guava tree and observe it under a microscope, it will be clearly seen that it has a cellular structure.

Cells are replaced after they die: Monkeys and guava trees are composed of millions of cells, which are regularly replaced by new cells. This never-ending process of formation of new cells by the division of pre-existing cells takes place inside the body.

Life starts with a single cell: A monkey's life starts from an egg, which arises from the ovary of the parent monkey. A guava's life starts from a seed, which comes from an embryo.

Cell theory was widely accepted, but after viruses were discovered and their characteristic of being living and non-living organisms was known, the first point of cell theory was no longer applicable. Thus, biologists believe that we are yet to produce a concrete definition of life.

Fun Fact!

It is believed that Virchow plagiarised the work of Robert Remak. That cells arise from pre-existing cells was initially stated by Robert Remak!

So, the **modified cell theory** is as follows:

- All living cells arise from pre-existing cells
- The cell is the smallest unit of function for all living things
- The cell is the smallest unit of structure for all living things
- Metabolism and biochemistry (energy flow) occurs within the cell
- Organisms of similar species have cells of the same chemical composition
- The total activity of cells results in the total activity of organisms
- Cells are made up of DNA (in chromosomes) and RNA (in the cytoplasm and nucleus)

TYPES OF CELLS

Have you ever observed a brick wall? Bricks when arranged in a sequence act as the basic structural unit of a building (Fig. 1.5). Similarly, cells when arranged in the same way act as the basic structural unit of a body.

(a) (b)

Fig. 1.5 (a) Brick wall and (b) tissue show how both the structures are formed using bricks and cells, respectively, as the building units
((a) Courtesy: PublicDomainPictures; https://pixabay.com/photos/wall-bricks-brick-wall-red-bricks-21534/)

Bricks are the structural units of buildings but buildings have different designs, sizes and shapes. Similarly, in the biotic environment, living organisms are made up of cells but differ in many aspects. Cells show variety in shape, size and number. Even when cells are placed under a microscope, they would not be visible as they need to be stained in order to be seen. Stains (dyes) are help us study the detailed structures of cells as they colour specific parts.

Number of Cells

A single organism is made up of billions and trillions of cells. The cells of organisms vary in size and shape and each category performs different functions. The number of cells in a body is directly related to the size of the organism. The larger the organism, the greater the number of cells.

On the basis of the number of cells (Fig. 1.6, Table 1.2), organisms can be categorised as:

- **Multi-celled/Multicellular:** Larger organisms like most plants and animals are all multicellular; that is, they are made up of billions of cells. For example, humans, monkey, guava, mango.
- **Few-celled:** Some plants and animals are made up of relatively few cells, ranging from a few hundreds to a few thousands. For example, *Spirogyra* and *Volvox*.
- **Single-celled/Unicellular:** The organisms that are made up of just one cell are known as single-celled organisms. For example, *Amoeba*, yeast and bacteria.

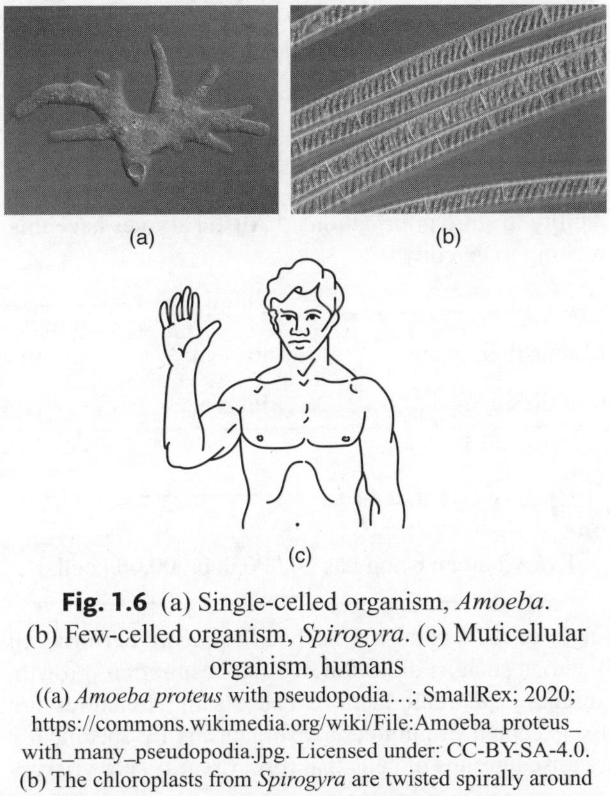

(a) (b) (c)

Fig. 1.6 (a) Single-celled organism, *Amoeba*. (b) Few-celled organism, *Spirogyra*. (c) Muticellular organism, humans
((a) *Amoeba proteus* with pseudopodia…; SmallRex; 2020; https://commons.wikimedia.org/wiki/File:Amoeba_proteus_with_many_pseudopodia.jpg. Licensed under: CC-BY-SA-4.0. (b) The chloroplasts from *Spirogyra* are twisted spirally around the cell…; Wiedehopf20; 2015; https://commons.wikimedia.org/wiki/File:The_freshwater_alga_Spirogyra.jpg. Licensed under: CC-BY-SA-4.0)

Smaller organisms may contain fewer cells than larger organisms, but their functionality is not affected in any way. Even multicellular organisms start their journey with a single cell, which is the fertilised egg. This one cell multiplies and increases in number to develop into a large organism.

Table 1.2 Differences between eukaryotic and prokaryotic cells

Characteristics	Eukaryotic cell	Prokaryotic cell
Definition	Any cell that contains a clearly defined nucleus and membrane-bound organelles	Any unicellular organism that does not contain a membrane-bound nucleus or organelles
Examples	Animal, plant, fungi and protist cells	Bacteria and Archaea
Nucleus	Present (membrane bound)	Absent (nucleoid region)
Cell size	Large (10–100 micrometres)	Small (less than a micrometre to 5 micrometres)
DNA replication	Highly regulated with selective origins and sequences	Replicates the entire genome at once
Organism type	Usually multicellular	Unicellular
Chromosomes	More than one	One long single loop of DNA and plasmids
Ribosomes	Large	Small
Growth rate/Generation time	Slower	Faster
Organelles	Present	Absent
Ability to store information relating to heredity	All eukaryotes have this ability	All prokaryotes have this ability
Cell wall	Simple: Present in plants and fungi	Complex: Present in all prokaryotes
Plasma membrane	Present	Present
Cytoplasm	Present	Present

Fun Fact!

Every human being has 30,000,000,000,000 cells!

Single-celled organisms such as *Amoeba* perform all functions such as digestion of food, respiration, growth, excretion and reproduction. These same functions are also executed in multicellular organisms by specifically organised groups of cells that form tissues. These tissues are organised to form organs.

SCIENCE IN DAILY LIFE

In the biology laboratory, specimens are available for single-celled organisms like *Amoeba* and *Paramecium*, few-celled organisms such as *Spirogyra* and multi-celled organisms like snake and frog. Study the differences between these organisms.

SCIENCE IN DAILY LIFE

Collect pond water and prepare a slide. Observe the different single-celled organisms under a microscope.

Fun Fact!

An adult human has around 10,000 million nerve cells in the cortex of the brain and about 5–6 million RBCs!

Size of Cells

The size of a living organism's cell can range from as small as a micron or micrometre to as large as a few centimetres. In general, most cells are microscopic in size, which means they cannot be seen by the unaided eye. They need to be magnified or enlarged using a microscope.

The smallest bacteria is 0.1–0.5 micrometres in diameter, whereas the largest cell is the egg of the ostrich which has a size of 170 × 130 mm. There is no relation between the size of the organism and the size of its cell. For instance, the cells in the body of an elephant may not be much bigger than the cells in a rat. Unlike the number of cells, the size of the cell is related to function. For example, nerve cells are long and branched, and perform the function of transferring messages from the brain to organs and vice versa.

On the basis of size (Fig. 1.7), here are some facts about cells:

- **Largest cell:** These are the cells of Aves. The eggs of birds, specifically, the ostrich, are some of the largest, even before development starts.
- **Longest cell:** The nerve cells are the longest. For example, a nerve cell starts from the fingertips and extends till the backbone.
- **Smallest cell:** Single-celled organisms like bacteria have a size of 0.3–0.5 micrometres. In multicellular organisms like humans, RBCs which have a size of 7 micrometres are the smallest cells.

Cells remain small for two reasons:

- To have a larger surface-area-to-volume ratio for better distribution of material in and out of the cell
- To communicate between different regions rapidly and effectively

The surface area to volume ratio is an aspect that needs to be examined in detail. Let us take an example of a cube with sides of 4 mm. The total surface area of this cube would be 96 mm. If we cut this cube into 8 equal cubes, we reduce its length to half. The total surface area would be 2 × 2 × 6 × 8 = 192 sq. mm. This is double that of the original size of the cube but the volume remains the same. This ratio is advantageous in many ways:

- Greater diffusion of metabolic waste from inside to outside the cell
- Greater diffusion of respiratory gases
- Greater diffusion of nutrients into the cell

Shape of Cells

Cells are present in a variety of shapes. The shape of the cell does not determine how it functions. For example, *Amoeba* has no definite shape and it keeps changing. Different projections, called pseudopodia ('*pseudo*' meaning 'feet' and '*podia*' meaning 'false'), that vary in length protrude out of the *Amoeba's* body. These projections appear and disappear according to the need for locomotion. In large animals, the single cell that can change its shape is the white blood cell (WBC). The

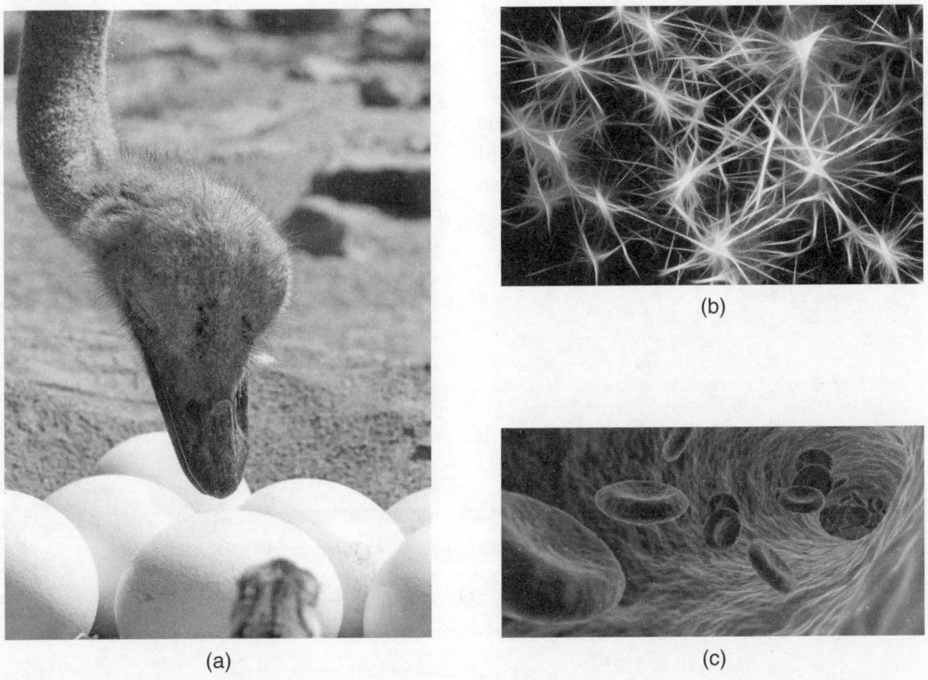

(a) (b) (c)

Fig. 1.7 (a) The largest cell, an ostrich egg. (b) The longest cells, neurons. (c) The smallest cells, red blood cells. ((a) Courtesy: Barbara Fraatz; https://pixabay.com/photos/ostrich-bird-strauss-bird-bouquet-4778261/. (b) Courtesy: Gerd Altmann; https://pixabay.com/photos/annoy-cells-stars-dendrites-sepia-2926087/. (c) Courtesy: Arek Socha; https://pixabay.com/illustrations/blood-cells-red-medical-medicine-1813410)

capability of independent existence is present in both WBCs and *Amoeba*.

On the basis of shape, cell are categorised as:

- Round
- Spherical
- Disc-shaped
- Polygonal
- Rectangular
- Cuboid
- Elongated
- Spindle-shaped
- Branched

Many components are present inside the cell enclosed in a membrane. The membrane provides a specific shape to the cell by providing rigidity. Along with the cell membrane, organisms like plants and bacteria have a cell wall. The shape of cells is related to their function. For example:

- White blood cells are amoeboid in shape so they can easily squeeze into and out of capillary walls.
- Muscle cells are long so that contractions or pull is easy.
- RBCs are biconcave and circular so that they can easily transport oxygen through narrow capillaries.
- In the leaves, the guard cells in the stomatal pores facilitate the opening and closing of the pore and so are bean-shaped.
- Nerve cells are long so as to conduct impulses to the brain from different parts and vice versa.

STRUCTURE AND FUNCTIONS OF CELLS

As we know, every organ performs different functions such as assimilation and digestion (in animals) and absorption of water and minerals (in plants). These organs are made up of tissues which is nothing but a group of similar cells that perform a specific function. This is why cells are known as the structural and functional units of life.

A cell may perform different functions, but its general structure remains the same. A generalised cell consists of three important parts:

- Nucleus
- Cytoplasm
- Cell membrane

Cells contain parts that have definite structure, shape and function, which are known as organelles (Figs 1.8 and 1.9, Table 1.3). The word 'organelle' comes from the word 'organ', and means 'little organs'.

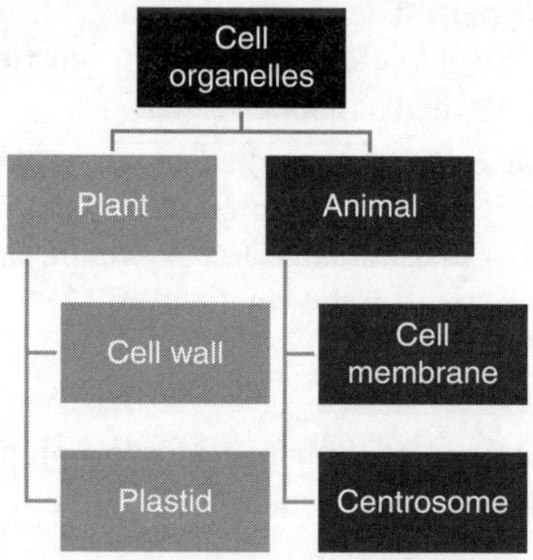

Fig. 1.8 Cell organelles that differ in plant and animal cells

Cell Membrane and Cell Wall

The cell membrane, also known as plasma membrane, encloses the cytoplasm and nucleus. It separates the cell from its surrounding environment and also one cell from another. Each cell is surrounded by a cell membrane

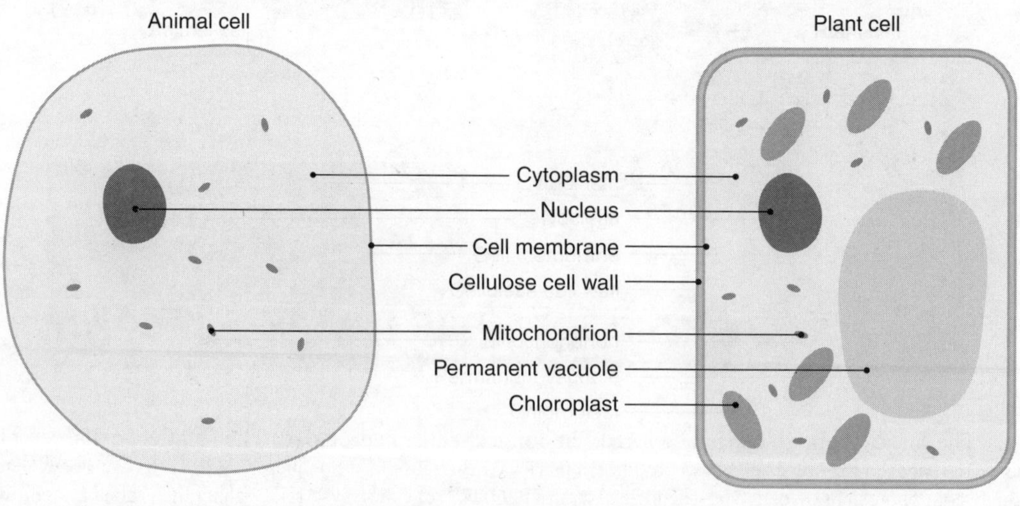

Fig. 1.9 Main differences between simple animal cell and simple plant cell

Table 1.3 Parts of a cell

Cell organelles
Cell wall
Cell membrane
Cytoplasm
Endoplasmic reticulum
Mitochondria
Golgi apparatus
Ribosome
Lysosome
Centrosome
Plastid
Granule
Vacuole
Fat droplets
Nucleus
Nuclear membrane
Chromatin fibres
Nucleoplasm

which has pores through which substances exit and enter the cell.

Did you know that the cell membrane is selectively permeable in nature? This means that it only allows specific substances to pass through while preventing others from entering. The cell membrane is porous in nature and gives the cell its shape. It is made up of lipids and proteins along with carbohydrates. The accepted model of the cell membrane is known as the **fluid mosaic model** (mosaic-like proteins embedded in the lipid-like fluid) proposed by Singer and Nicolson (Fig. 1.10).

Outside the cell membrane, there is a thick layer known as the cell wall, which is present in prokaryotes and plants (Fig. 1.11). This layer provides protection to the plant cell against high-speed winds, variations in temperature and atmospheric moisture. While the cell membrane is selectively permeable, the cell wall is freely permeable and allows all substances to move in and out without any hinderance. The cell wall provides a certain rigidity to the cell and a definite shape without affecting the function of the cell membrane. In plant cells, the lining of the cell wall is easily visible. The cell wall is made up of cellulose, hemicellulose and pectin in plants. In bacteria, the composition of the cell wall is a little different, as it is made up of N-acetyl-glucosamine and N-acetyl murein. A middle layer between the cell membrane and cell wall serves as a barrier.

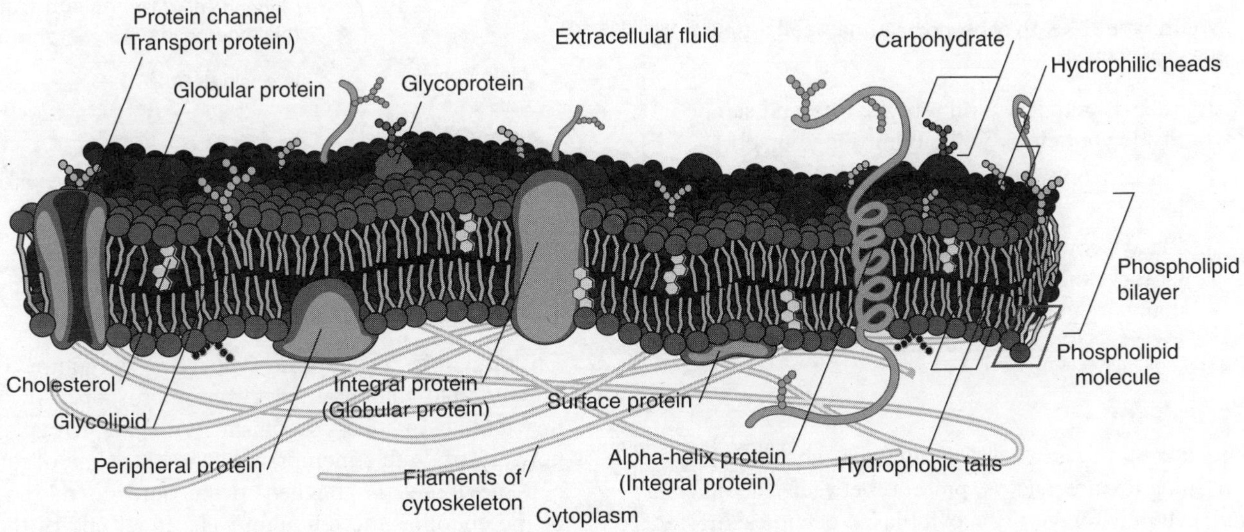

Fig. 1.10 Fluid mosaic model

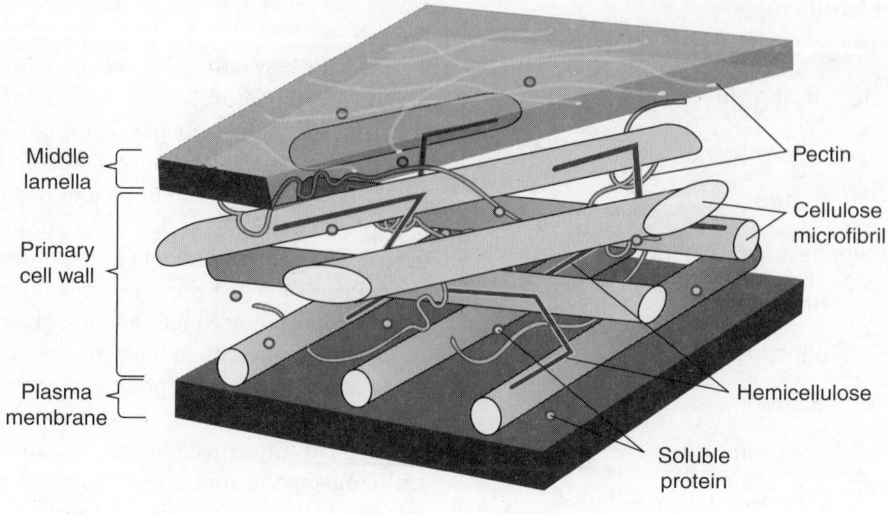

Fig. 1.11 Plant cell wall

> **SCIENCE IN DAILY LIFE**
>
> 1. Take an onion bulb and peel its dry pink covering. The fleshy white layers of the bulb can be removed using forceps.
> 2. Place it on a glass slide on a drop of water. Cut this thin layer into a smaller square with a blade and add methylene blue dye.
> 3. Place a coverslip on it ensuring that there is no air bubble. Observe it under a microscope and draw and label the structures you see.

> **SCIENCE IN DAILY LIFE**
>
> Would you like to observe animal cells under a microscope?
>
> 1. Take a toothpick with a broken tip and scrape it inside your cheek without hurting yourself.
> 2. Place it on a slide in a drop of water and add iodine.
> 3. Place a coverslip over it and add methylene blue. Observe the absence of cell wall and presence of many organelles.

Cytoplasm

The cytoplasm is a semi-liquid substance that has a jelly-like texture and is present between the nucleus and cell membrane. The cytoplasm occupies most of the cell within the membrane. The majority of the cell components and organelles are present in the cytoplasm. It appears partly transparent, colourless and watery in nature. The cytoplasm remains in a dynamic state and in a state of movement; it is the site of most chemical reactions. The organelles present in the cytoplasm are discussed below:

1. **Ribosomes:** Ribosomes are multiple in number and granular in nature. They are attached to the membrane or scattered freely in the cytoplasm. They are used for the synthesis of proteins (Fig. 1.12).

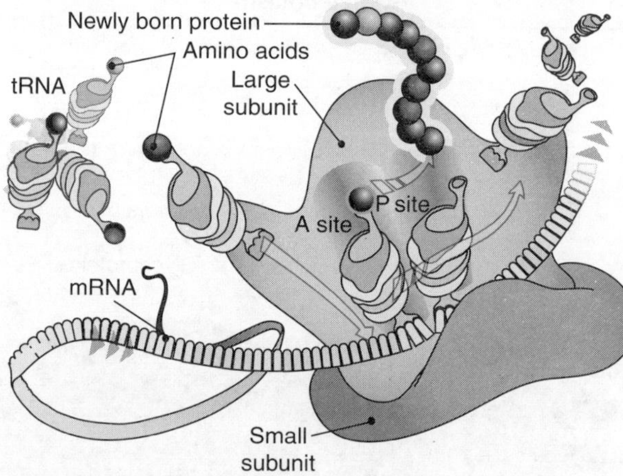

Fig. 1.12 Ribosome

2. **Endoplasmic reticulum:** This fine structure is only visible under an electron microscope. It is distributed in the cytoplasm with an irregular network of double membranes. The outer end of the cell membrane is attached to the outer membrane of the endoplasmic reticulum. The inner end of the endoplasmic reticulum is attached to the nuclear membrane. Endoplasmic reticulum is of two types:

rough (ribosomes are attached to it) and smooth (no particles are attached to it). It serves as a pathway for material distribution from one part of the cell to another and forms a supporting framework for the cell (Fig. 1.13).

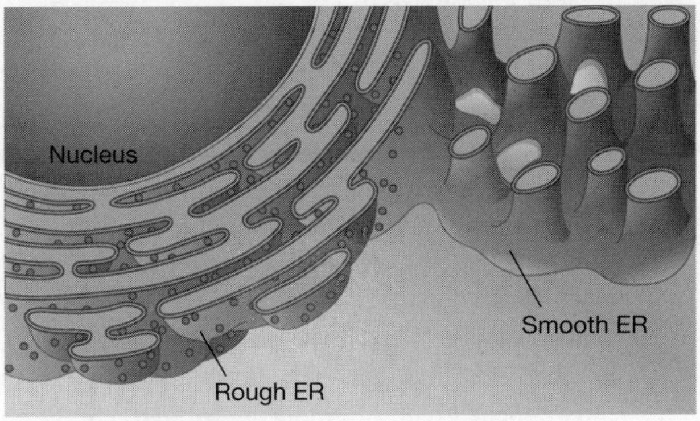

Fig. 1.13 Endoplasmic reticulum

3. **Lysosomes:** These are small vesicles of various shapes that contain digestive enzymes which digest and destroy foreign substances in the cell. During starvation of the cell, food is digested by the enzymes present in the lysosomes. They are also called 'suicide bags' as they rapidly dissolve and destroy damaged cells (Fig. 1.14).

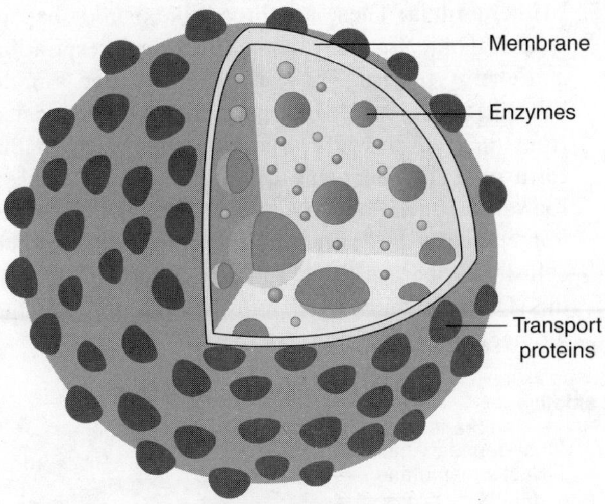

Fig. 1.14 Lysosome

4. **Centrioles and Centrosome:** They are only found in animal cells and appear as a clear area of cytoplasm close to the nucleus. They help in spindle fibre development during meiosis and mitosis.

   The centrosome (Fig. 1.15) has two centrioles arranged at right angles to each other. These centrioles are short bundles of microfilaments which are only visible under a microscope. There are no centrioles or centrosomes in plant cells.

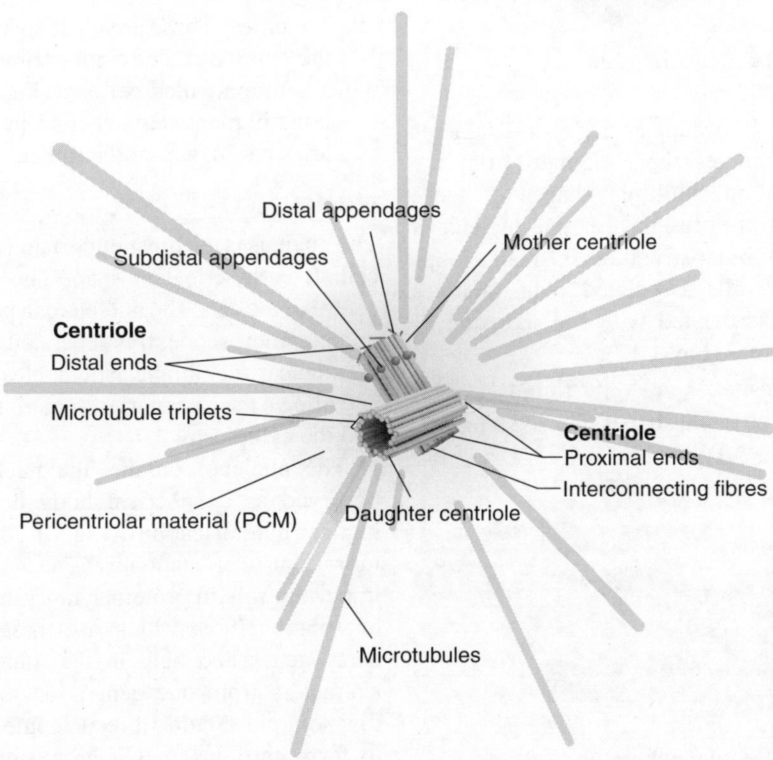

Fig. 1.15 Centrosome

5. **Mitochondria:** These are thread-like or rod-shaped, spherical objects which are the sites of cell respiration and energy release. They are double-walled minute bags that have finger-like projections called cristae from their inner walls. Energy is produced in the form of ATP (adenosine triphosphate) and is used for various metabolic functions of the body; hence, mitochondria are known as the powerhouse of the cell. It has its own DNA and can divide on its own; this DNA is transferred to offspring only by the mother (Fig. 1.16).

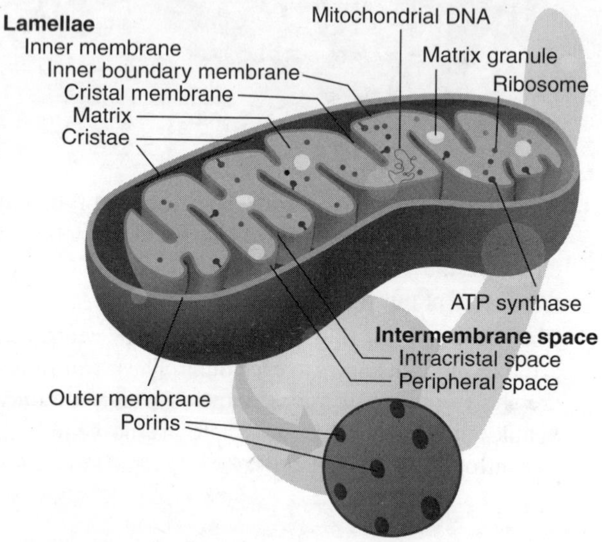

Fig. 1.16 Mitochondria

6. **Golgi Apparatus:** It occurs as rods or filaments in the form of granules that originate from the endoplasmic reticulum. Multiple shapes of these very small vesicles are present near the nucleus. It consists of hollow, tubular structures in small groups with membranous walls associated with vacuoles and vesicles. It is associated with the secretion of hormones and enzymes (Fig. 1.17).
7. **Plastids:** These organelles are only found in plant cells. They occur in different shapes such as spherical, disc and oval. On the basis of colour, plastids can be classified as:
   - **Leucoplasts** ('*leuco*' means 'white'): They are colourless with no pigment and perform the important function of storing starch. Potatoes have a lot of leucoplasts.
   - **Chromoplasts** ('*chromo*' means 'colour'): They exist in many colours such as red, orange and yellow. They are generally present in fruits and flowers along with pigments like xanthophyll and carotene. They provide colour to flowers in order to attract pollinators. Colours like blue, violet and purple come from the anthocyanin present in cell sap and not plastids.
   - **Chloroplasts** ('*chloro*' means 'green'): They have a green colour pigment, called chlorophyll, and thus appear green in colour and impart the same to the leaves. Leaves when exposed to light show abundant chlorophyll but also contain pigments that give yellow and orange colours. Chlorophyll traps sunlight and absorbs carbon dioxide to produce starch and sugar during photosynthesis. Chloroplasts have their own DNA and can divide by themselves (Fig. 1.18).

Fig. 1.17 Golgi apparatus

8. **Granules:** These are particles present in the cytoplasm that contain food materials like fats, glycogen and starch.
9. **Vacuoles:** These are clear spaces filled with water in the cytoplasm and with various substances mixed in a solution called cell sap (Fig. 1.19). They are quite large in plant cells whereas in animal cells, vacuoles are small in size and number.

Nucleus

The nucleus is the most important component of the living cell. It is spherical in shape and located at almost the centre of the cell. The nucleus can be stained and observed under a microscope. It is separated from the cytoplasm by the nuclear membrane. The nuclear membrane is porous and allows the transport of materials between the nucleus and the cytoplasm.

The nucleus contains the nucleolus, which can be observed as a spherical body under a microscope of higher magnification (Fig. 1.20). Some cells have more than one nucleolus. The nucleolus also plays an important role in protein synthesis. The nucleus contains chromatin fibres which are thread-like structures that carry genes and help in the transfer or inheritance of characters from one generation to another. During cell division, chromatin fibres become thick and ribbon-like to form chromosomes. Chromosomes are visible under

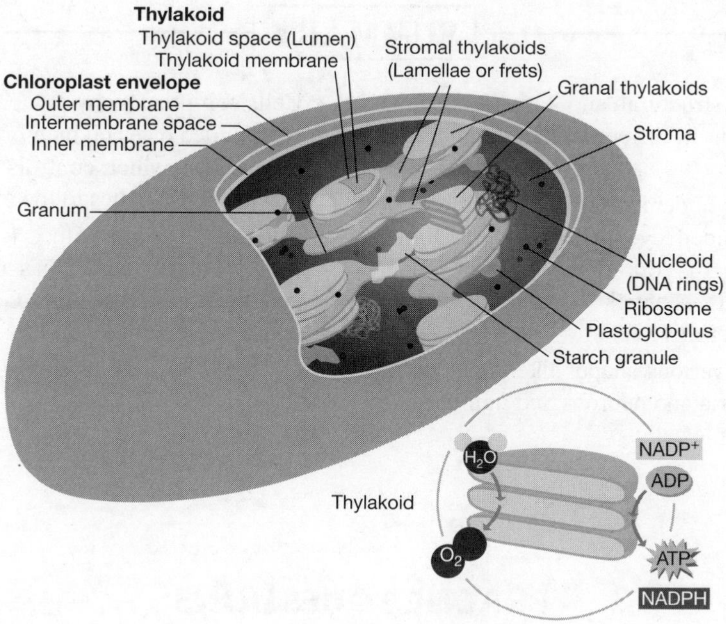

Fig. 1.18 Chloroplast

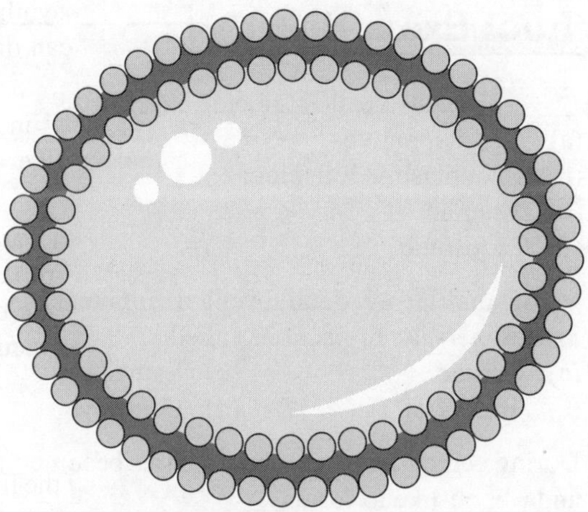

Fig. 1.19 Vacuole

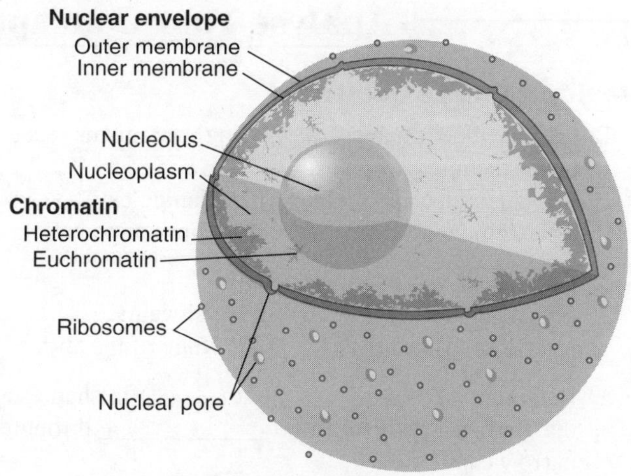

Fig. 1.20 Nucleus

the microscope during cell division. The nucleus acts as a control centre for the cell's activities in addition to playing a role in inheritance. Protoplasm is the content of the living cell that includes the nucleus and cytoplasm.

The nuclear membrane is absent in prokaryotic cells and chromatin fibres are freely suspended in the cytoplasm, like in bacteria. A double nuclear membrane is present in eukaryotic cells. The number of chromosomes is fixed for each species. Every human cell has 46 (23 pairs of) chromosomes. Chromatins are made up of genes, which are the hereditary units of the organism. Genes are composed of a complex called DNA. Genes define the characteristics of a species.

Fun Fact!

Did you know that DNA helps in identifying a person and is specific and unique to that person? The technique based on this principle is called DNA fingerprinting and it helps us find criminals and solve disputes relating to paternity!

SUMMARY

- Cells are the basic structural and functional units of life and are present in different sizes, shapes and number.
- Cells are only visible under a microscope and cannot be seen by the unaided eye. After Robert Hooke discovered the cells of cork under a microscope, Antonie van Leeuwenhoek invented better microscopes to observe live cells.
- Cells are present in various shapes like oval, rod or disc; sizes like bacteria and neurons and number like a single cell or multiple cells.
- Cells comprise many organelles and parts. The cell is surrounded by a cell membrane or cell wall that holds the nucleus which controls the activities of the cells and also pass genes to the offspring.
- The cell is filled with a jelly-like substance called the cytoplasm which has many organelles such as endoplasmic reticulum, Golgi apparatus, plastids, centrosome, mitochondria and vacuoles. The cells parts are different in plant and animal cells.

PRACTICE QUESTIONS

Analyse Your Concepts (School Exam Based)

Multiple Choice Questions

1. Leeuwenhoek named the living organisms seen under the microscope as:
(a) Animalcules (b) Plantcules
(c) Cellulae (d) Micrographia

2. Cell theory was hypothesized by:
(a) Schleiden (b) Schwann
(c) Both (a) and (b) (d) None of the above

3. Red blood cells are ______ and ______ in shape so that they can easily transport ______ ______ through narrow capillaries.
(a) Biconvex, oval, oxygen
(b) Biconcave, oval, carbon dioxide
(c) Biconcave, circular, carbon dioxide
(d) Biconcave, circular, oxygen

4. The accepted model of the cell membrane is known as ______ proposed by ______ and ______.
(a) Solid mosaic model, Singer, Nicolson
(b) Fluid mosaic model, Schleiden, Schwann
(c) Fluid mosaic model, Singer, Nicolson
(d) Solid mosaic model, Schleiden, Schwann

5. ______serves as a supporting framework for the cell.
(a) Cytoplasm
(b) Endoplasmic reticulum
(c) Lipid
(d) Nucleus

6. _______are also called 'suicide bags'.
(a) Golgi apparatus
(b) Endoplasmic reticulum
(c) Vacuole
(d) Lysosome

7. Mitochondria are double-walled minute bags that have finger-like projections called:
(a) Cristae (b) Lamella
(c) Tonoplast (d) Centriole

8. During cell division, chromatin fibres become _____ and ribbon-like to form ______.
(a) Thick, chromosome
(b) Thick, DNA
(c) Thin, chromosome
(d) Thin, DNA

9. _______are colourless and store starch.
(a) Chromoplasts (b) Leucoplasts
(c) Mitochondria (d) Vacuoles

10. _______are short bundles of microfilaments.
(a) Centrosomes (b) Centrioles
(c) Tonoplasts (d) Membranes

Match the Following

Each question contains statements given in two columns which have to be matched. Statements in column I have to be matched with statements in column II.

1. Choose the correct option:

Column I	Column II
i. Ribosome	a. Chlorophyll
ii. Genes	b. Protein synthesis
iii. Endoplasmic reticulum	c. DNA
iv. Plastid	d. SER and RER

2. Choose the correct option:

Column I	Column II
i. *Amoeba*	a. Unbranched
ii. Neuron	b. Multicellular
iii. Muscle cell	c. Longest cell
iv. Human	d. Unicellular

3. Choose the best match:

Column I	Column II
i. *Amoeba*	a. Gate of cell
ii. Red blood cells	b. Lysosome
iii. Cell membrane	c. Oval
iv. Suicide bag	d. Irregular

4. Choose the best match:

Column I	Column II
i. Chromoplast	a. Stores starch
ii. Chloroplast	b. Anthocyanin
iii. Leucoplast	c. Chlorophyll

5. Choose the correct option:

Column I	Column II
i. Antonie van Leeuwenhoek	a. Cell theory
ii. Robert Hooke	b. Animalcules
iii. Schleiden and Schwann	c. Modified cell theory
iv. Virchow	d. *Micrographia*

True or False

Read the following statements and write your answer as true or false.

1. All animals and plant cells contain cell wall.
2. Large vacuoles are present in the plant cell.
3. Cell walls are only made of protein.
4. Chloroplasts are present in plant cells that contain chlorophyll.
5. Genes are located inside the nucleus in the chromosomes.
6. The chromoplast contains green-colour-imparting chlorophyll.
7. *Micrographia* is the book written by Antonie van Leeuwenhoek.
8. Cell theory was stated by Schleiden and Schwann.
9. Yeast is an example of a multicellular organism.
10. In bacteria, the composition of the cell wall is a little different as it is made up of N-acetyl guanine and N-acetyl uracil.
11. Ribosomes are used for the synthesis of protein.
12. Centrosomes are also known as suicide bags.
13. Mitochondria have finger-like projections called cristae.
14. Genes are the hereditary units of an organism.
15. There are no centrioles or centrosomes in animal cells.

Fill in the Blanks

Complete the following statements with an appropriate word/term to be filled in the blank spaces.

1. ________ is the cell organelle that is involved in the secretion of hormones and enzymes.
2. The double-walled minute projections present in mitochondria are known as ______.
3. *Amoeba* is a/an ________ celled organism.
4. One of the longest cells present in the human body is ________.
5. _______ is a type of plastid that stores starch.
6. ________ was written by Robert Hooke.
7. Chloroplast contains _______ that imparts a green colour.
8. ________ are known as the hereditary units.
9. __________ is the coating around the cytoplasm and is semi-permeable.
10. The centrosome is surrounded by _______, located near the nucleus.

Assertion–Reason Questions

The following questions consist of two statements: Assertion (A) and Reason (R). Answer these questions selecting the appropriate option given below:

(a) Both A and R are true and R is the correct explanation of A
(b) Both A and R are true but R is not the correct explanation of A
(c) A is true but R is false
(d) A is false but R is true

1. Assertion (A): Cells are the smallest units of life.
   Reason (R): Cells are microscopic.
2. Assertion (A): Cell membrane is selectively permeable in nature.
   Reason (R): Cell membrane allows all substances to pass through.
3. Assertion (A): Lysosomes are also called suicide bags.
   Reason (R): They rapidly dissolve and destroy damaged cells.
4. Assertion (A): Chromoplasts provide colour to flowers.
   Reason (R): They do so in order to attract pollinators.
5. Assertion (A): Vacuoles are clear spaces filled with water in the cytoplasm.
   Reason (R): They are quite small in plant cells.

One-Word Answers

1. Who stated cell theory?
2. Who was the first scientist to observe cells under a microscope?
3. Which organelle performs photosynthesis?
4. Which organelle is known as the powerhouse of the cell?
5. What are the two types of endoplasmic reticulum?
6. Which cells are devoid of centrioles and centrosomes?
7. During starvation of the cell, food is digested by which organelle?
8. What is the composition of the cell wall in plants?
9. What are chromatins made of?
10. Which type of plastid is present in fruit and vegetables?

Define the Term

1. Endoplasmic reticulum
2. Centrioles and centrosomes
3. Lysosomes
4. Golgi apparatus
5. Vacuole

Groups of Terms

Given below are groups of terms. In each group the first pair indicates the relationship between the two terms. Rewrite and complete the second pair on a similar basis.

1. Schleiden: Cell theory :: Nicolson: ________
2. Chloroplast: Photosynthesis :: Ribosome: ____
3. Chromatin: Genes :: Cell membrane:_____
4. Chloroplast: Lamellae :: Mitochondria:______
5. Microfilament: Centriole :: N-acetyl murein: ______

Identify the Odd One Out

1. Chromoplast, chloroplast, mitochondria, leucoplast
2. Protein, lipid, cell membrane, N-acetyl murein
3. Robert Hooke, Antonie van Leeuwenhoek, Virchow, Newton
4. Mitochondria, chloroplast, bacteria, ribosome
5. Centriole, cell wall, chloroplast, chromoplast

Image-Based Questions

1. What is the name of the projections in this organelle?

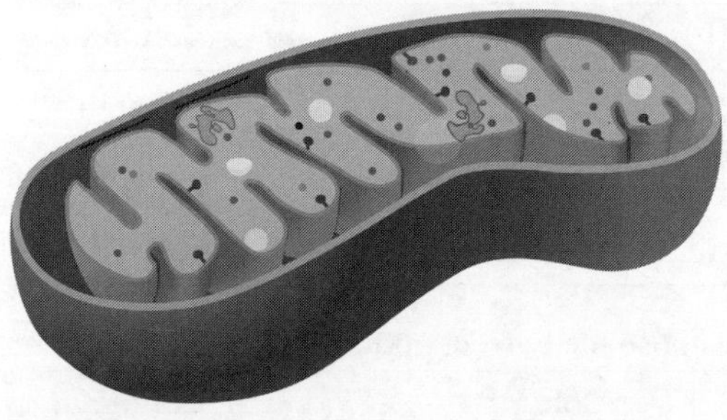

(a) Cristae (b) Lysosome
(c) Centrosome (d) Lamella

2. To which category of cell does this microbe belong?

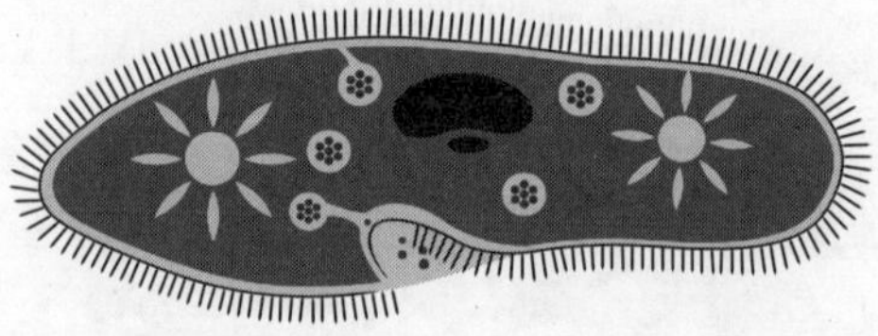

(a) Few celled (b) Unicellular
(c) Multicellular (d) None of the above

3. What is the role of this organelle?

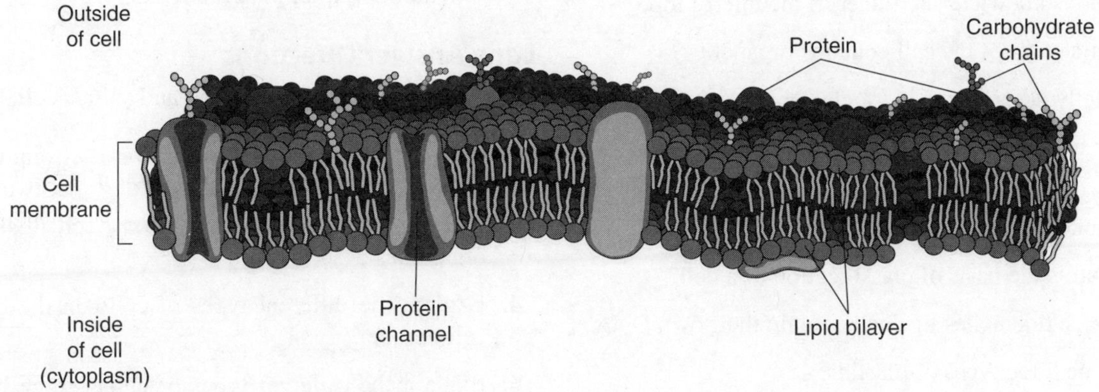

(a) Control of cell (b) Protection
(c) Gate of cell (d) Synthesis

4. What is the name of this scientist?

(a) Robert Brown
(b) Schleiden
(c) Schwann
(d) Antonie van Leeuwenhoek

5. Choose the correct term for this structure:

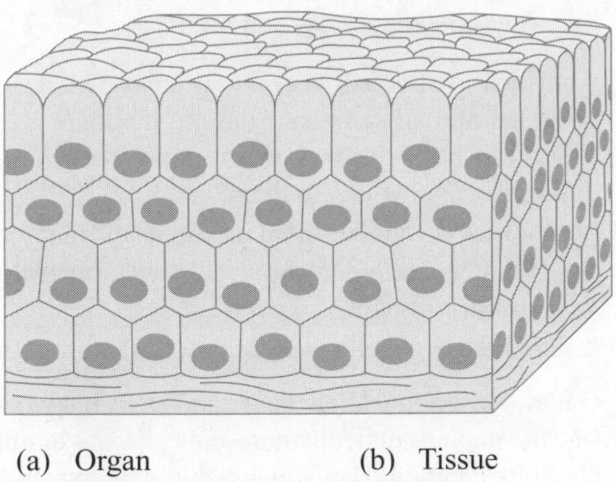

(a) Organ (b) Tissue
(c) Organ system (d) Cell

6. What is the role of this organelle?

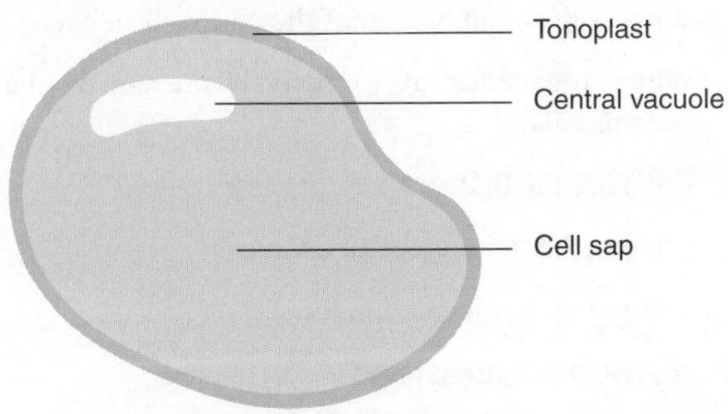

(a) Maintain osmotic pressure
(b) Store food
(c) Synthesize protein
(d) (a) and (b)

7. Choose the correct distinguishing feature of this organelle:

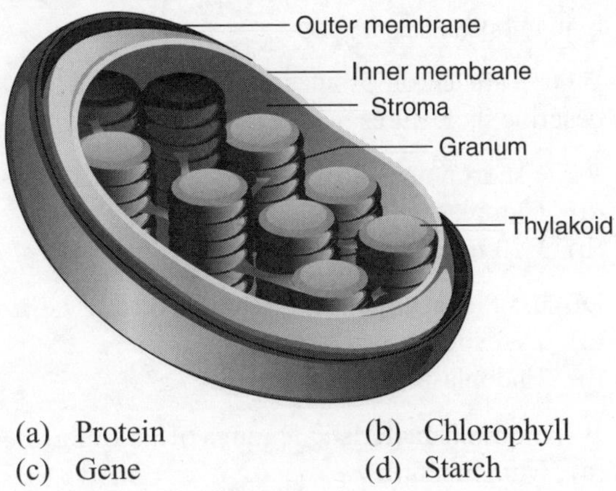

(a) Protein (b) Chlorophyll
(c) Gene (d) Starch

Very Short Answer Questions

1. Who is known as the father of the microscope?
2. Which part of the cell contains organelles?
3. Briefly describe the hierarchy from cell to body.
4. What is cell membrane made of?
5. What is the name of the book written by Robert Hooke?
6. What is the basis of classification of a cell?
7. Which organelles in a cell contain their own DNA?
8. Name three types of plastids.
9. What is the role of the centrosome and centriole in a cell?
10. What are the two types of endoplasmic reticulum?
11. What is the composition of the cell wall in bacteria?
12. Which organelles are exclusively present in plants and animals?
13. What are the three important parts of cells?
14. What is the function of ribosomes?
15. Where is DNA located in a cell?

Short Answer Questions

1. Draw a well-labelled diagram of a nerve cell, explaining its functions.
2. Draw the cell membrane fluid mosaic model.
3. What is DNA fingerprinting?
4. What are the three types of plastids?
5. What are the different shapes of cells observed in the human body?
6. Where are chromosomes found in a cell? Briefly describe its features.
7. Write short notes on:
   (a) Cytoplasm
   (b) Cell membrane
8. Discuss the distinguishing features of:
   (a) Lysosome
   (b) Endoplasmic reticulum
9. Write two characteristic features of:
   (a) Mitochondria
   (b) Golgi apparatus
   (c) Ribosomes
10. What are the different types of endoplasmic reticulum? Explain the functions.

Long Answer Questions

1. Make a sketch of a plant cell and animal cell. Discuss the differences between them.
2. Why are chloroplasts only present in a plant cell?
3. Draw a well-labelled diagram of a cell with all the organelles described.
4. Explain the different types of cells on the basis of cell number.
5. Explain the different types of cells on the basis of their size.
6. Explain the functions of the cell in detail.
7. Explain cell theory.
8. Describe why the cell is known as the structural and functional unit of life.
9. Explain the role of:
   (a) Nucleus
   (b) Mitochondria
   (c) Endoplasmic reticulum
   (d) Golgi apparatus
10. Write a short note on the cell membrane and cell wall.

Case Study-Based Questions

1. Nisha has been taught that cells can only be observed under a microscope, but Rahul disagrees with her. Can you help Rahul teach Nisha more about cells?
   (a) What are cells?
   (b) What are the different types of cells?
   (c) Which is the largest cell?
2. Abhishek collected leaves whereas Richa was provided animal samples from the laboratory. After observing both under a microscope, they found a difference. Can you help them?
   (a) Which organelles differ in animal and plantcells?
   (b) What is different about cell wall presence or absence in both?
   (c) Which organelle is present or absent in both?
3. David was reading about genes and DNA but when he saw the nucleus under a microscope, he was confused about its location. Can you help him understand?
   (a) How does the nucleus store genes and genetic material?
   (b) What are the different parts of the nucleus?

4. While taking a stroll in a garden, Sweety noticed her little sister keenly looking at leaves and flowers. Later she started asking Sweety different questions about them; help her to answer the questions.
(a) Why are the leaves green in colour?
(b) What is the role of leaves in plants?
(c) Why are flowers and fruits colourful?

5. While observing blood cells under a microscope, Shahid saw cells of different shapes. Can you answer his queries?
(a) Why are WBCs amoeboid in shape?
(b) Why are RBCs circular in shape?
(c) What is the role of WBC and RBC?

NCERT Corner

1. Select the correct option:
(a) Cell organelles are present in:
  (i) Nucleoplasm
  (ii) Cell membrane
  (iii) Cytoplasm
  (iv) Cell wall
(b) The following has its own DNA:
  (i) Golgi apparatus
  (ii) Endoplasmic reticulum
  (iii) Ribosome
  (iv) Mitochondria
(c) Genes are present in:
  (i) Cell wall (ii) Vacuole
  (iii) Chromosome (iv) Cell wall
(d) The following is also known as suicide bag:
  (i) Lysosomes (ii) Ribosomes
  (iii) Vacuole (iv) Cell wall
(e) The following is only present in plant cells:
  (i) Lysosome (ii) Chromoplast
  (iii) Centrosome (iv) Ribosome
(f) The finger-like projections in mitochondria are called:
  (i) Anthocyanin (ii) Nucleolus
  (iii) Cristae (iv) Lamella

2. True or False
(a) Unicellular organisms have a single-celled body.
(b) Muscle cells are branched.
(c) Organ is the basic living unit.
(d) *Amoeba* is irregular in shape.

3. Complete the crossword with the help of clues:
Across
1. This is necessary for photosynthesis
3. Term for component present in the cytoplasm
6. The living substance in the cell
8. Units of inheritance present on the chromsomes

Down
1. Green plastids
2. Formed by collection of tissues
4. It separates the contents of the cell from the surrounding medium
5. Empty structure in the cytoplasm
7. A group of cells

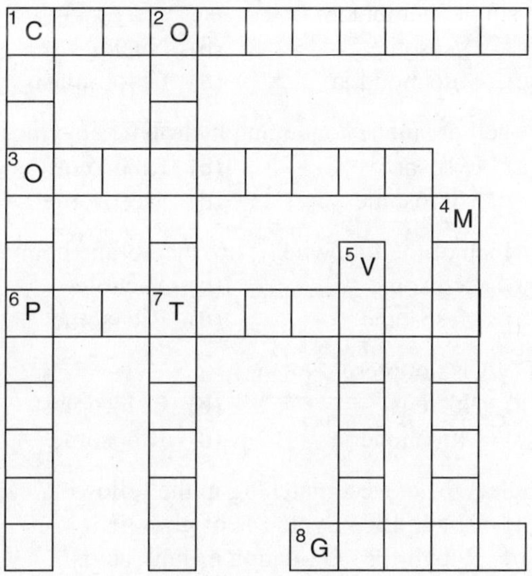

4. Match the organisms in Column A with their action in Column B.

Column A	Column B
i. Genes	a. Animalcules
ii. Centrosome	b. Longest cell
iii. *Amoeba*	c. Powerhouse of cell
iv. Mitochondria	d. Mitotic spindles
v. Schwann and Schleiden	e. *Micrographia*
vi. Neuron	f. DNA
vii. Robert Hooke	g. Irregular shape
viii. Antonie van Leeuwenhoek	h. Cell theory

COMPETITION WINDOW (OBJECTIVE TYPE)

1. Which of the following cell organelles is responsible for converting energy from carbohydrate to ATP?
(a) Ribosome (b) Chloroplast
(c) Mitochondria (d) Lysosome

2. Which of the following cell organelles is enclosed by a single membrane?
(a) Lysosome (b) Nuclei
(c) Mitochondria (d) Chloroplasts

3. A cell organelle containing hydrolytic enzymes is:
(a) Lysosome (b) Ribosome
(c) Microsome (d) Mesosome

4. Which of the following is not membrane bound?
(a) Lysosome (b) Vacuole
(c) Mesosome (d) Ribosome

5. DNA is not present in:
(a) Nucleus (b) Chloroplast
(c) Mitochondria (d) Ribosome

6. Select the correct matching in the following pairs:
(a) Rough ER - Synthesis of glycogen
(b) Rough ER - Oxidation of fatty acids
(c) Smooth ER - Oxidation of phospholipids
(d) Smooth ER - Synthesis of lipids

7. The osmotic expansion of a cell kept in water is chiefly regulated by:
(a) Mitochondria (b) Vacuole
(c) Plastid (d) Ribosome

8. Match the following and select the correct order:

Column A	Column B
A. Centriole	i. Infolding in mitochondria
B. Chlorophyll	ii. Thylakoid
C. Cristae	iii. Nucleic acid
D. Ribozyme	iv. Basal body of cilia

(a) A(iv), B(ii), C(i), D(iii)
(b) A(i), B(ii), C(iv), D(iii)
(c) A(i), B(iii), C(ii), D(iv)
(d) A(iv), B(iii), C(i), D(ii)

9. A major site for the synthesis of lipids is:
(a) Symplast (b) Nucleoplast
(c) SER (d) RER

10. Peptide synthesis inside a cell take place in:
(a) Chloroplast (b) Chromoplast
(c) Mitochondria (d) Ribosome

11. Lysosomes contain:
(a) Oxidative enzyme (b) Hydrolytic enzyme
(c) Anabolic enzyme (d) Reductive enzyme

12. Lysosomes are rich in:
(a) Nucleic acid (b) Hydrolytic acid
(c) Carbohydrate (d) Hormone

13. Protein are synthesised in:
(a) Golgi apparatus (b) Ribosome
(c) Mitochondria (d) Centrosome

14. Which of the following contains digestive enzymes?
(a) Ribosome (b) Plastid
(c) Polysome (d) Lysosome

15. Schleiden and Schwann are associated with:
(a) Protoplasm as the physical basis of life
(b) Cell theory
(c) Theory of cell lineage
(d) Nucleus function as control centre of cell

16. Which of the following are multicellular organisms?
I. Fish II. Fungi
III. Gymnosperm IV. Angiosperm
(a) I, II, III, IV (b) II, III, IV
(c) III, IV (d) I, II

17. Which part of the cell contains organelles?
(a) Protoplasm (b) Cytoplasm
(c) Nucleoplasm (d) Cell membrane

18. Match column I and column II:

Column A	Column B
A. Nucleus	I. Control centre
B. Cell membrane	II. Gate of the cell
C. Vacuoles	III. Food producers of the cell
D. Cell wall	IV. Storage tanks
E. Chloroplast	V. Supporter and protector

(a) A(V), B(IV), C(I), D(II), E(III)
(b) A(III), B(V), C(I), D(IV), E(II)
(c) A(I), B(II), C(IV), D(V), E(III)
(d) A(I), B(III), C(II), D(V), E(IV)

19. Fluid mosaic model of cell membrane explains:
(a) Structural aspects
(b) Cellular aspects
(c) Functional aspects
(d) Structural and functional aspects

Answer Keys

Multiple Choice Questions

1. (a) 2. (c) 3. (d) 4. (c) 5. (a)
6. (d) 7. (a) 8. (a) 9. (b) 10. (b)

Match the Following

1. i (b), ii (c), iii (d), iv (a)
2. i (d), ii (c), iii (a), iv (b)
3. i (d), ii (c), iii (a), iv (b)
4. i (b), ii (c), iii (a)
5. i (b), ii (d), iii (a), iv (c)

True or False

1. F 2. T 3. F 4. T 5. T
6. F 7. F 8. T 9. F 10. F
11. T 12. F 13. T 14. T 15. F

Fill in the Blanks

1. Golgi apparatus 2. Cristae
3. Single 4. Neurons
5. Leucoplast 6. *Micrographia*
7. Chlorophyll 8. Genes
9. Cell membrane 10. Microtubules

Assertion–Reason Questions

1. (b) 2. (c) 3. (a) 4. (a) 5. (d)

One-word Answers

1. Schleiden and Schwann; modified by Rudolf Virchow
2. Robert Hooke
3. Chloroplast
4. Mitochondria
5. Smooth and rough endoplasmic reticulum
6. Plant cells
7. Lysosome
8. Cellulose, hemi-cellulose, pectin
9. Genes
10. Chromoplast

Define the Term

1. These fine structures are distributed in the cytoplasm with an irregular network of double membranes. The outer end of the cell membrane is attached to the outer membrane of the endoplasmic reticulum. The inner end of the endoplasmic reticulum is attached to the nuclear membrane. Endoplasmic reticulum is of two types: rough (ribosomes are attached to it) and smooth (no particles are attached to it). It serves as a pathway for material distribution from one part of the cell to the other and is the supporting framework of the cell.

2. They are only found in animal cells and appear as a clear area of cytoplasm close to the nucleus. They are the origin of spindle fibre development during meiosis and mitosis. Centrosomes have two centrioles arranged at right angles to each other. Centrioles are short bundles of microfilaments which are only visible under a microscope.
3. Small vesicles of various shapes that contain enzymes that digest and destroy foreign substances in the cell. During starvation of the cell, food is digested by the enzymes present in the lysosome. These are also called suicide bags as they rapidly dissolve and destroy many damaged cells.
4. It occurs as rods or filaments in the form of granules which originate from the endoplasmic reticulum. Multiple shapes of these very small vesicles are present near the nucleus. It consists of hollow, tubular structures in small groups with the walls of the membrane associated with vacuoles and vesicles. It is associated with secretion of hormones and enzymes.
5. These are clear spaces filled with water in the cytoplasm with various substances mixed in a solution, called cell sap. They are quite large in plant cells whereas in animals, vacuoles are small in size and number.

Groups of Terms

1. Fluid mosaic model
2. Protein synthesis
3. Phospholipids
4. Cristae
5. Cell membrane

Identify the Odd One Out

1. Mitochondria
2. N-acetyl murein
3. Newton
4. Ribosome
5. Centriole

Image-Based Questions

1. (a) 2. (b) 3. (c) 4. (d) 5. (b)
6. (d) 7. (b)

Very Short Answer Questions

1. Antonie van Leeuwenhoek is known as the father of the microscope.
2. The cytoplasm contains organelles in cells.
3. Cell—>Tissue—>Organ—>Organ system—>Body
4. The cell membrane is made up of lipid and protein.
5. The book written by Robert Hooke is *Micrographia*.
6. Cells are classified according to number, shape and size.
7. Mitochondria and chloroplasts contain their own DNA.
8. The three types of plastids are chloroplast, leucoplast and chromoplast.
9. They act as the origin of spindle fibre development during meiosis and mitosis.
10. There are two types of endoplasmic reticulum: smooth and rough.
11. The cell wall in bacteria is made up of N-acetylglucosamine and N-acetyl murein.
12. Plastids are exclusively present in plants while centrioles and centrosomes are exclusively present in animals.
13. The three important parts of the cell are the nucleus, cell membrane and cytoplasm.
14. Ribosomes are responsible for the synthesis of protein.
15. DNA is the genetic material which is present in the nucleus of the cell.

Short Answer Questions

1. Hint: Draw a well-labelled diagram, as given in the chapter.
2. Hint: Draw a well-labelled diagram of the cell membrane, as given in the chapter.
3. DNA is specific and unique to a person and so it helps in identifying the person. The technique based on this principle is called DNA fingerprinting and it helps to find criminals, solve parental disputes and so on.
4. On the basis of colour, plastids are classified into three types:
   (a) **Leucoplasts:** They are colourless with no pigment and have the important function of storing starch. Potatoes have a lot of leucoplasts.
   (b) **Chromoplasts:** They exist in many colours like red, orange and yellow. They are generally present in fruits and flowers along with pigments like xanthophyll and carotene. They provide colour in flowers in order to attract pollinators. The colours like blue, violet and purple come from anthocyanin present in cell sap and not plastids.
   (c) **Chloroplasts:** They have a green pigment, called chlorophyll, and thus appear green in colour; this the same in leaves. Leaves when exposed to light show abundant chlorophyll but also contain pigments giving yellow and orange colours.

Chlorophyll traps the sunlight and absorbs carbon dioxide to produce starch and sugar during photosynthesis. Chloroplasts have their own DNA and can divide on their own.

5. The functions of cells is the reason behind some of their shapes. For example:
   (a) White blood cells are amoeboid in shape so that they can easily squeeze into and out of capillary walls.
   (b) Muscle cells are long in shape so that the contractions or pull of parts is easy
   (c) RBCs are biconcave and circular so that they can easily transport oxygen through narrow capillaries
   (d) In the leaves, the guard cells in the stomatal pore facilitate the opening and closing of pores and so are bean-shaped.
   (e) To conduct impulses to the brain from different parts or vice versa, nerve cells are long.
6. During cell division, chromatin fibres become thick and ribbon-like to form chromosomes that are visible under the microscope. The nucleus contains chromosomes which are responsible for inheritance. In prokaryotic cells, chromatin fibres are freely suspended in the cytoplasm due to lack of a nucleus, like in bacteria. The number of chromosomes is definite in each species. Every human cell has 46 (23 pairs) chromosomes. Chromatins are made up of genes, which are the hereditary units of the organism and define the characteristics of a species. Genes are made up of DNA.
7. (a) **Cytoplasm:** This semi-liquid substance which has a jelly-like texture is present between the nucleus and cell membrane. The cytoplasm occupies most of the cell within the membrane. Most cell components and organelles are present in the cytoplasm. The cytoplasm appears partly transparent, colourless and watery in nature. It remains in dynamic state and in a state of movement with the majority of chemical reactions taking place. The organelles that are present are endoplasmic reticulum, mitochondria, ribosomes, Golgi bodies, plastids and lysosomes.
   (b) **Cell membrane**: It is also known as plasma membrane and encloses the cytoplasm and nucleus. The cell membrane separates the cell from the surrounding environment and one cell from another. Each cell is surrounded by a cell membrane which has pores through which substances exit and enter the cell. The cell membrane is selective permeable in nature. This means that it only allows specific substances to pass through while it prevents others from entering. The membrane is porous in nature and gives a shape to the cell. It is made up of lipid and protein along with carbohydrate. The correct model of the cell membrane is known as the fluid mosiac model given by Singer and Nicolson.
8. (a) **Lysosome:** These are small vesicles of various shapes that contain digestive enzymes that digest and destroy foreign substances in the cell. During starvation of the cell, the food is digested by the enzymes present in lysosomes. These are also called 'suicide bags' as they rapidly dissolve and destroy many damaged cells.
   (b) **Endoplasmic reticulum:** These fine structures are only visible under the electron microscope. It is distributed in the cytoplasm with an irregular network of double membranes. The outer end of the cell membrane is attached to the endoplasmic reticulum outer membrane. The inner end of the endoplasmic reticulum is attached to the nuclear membrane. The endoplasmic reticulum is of two types: rough endoplasmic reticulum: ribosomes are attached to it and smooth endoplasmic reticulum: no particles are attached to it. It serves as a pathway for material distribution from one part of the cell to another, and forms a supporting framework for the cell.
9. (a) **Mitochondria:** They are thread-like or rod-shaped structures, which are sites of cell respiration to release energy. They are double-walled minute bags that have finger-like projections in their inner walls, called cristae. Energy is produced as ATP (adenosine triphosphate). This ATP is used in various metabolic functions of the body and cells; hence mitochondria are known as the powerhouse of the cell. It has its own DNA and can divide on its own and is the only DNA which is transferred to offspring solely by the mother.
   (b) **Golgi bodies:** It occurs as rods or filaments in the form of granules which originate from the endoplasmic reticulum. Multiple shapes of these very small vesicles are present near the nucleus. It consists of hollow, tubular structures in small groups with the walls of the membrane associated with vacuoles and vesicles. It is associated with secretion of hormones and enzymes.
   (c) **Ribosomes:** They are multiple in number and granular in nature which are either attached to the

membrane or scattered freely in the cytoplasm. Ribosomes are used for the synthesis of protein.

10. Endoplasmic reticulum is distributed in the cytoplasm with an irregular network of double membranes. The outer end of the cell membrane is attached to the endoplasmic reticulum outer membrane. The inner end of the endoplasmic reticulum is attached to the nuclear membrane. The endoplasmic reticulum is of two types: rough endoplasmic reticulum: ribosomes are attached to it and smooth endoplasmic reticulum: no particles are attached to it. It serves as a pathway for material distribution from one part of the cell to another, and forms a supporting framework for the cell.

Long Answer Questions

1. Hint: Draw diagrams of plant and animal cells and refer to Fig. 1.9 in this chapter.

2. Hint: Refer to the chloroplast section in this chapter.

3. Hint: Refer to the figures given in the chapter and draw a well-labelled diagram.

4. Hint: Write about the classification of cells on the basis of number.

5. Hint: Write about the classification of cells on the basis of size.

6. Hint: Refer to the section on functions of cells in this chapter.

7. Hint: Explain cell theory and modified cell theory.

8. Hint: Define the cell, its functions, types and structure along with a few organelles.

9. Hint: Refer to the topics mentioned in this chapter and explain them with the help of a well-labelled diagram.

10. Hint: Explain the theory, functions, composition and structure of the cell membrane and cell wall in different organisms.

Case Study-Based Questions

1. (a) The smallest unit of life is known as cells. Cells are the basic functional, structural and biological units of all organisms. They are also called the building blocks of life.

(b) Cells can be classified based on size, number and shape.

(c) Ostrich's egg is the largest cell.

2. (a) Plant cells have a cell wall and chloroplasts while animal cells have centrosomes and centrioles.

(b) The cell membrane is present in both animal and plant cells while cell wall is only present in plants.

(c) Ribosomes, mitochondria, nucleus, Golgi apparatus, endoplasmic reticulum, cell membrane and, vacuole are present in both, the while cell wall and plastid are present only in plants and centrosome and centriole are present only in animals.

3. (a) Nucleus also contains chromatin fibres that carry genes (DNA) and help in the transfer or inheritance of characters from one generation to another.

(b) Nucleus has nucleolus, nucleopore and nuclear membrane.

4. (a) Due to the presence of chloroplasts that carry chlorophyll which makes it green

(b) It performs photosynthesis

(c) Chromoplasts present in cells come in different colours

5. (a) White blood cells are amoeboid in shape so that they can easily squeeze into and out of capillary walls.

(b) RBCs are biconcave and circular so that they can easily transport oxygen through narrow capillaries.

(c) WBCs are involved in immunity while RBCs have haemoglobin which carries and transports oxygen.

NCERT Corner

1. (a) (iii) (b) (iv)
(c) (iii) (d) (i)
(e) (ii) (f) (iii)

2. (a) T (b) F
(c) F (d) T

3.

Across	**Down**
1. Chlorophyll	1. Chloroplasts
3. Organelle	2. Organ
5. Protoplasm	4. Membrane
8. Genes	5. Vacuole
	7. Tissue

4. i-(f), ii-(d), iii-(g), iv-(c), v-(h),vi-(b), vii-(e), viii-(a)

Competition Window

1. (c)	**2.** (a)	**3.** (a)	**4.** (d)	**5.** (d)
6. (d)	**7.** (b)	**8.** (a)	**9.** (c)	**10.** (d)
11. (b)	**12.** (b)	**13.** (b)	**14.** (d)	**15.** (b)
16. (a)	**17.** (b)	**18.** (c)	**19.** (a)	

CHAPTER 2

Adolescence

PREREQUISITES

You must have knowledge of:

- Human body parts
- Stages of human life
- Glands of the human body
- Basics of reproduction

LEARNING OBJECTIVES

At the end of this chapter, you will be able to:

- Define puberty and describe the changes
- List the secondary sexual characters
- Explain the role of hormones in reproductive functions
- Discuss reproductive health

INTRODUCTION

In the previous chapters, you learned about how animals and humans reproduce. Reproduction is an essential process to ensure continuity of a species. Human beings are capable of reproduction only after they reach a certain age.

Have you ever wondered what changes occur when a person attains an age where she is capable of reproduction? Are these changes taking place inside the body or externally? How long does the reproductive ability last?

In this chapter, we will learn about the changes that take place in a boy's body and a girl's body which make them capable of reproduction.

ADOLESCENCE AND PUBERTY

Growing up is a normal process. You may have observed that when a boy grows into an adult, several changes take place. Some of these changes are growth of hair on the face, change in voice and body shape.

There are four stages of growth in life. **Infancy** is the earliest period of growth, from birth up to the age of 2 years. **Childhood** is the state between infancy and adolescence. It occurs between 2 and 11 years. The age group of 10–19 years is defined as **adolescence**. Since this period also covers the teens (13–19 years), adolescents are also known as teenagers. Adolescence is a transition phase of growth and development between childhood and adulthood. It begins at the age of 10–11 years and ends at the age of 18–19 years (Table 2.1). The last stage of growth is **adulthood**; it begins after 20 years and continues till death.

During adolescence, the human body undergoes many changes such as growth and maturation of the reproductive system. As a result, adolescents attain sexual maturity and become capable of reproduction. These changes constitute puberty. The transformation that takes place during adolescence is known as **puberty**. The most important change which marks the onset of puberty is that boys and girls become capable of reproduction.

Table 2.1 Time period of age of adolescence

Adolescence in	Begins at (age)	Ends at (age)
Boys	11 years	18 or 19 years
Girls	10 or 11 years	18 or 19 years

Changes at Puberty

Adolescence is a time of growth in physical and mental health. During adolescence, the human body undergoes changes that mark the onset of puberty. A baby is born with a full set of reproductive organs. However, these organs do not function during the first 10–12 years of life. Under the influence of hormones produced in the body, the reproductive organs become functional at the time of puberty.

Let us discuss the changes that mark the onset of puberty and reproductive maturity.

Fun Fact!

The brain reaches its biggest size early in adolescence. For girls, it takes place around the age of 11 years. For boys, it occurs around the age of 14 years.

Increase in Height

The most visible change during puberty is a sudden increase in height. Some children may grow quickly during puberty and then slow down, while others may grow gradually. The bones and arms elongate and the individual becomes tall and lanky. Girls grow faster than boys; at around 18 years of age, both reach their maximum height. All the parts of the body do not grow at the same rate. For example, the arms, hands or legs may appear oversized and out of proportion with the body. The rate of growth in height also varies. Height mainly depends on the genes inherited from parents. It is very important to eat the right kind of food during the growing years. Consuming a balanced diet helps the bones, muscles and other parts of the body get adequate nourishment for growth.

The percentage of full height in boys and girls along with the age is given in Table 2.2 and the graph is presented in Fig. 2.1.

Table 2.2 Percentage of full height of boys and girls

Age	Percentage of full height in boys	Percentage of full height in girls
8 years	72%	77%
9 years	75%	81%
10 years	78%	84%
11 years	81%	88%
12 years	84%	91%
13 years	88%	95%
14 years	92%	98%
15 years	95%	99%
16 years	98%	99.5%
17 years	99%	100%
18 years	100%	100%

The following formula is used to calculating the full height:

$$\frac{\text{Present height}}{\text{\% of full height (from the table)}} \times 100$$

For example, if a boy is 9 years old and 120 cm tall, at the end of the growth period, he is likely to be:

$$\frac{120}{75} \times 100 = 160 \text{ cm}$$

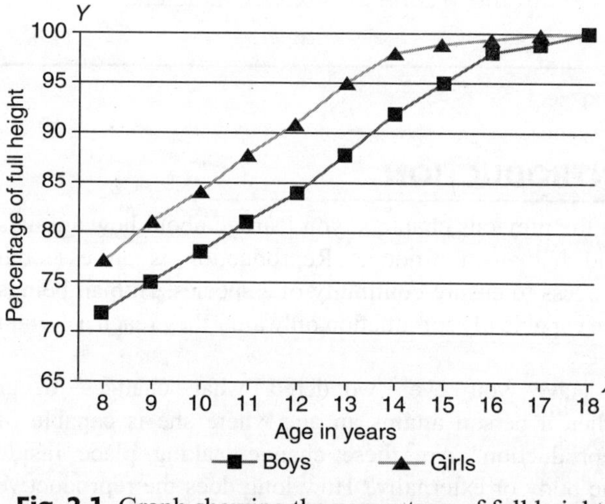

Fig. 2.1 Graph showing the percentage of full height with age in boys and girls

Changes in the Shape of the Body

Change in body shape is different in girls and boys. In boys, the body becomes more muscular, the shoulders become broader and the chest grows wider. In girls, the body shows curves in certain parts and the region below the waist becomes wider and broader.

Changes in Voice

Change in voice is caused by the voice box or larynx. At puberty, the voice box begins to grow and because boys develop larger voice boxes, it is seen as a protruding part of the throat, called the Adam's apple (Fig. 2.2). Sometimes in boys, as the larynx begins to grow, the muscles get enlarged due to which their voice becomes hoarse. In girls, the larynx is small and is hardly visible from the outside. Girls have a higher pitched voice than boys.

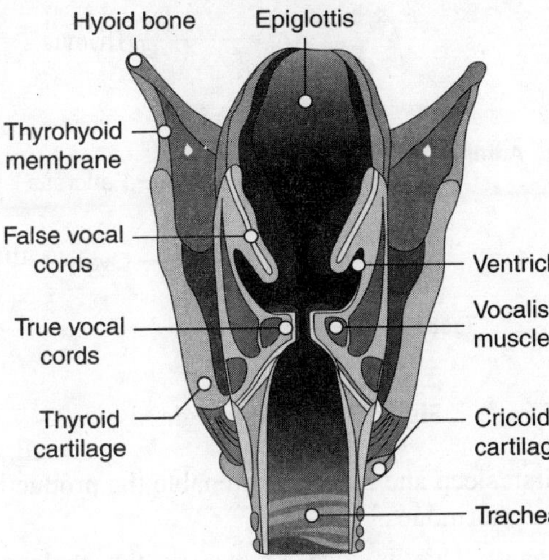

Fig. 2.2 Adam's apple in male voice box

Activity of the Sweat and Sebaceous Glands

In adolescents, the sweat glands become more active and secrete more sweat. The sebaceous glands secrete oil and many young girls and boys develop acne and pimples. These secretions are released through ducts by the exocrine glands.

Mental, Intellectual and Emotional Maturity

Adolescents start thinking independently and become self-conscious. Intellectual development takes place and they spend considerable time in contemplation. At this time of life, the brain has the greatest capacity for learning and thinking. Sometimes, adolescents may feel insecure about their elder siblings while trying to adjust to the changes in their body and mind. These changes are a part of the natural growth that takes place in all individuals.

Development of Sex Organs

As a child reaches the age of adolescence, the reproductive parts develop completely. In females, the ovaries develop completely and release the egg or ovum. In males, the testes and penis are completely developed. The testes become capable of producing sperms and testosterone.

Secondary Sexual Characters: Secondary sexual characters are those that develop in males and females at the time of attaining sexual maturity. These characters help to distinguish a male from a female. In girls, breasts begin to develop at puberty while boys grow facial hair and hair on the chest. In both boys and girls, hair grows under the arms and in the pubic region.

Secondary sexual characters in girls:

- Breasts develop and become enlarged
- The ovaries start to release eggs; the ovaries, oviducts, uterus and vagina become enlarged
- Menstruation begins
- Hips broaden; extra fat may be deposited on the hips and thighs
- Hair grows in the armpits and in the pubic region between the thighs

Secondary sexual characters in boys:

- Hair grows on the face as moustache and beard, and on other parts of the body like the chest and armpits
- The testes begin to make sperm; the testes and penis become larger
- The chest and shoulders broaden and the body becomes muscular
- Hair grows in armpits and in the pubic region between the thighs
- Sexual drive associated with adulthood begins to develop

The changes that take place in adolescence are summarised in Table 2.3.

Table 2.3 Changes that take place during puberty

Change	Explanation
Height increases suddenly	• The bones of the legs and the arms start to grow longer; during puberty, boys and girls become taller. • Occurs till the age of 18 years in both boys and girls.
Change in the shape of the body	• Boys generally experience a change in the chest and shoulders, as they become wider. • In girls, the area below the waist becomes wider.
Change in voice	• The voice box or larynx develops in boys. • It can be seen as an enlarged part of the throat in boys and is called the Adam's apple. • Boys experience a hoarse voice for some time due to changes in the larynx. • The larynx in girls is not visible because of its smaller size.

(Continued)

Change	Explanation
Increased sweat and oil production in the skin	• The sweat glands and oil glands (called sebaceous glands) start secreting more oil and sweat during puberty.
Development of sex organs	• In males, sperm production begins. • In females, the size of the ovaries increases and it starts releasing eggs.
Mental, emotional and intellectual maturity	• During adolescence, the brain attains its maximum learning power. • The way of thinking starts to change. • Adolescents may feel a little uncomfortable and insecure due to sudden changes in the body. However, all these changes are completely normal.

HORMONES

The changes that take place in the human body during adolescence occur due to chemical substances called hormones secreted by the endocrine glands. The endocrine glands are distributed throughout the body. Hormones control many physiological functions as well as psychological health. They act as chemical messengers and are responsible for transporting signals from one cell to another. Hormones have certain **properties** or characteristics:

- They are secreted by the endocrine glands.
- They are responsible for controlling the behaviour of the target cells.
- They cause long-term effects in the body such as changes in behaviour, growth, development of organs and puberty.

The endocrine glands are also known as **ductless glands** because they secrete hormones directly into the bloodstream (Fig. 2.3). In males, the testes secrete testosterone and this results in changes such as the growth of facial hair, deepening voice and hairy chest. In females, the ovaries secrete oestrogen, which results in breast development.

The different types of glands and the hormones they secrete are given below:

- **Hypothalamus:** Secretes hormones that help control the body temperature, regulate emotions, hunger, thirst, sleep and moods and enable the production of other hormones.

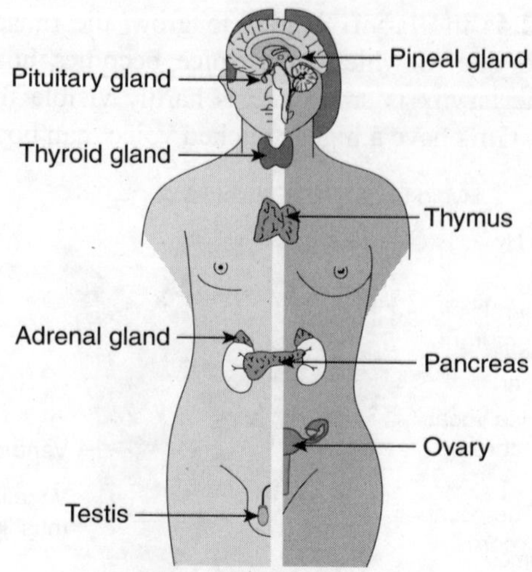

Fig. 2.3 Endocrine glands

- **Pineal:** This is also known as the thalamus. It produces serotonin derivatives of melatonin, which affect sleep patterns in the human body.
- **Thyroid:** It produces hormones that affect the heart rate and how calories are burnt in the body.
- **Adrenal:** This gland produces hormones that control the sex drive and stress.
- **Pituitary:** It is known as the 'master control gland'. It helps in controlling other glands. Moreover, it develops hormones that trigger growth and development.
- **Pancreas:** This gland is involved in the production of insulin, which plays an important role in maintaining blood sugar levels.
- **Testosterone:** In men, the testes secrete testosterone. It plays an important role in sperm production.
- **Oestrogen:** The ovaries release oestrogen, progesterone and other female sex hormones.

All these glands work together to produce and manage the hormones of the body.

Role of Hormones in Initiating the Reproductive Function

The part of the body for which a hormone is intended is called the **target site**. The endocrine glands secrete hormones into the blood and it reaches the target site. The target site then responds accordingly (Table 2.4).

Table 2.4 Endocrine glands and their hormones

Gland	Hormone	Location and Function
Thyroid 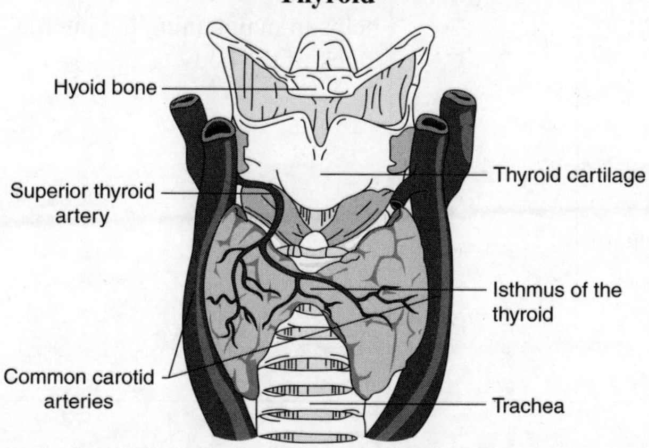	Thyroxine	Found at the front of the neck. Secretes a collection of hormones called the thyroid hormones. Controls the rate of metabolism in the body. Lack of thyroxine can lead to goitre.
Pancreas 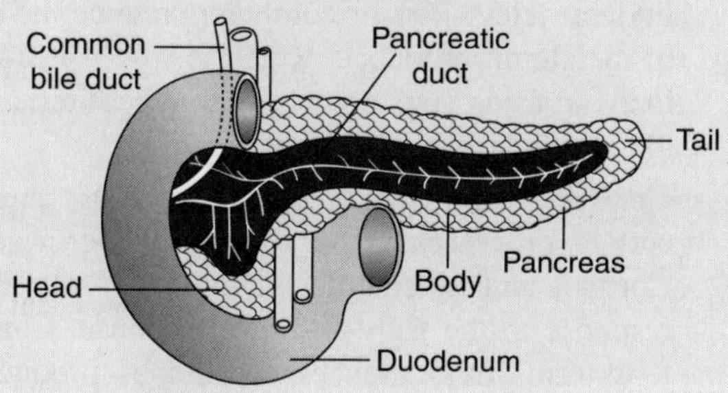	Insulin, Glucagon	Located behind the stomach. Produces some digestive enzymes, glucagon and insulin. Insulin maintains the blood glucose level. Lack of insulin leads to diabetes.
Adrenal 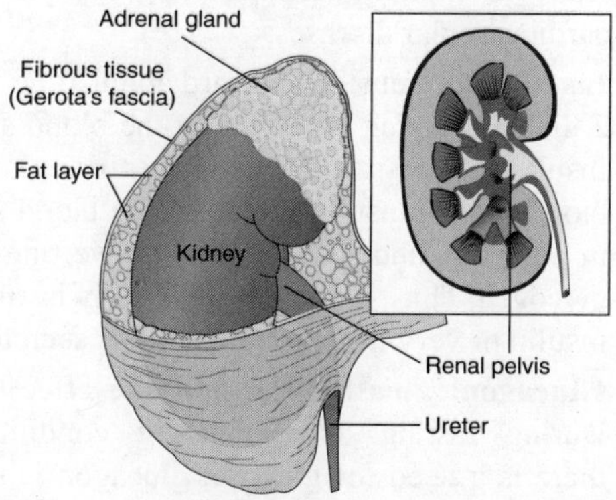	Adrenaline	Located above the kidneys. Adrenaline helps in managing stress and also maintains salt balance in blood.
Pituitary 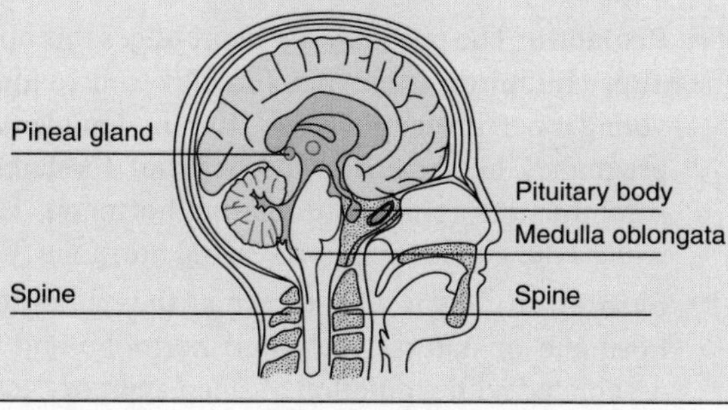	Growth hormone	Located inside the skull. Also called the 'master gland' as it controls the functions of all other glands in body.

Gland	Hormone	Location and Function
Pineal 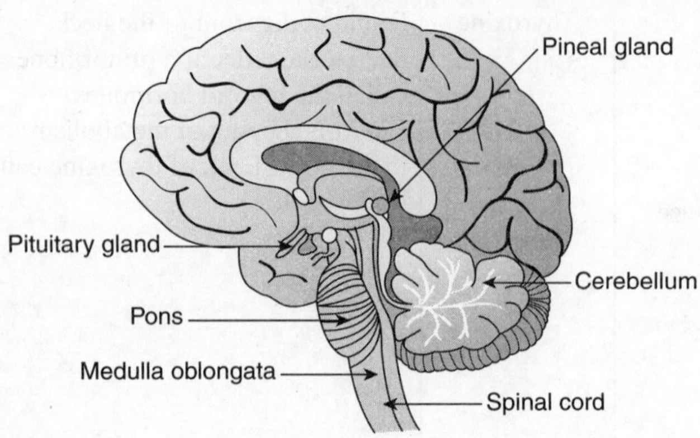	Melatonin	Known as the 'third eye' and located in the brain. Melatonin helps in maintaining the internal clock of the body.

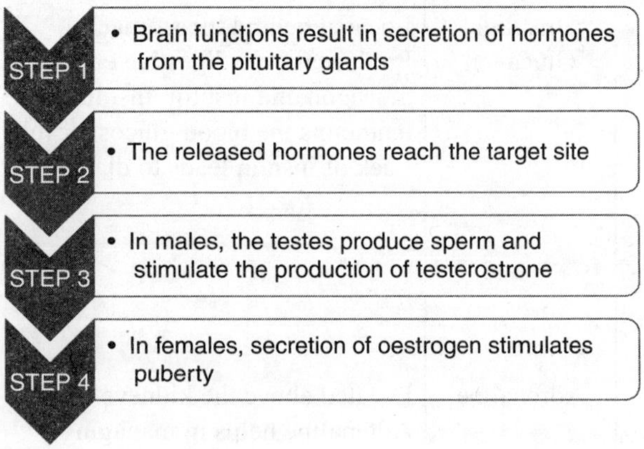

Fig. 2.4 Hormones initiate puberty

Secretion of hormones from the endocrine glands into the bloodstream stimulates the male and female reproductive organs, that is, the testes and ovaries (Fig. 2.4). These organs release testosterone (in males) and oestrogen (in females). The released hormones then stimulate changes in the body.

Apart from the sex hormones, there are different types of hormones present in our body:

- **T3 and T4:** These are the two main hormones produced by the thyroid glands. These hormones regulate metabolism, which plays an important role in overall energy level and digestion. If the thyroid secretes too much of these hormones, the condition is called hyperthyroidism. If it makes less than normal, it is known as hypothyroidism, which is a more common problem.
- **Melatonin:** The circadian rhythm or sleep/wake cycle of our body is controlled by melatonin. Sunlight prevents melatonin production and it is secreted by the pineal gland. More melatonin is secreted as it gets dark at night and this increases the urge to sleep. Phones, televisions or computers reduce the amount of melatonin produced, which is why it is advisable to avoid using such devices before bedtime or to use glasses that block blue light at night.
- **Cortisol:** Long-term or short-term stress can trigger certain processes in our body's endocrine system. Cortisol and adrenaline are made by our body in response to the fight-or-flight response. Chronic or long-term stress induces continuous production of cortisol and other stress-related hormones, which leads to anxiety, depression, obesity as well as cardiovascular diseases.
- **Insulin:** You must have heard about Type 1 and Type 2 diabetes which is related to the blood sugar level. Insulin is necessary for cells to utilise glucose in the bloodstream. Insulin regulates the blood sugar level in our body and reduces it by converting sugar into glycogen. Thus, in diabetes, the body is unable to use insulin or very little or no insulin is secreted.
- **Glucagon:** This has the opposite effect to insulin. During fasting or flight/fight conditions, when there is a need for glucose, glucagon is secreted so that glycogen is broken down to form glucose to be utilised by the body for the production of energy.
- **Prolactin:** The pituitary gland produces this hormone after childbirth; it enables females to breastfeed the young one. Prolactin levels rise in females during pregnancy because it is also crucial for fertility as it inhibits the follicle stimulating hormone (FSH) as well as the gonadotropin-releasing hormone (GnRH).
- **Serotonin:** This is also known as the mood-boosting hormone or natural feel-good hormone and is also associated with regulation of sleep, memory, learning,

mood regulation as well as muscular functions. When the brain does not produce enough of this hormone to regulate stress levels or mood, anxiety, weight gain, carbohydrate craving and depression result. Excess of this hormone causes confusion, agitation and sedation.

- **Adrenaline:** This hormone is secreted in the adrenal gland's medulla region as well as in the neurons of the central nervous system (CNS). It is also known as the emergency hormone as it induces and initiates a quick reaction to respond to stress by increasing the metabolic rate and dilating blood vessels to increase blood flow to the brain and heart.
- **Growth hormone/Somatotropin:** This protein hormone is synthesised and secreted by the cells of somatotrophs in the anterior pituitary gland. Cell reproduction, growth stimulation, cell regeneration and metabolism is boosted by this hormone as it plays a crucial role in human development.

MENSTRUAL CYCLE IN FEMALES

In females, the reproductive age begins at 10–12 years and ends at the age of 45–50 years. At birth, the ovaries contain thousands of immature eggs. When a female reaches puberty, maturation of the eggs takes place. The release of an egg from an ovary is called **ovulation**. It takes place 14 days after the start of a menstrual cycle.

Before every ovulation, the inner lining of the uterus become thick and soft with a lot of blood vessels in it. If the egg is fertilised by a sperm, pregnancy occurs. If the egg is not fertilised by a sperm, the thick and soft uterus lining breaks down. It results in vaginal bleeding, called menstrual flow, **menstruation** or periods. Menstruation occurs when an egg released by the ovary does not get fertilised.

In females, menstruation begins in adolescence and the event is called **menarche**. The cessation of menstrual flow at the age of 45–50 years is called **menopause**.

The menstrual cycle includes the following steps:

1. Maturation of eggs
2. Release of mature eggs
3. Thickening of the wall of the uterus
4. Break down of the thickened wall if pregnancy does not occur

SEX DETERMINATION

Every human being has 23 pairs of chromosomes. Of these, 22 are similar in all respects and are known as autosomes. Only one pair is different; it is known as the 'sex chromosome', and it plays an active role in sex determination.

Both males and females carry two sets of sex chromosomes. Males hae one X and one Y (XY) sex chromosomes. Females have two X (XX) sex chromosomes. The XY sex determination system is found in humans, mammals and in some plant species. A child will always inherit an X chromosome from its mother. Thus, the sex of the child will be determined by the chromosome inherited from the father.

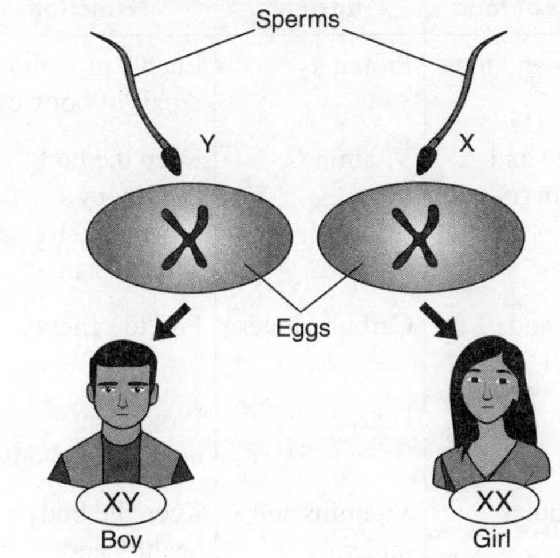

Fig. 2.5 Mechanism of sex determination

The sex of an individual can be genetically determined as follows (Fig. 2.5):

- If a sperm carrying an X chromosome fertilises an ovum (or egg) which carries an X chromosome, the child will be female. This is because the child will have an XX combination of sex chromosomes.
- If a sperm carrying a Y chromosome fertilises an ovum (or egg) which carries an X chromosome, the child will be male. This is because the child will have an XY combination of sex chromosomes.

REPRODUCTIVE HEALTH

According to the World Health Organization (WHO), reproductive health refers to "the sum total of wellbeing in all aspects of reproduction, that is, physical, emotional, behavioural and social."

Factors that lead to wellbeing in adolescents:

1. **Nutritional Health:** As a child reaches puberty, his or her body undergoes several changes and transformations. A balanced diet plays an important role in puberty. It includes sufficient amount of proteins, carbohydrates, fats, vitamins and minerals.

In a balanced diet, pulses, fruits, vegetables, milk, iron-rich food and other nutrients are present in the requisite proportions (Table 2.5).

Healthy and nutrient-rich food is essential for the growth and repair of cells. Junk food such as fried food, soft drinks and sweets should be avoided as they are of low nutritional value.

Table 2.5 Food and their nutrients

Types of food	Major nutrient	Function
Pulses and nuts	Proteins	Enable growth and repair of body cells
Orange and gooseberry	Vitamin C	Keep the body healthy by providing strength and resistance
Sugar and chapati	Carbohydrates	Provide energy
Oils	Fats	Provide energy and nourish the body
Vegetables	Vitamins and minerals	Keep the body healthy and disease-free

2. **Personal Hygiene:** Personal hygiene is about caring for one's body. It plays an important role in maintaining the mental and physical wellbeing of a person. In a poor hygienic condition, the body provides an ideal environment for germs to grow, developing microbial infections. Here are a few ways in which we can maintain our hygiene:
   - Have a bath daily as sweat and oil glands secrete large amounts of sweat and oil into the skin which can lead to bad odour and skin infections.
   - Maintain the cleanliness of the body by bathing, washing hands and face and wearing neat and clean clothes to reduce this risk of a microbial infection.
   - Keep the surroundings clean and maintain hygiene standards to live a healthy and disease-free life.
3. **Physical Exercise:** One way to keep the body healthy and disease-free is by exercising regularly. Exercise keeps the body fit and helps relieve mental stress. Adolescents should be involved in sports, exercises, walking, cycling and other physical activities. Being physically fit does not mean a rigorous schedule of workouts. Various benefits are involved in living a physically fit and active lifestyle:
   - **Prevents chronic diseases:** Being physically fit helps lower blood sugar levels and controls the blood pressure. It also reduces the risk of stroke or heart disease.
   - **Controls weight:** The present generation is highly vulnerable to diseases owing to extra body weight and increased cholesterol levels. Being fit allows you to get rid of fat, which in turn supports a healthy lifestyle.
   - **Ensures strong bone, muscle and joint development:** Adolescence is the time to invest in your body. Physical exercise ensures that you have strong bones, muscles and joints.
   - **Reduces stress:** Stress is one of the most important health hazards among youth. Being unfit results in loss of confidence and is one of the factors causing stress. Being fit makes a person physically smart and helps in better inter-personal relationships.
   - **Increases energy levels and confidence:** Laziness is associated with an unfit body. Being fit makes one active and energetic, thereby making a person more competent and brisk. This ultimately reflects in higher confidence levels.
4. **Say 'No' to Drugs:** Adolescence incorporates different changes in the mind and body of a person. Drugs should be taken only if prescribed by a doctor. This is because drugs are highly addictive and can be habit-forming. They can have deleterious effects on the health of a person. The mental and physical health of a person is affected by drugs. Drugs can also cause long-term side effects on the organs and systems in your body, such as throat, stomach, lungs, liver, pancreas, heart, brain and nervous system. Taking drugs through syringes can cause AIDS (Acquired Immunodeficiency Syndrome). This disease is caused by HIV (Human Immunodeficiency Virus) that can be transmitted from an infected person to a healthy person through sharing of syringes. It can also be transmitted through sexual contact or an infected mother's milk.

SUMMARY

- Human beings become capable of reproduction after attaining puberty.
- The onset of puberty results in the development of reproductive organs.
- There is an increase in the height of males and females.
- Development of breasts in girls and facial hair (moustache and beard) in boys is a significant feature of puberty. The voice of boys also becomes hoarse as the voice box enlarges.
- The onset of puberty and maturity of reproductive parts are controlled by special chemical substances known as hormones.
- Hormones are secreted from the endocrine glands and they flow directly into the bloodstream.
- The pituitary gland is also known as the master gland as it secretes hormones which include growth hormones and those that control other glands such as the testes, ovaries, thyroids and adrenals.
- The pancreas secretes insulin which controls blood sugar, while the thyroid produces thyroxine and the adrenals produce adrenaline.
- Testosterone is the male sex hormone and oestrogen is the female sex hormone; they help in controlling secondary sexual characters in males and females, respectively.
- The uterine wall in females prepares to receive the developing fertilised egg. In case there is no fertilisation of gametes, the thickened lining of the uterine wall breaks down and flows out of the body along with blood. This is called menstruation.
- The sex of the unborn child is determined by the chromosomes inherited from the father.
- It is important to eat a balanced diet, maintain personal hygiene and exercise regularly during adolescence.

PRACTICE QUESTIONS

Analyse Your Concepts (School Exam Based)

Multiple Choice Questions

1. Change in voice is marked by:
(a) Larynx (b) Pharynx
(c) Oesophagus (d) Epiglottis

2. Girls have a:
(a) Low-pitched voice
(b) High-pitched voice
(c) Low-pitched and baritone voice
(d) High-pitched and baritone voice

3. Among the following, which is the exocrine gland?
(a) Pituitary gland (b) Hypothalamus
(c) Thyroid (d) Sebaceous gland

4. Which gland produces and releases serotonin?
(a) Pituitary (b) Pineal
(c) Pancreas (d) Adrenal

5. The master control gland in the human body is:
(a) Pituitary (b) Pineal
(c) Pancreas (d) Adrenal

6. If the thyroid gland makes less hormone than normal, it is known as:
(a) Hypothyroidism
(b) Hyperthyroidism
(c) Both (a) and (b)
(d) None of the above

7. Prolactin levels rise in females during pregnancy as it inhibits:
(a) Follicle stimulating hormone (FSH)
(b) Gonadotropin-releasing hormone (GnRH)
(c) Both (a) and (b)
(d) None of the above

8. Before every ovulation, the inner lining of the uterus becomes:
(a) Thin and soft with a lot of blood vessels
(b) Thick and soft with a lot of blood vessels
(c) Thick and hard with a lot of blood vessels
(d) Thick and soft with fewer blood vessels

9. Every human being has 23 pairs of chromosomes. Males have ____sex chromosomes and females have ____ sex chromosomes.
(a) XY, XX (b) XX, XY
(c) XX, XX (d) XY, XY

10. The juvenile hormone (JH) is secreted from the ________ during insect development.
(a) Prothoracic gland (b) Corpora allatum
(c) Both (a) and (b) (d) None of the above

Match the Following

Each question contains statements given in two columns which have to be matched. Statements in column I have to be matched with statements in column II.

1. Choose the correct option:

Column I	Column II
i. Pineal gland	a. Stress hormone
ii. Adrenal gland	b. Thalamus
iii. Pituitary gland	c. Effect on heart rate
iv. Thyroid gland	d. Master control gland

2. Choose the correct option:

Column I	Column II
i. X chromosome	a. Endocrine gland
ii. Sebaceous gland	b. Developed larynx
iii. Pancreas	c. Sex determinant
iv. Adam's apple	d. Exocrine gland

3. Choose the best pair:

Column I	Column II
i. Adolescence	a. Sperm
ii. Menstrual cycle	b. Frog
iii. Metamorphosis	c. Vaginal bleeding
iv. Testes	d. 13–19 years

True or False

Read the following statements and write your answer as true or false.

1. The period from 19 years of age to 29 years of age is called teenage.
2. Adolescents attains sexual maturity and become capable of reproduction.
3. Under the influence of hormones produced in the body, the reproductive organs become functional at the time of puberty.
4. At puberty, the voice box or pharynx begins to grow in boys.
5. Girls have a high-pitched voice.
6. Oil and sweat secretions are released through ducts by the endocrine glands.
7. In males, the testes and penis become completely developed in adolescence.
8. In females, the ovaries develop completely in infancy and release an egg or ova.
9. One of the secondary sexual characters which occurs in girls during puberty is the beginning of the menstrual cycle.
10. Hormones are secreted by the endocrine glands.
11. In males, the testes secrete the male hormone which is known as progesterone.
12. The pituitary gland is also known as the thalamus.
13. The pancreas are involved in the production of insulin.
14. The part of the body for which a particular hormone is intended is called the target site.
15. The release of an egg from an ovary is called ovulation.

Fill in the Blanks

Complete the following statements with an appropriate word/term to be filled in the blank spaces.

1. The process of transformation that takes place during adolescence is known as ________.
2. Boys develop larger voice boxes and it is seen as a protruding part of the throat called ________.
3. The ________ secrete oil.
4. The features that help distinguish between male and female are called ________ sexual characters.
5. The breakdown of the thick and soft lining of the uterus along with its blood vessels results in vaginal bleeding, which is called ________.
6. The child will inherit a/an ________ chromosome from the mother, regardless of gender.
7. The changes that occur from larva to adult in insects is called ________.
8. Taking drugs through syringes can cause ________.

9. The ________ gland produces hormones that control the sex drive and the stress hormone.

10. The endocrine glands are also known as ________ because they secrete hormones directly into the bloodstream.

Assertion-Reason Questions

The following questions consist of two statements: Assertion (A) and Reason (R). Answer these questions selecting the appropriate option given below:

(a) Both A and R are true and R is the correct explanation of A
(b) Both A and R are true but R is not the correct explanation of A
(c) A is true but R is false
(d) A is false but R is true

1. Assertion (A): The XY sex-determination system is found in humans.
Reason (R): Males have one X and one Y chromosome that determines the sex of the child.

2. Assertion (A): We should have a bath daily.
Reason (R): Sweat and oil glands secrete large amounts of sweat and oil, leading to skin infections.

3. Assertion (A): When a female reaches puberty, maturation of eggs takes place.
Reason (R): The release of an egg from an ovary is called ovulation.

4. Assertion (A): The pituitary gland produces prolactin after childbirth.
Reason (R): It enables females to breastfeed the young one.

5. Assertion (A): When there is a need for glucose, insulin is secreted.
Reason (R): Insulin regulates the blood sugar level in our body.

One-word Answers

1. Which hormone is secreted to break down glycogen into glucose?
2. Excess of which hormone causes confusion, agitation and sedation?
3. Which hormone is known as the emergency hormone?
4. What will be the sex of a child if the sperm carries an X chromosome?
5. On which day does ovulation start in a menstrual cycle?
6. Which hormones play a crucial role in human development?
7. Hyperthyroidism is caused by excess secretion of which hormone?
8. The testes secrete which hormone that is responsible for sperm production?
9. Which gland is responsible for acne and pimples in many young girls and boys?

Define the Term

1. Secondary sexual characters
2. Properties of hormones
3. Melatonin
4. Insulin

Groups of Terms

Given below are groups of terms. In each group the first pair indicates the relationship between the two terms. Rewrite and complete the second pair on a similar basis.

1. Pituitary: endocrine :: Sebaceous: _____
2. Pancreas: insulin :: Testes: ______
3. Serotonin: agitation :: Thyroid (T3, T4): _____
4. Female: XX :: Male: ______
5. Frog: thyroxine :: Insect: ______

Identify the Odd One Out

1. Pituitary, pineal, thyroid, sebaceous
2. XX, XY, XA, XO
3. Insulin, sweat, testosterone, oestrogen
4. Humans, hen, frog, mice
5. Moustache, armpit hair, breast development, voice change, grey hair

Image-Based Questions

1. What is the Y-axis of the graph?

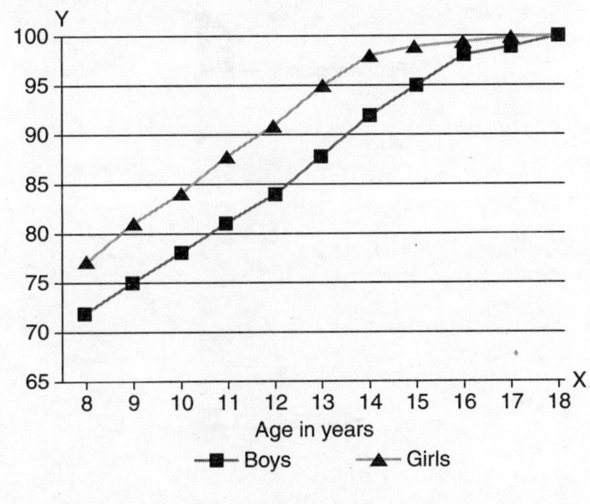

(a) Weight (b) Percentage of weight
(c) Full height (d) Percentage of full height

2. What does a and b stand for?

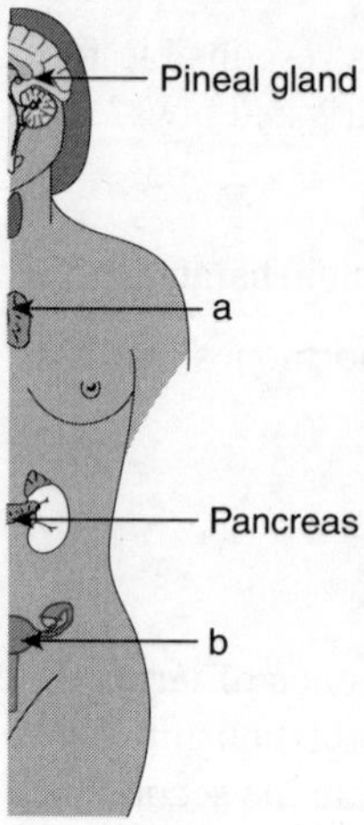

(a) a - pituitary, b - thymus
(b) a - thymus, b - ovary
(c) a - pineal, b - testes
(d) a - pineal, b - thymus

3. What is the name of this organ found in males?

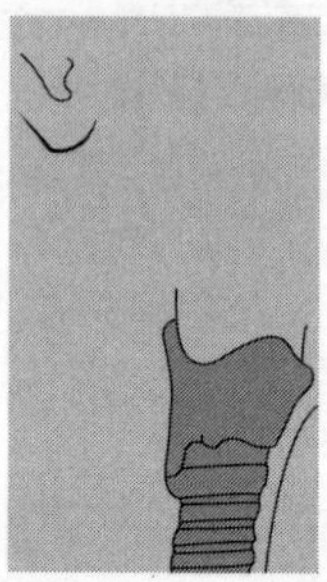

(a) Axis (b) Pharynx
(c) Adam's apple (d) Bronchioles

4. What is the name of glands a and b?

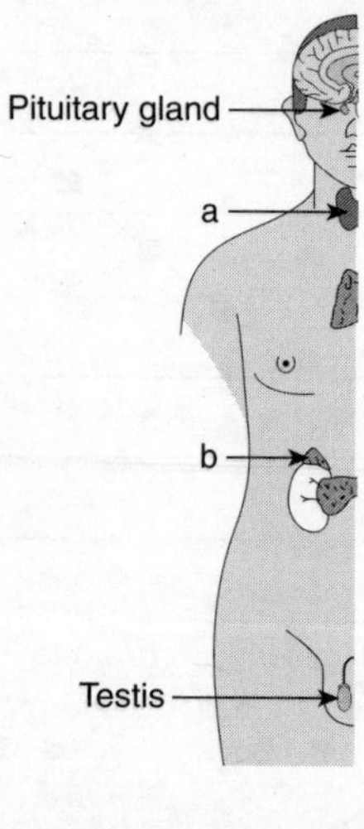

(a) a - adrenal, b - thyroid
(b) a - thymus, b - ovary
(c) a - thyroid, b - adrenal
(d) a - pineal, b - adrenal

5. What is another name for this endocrine gland?

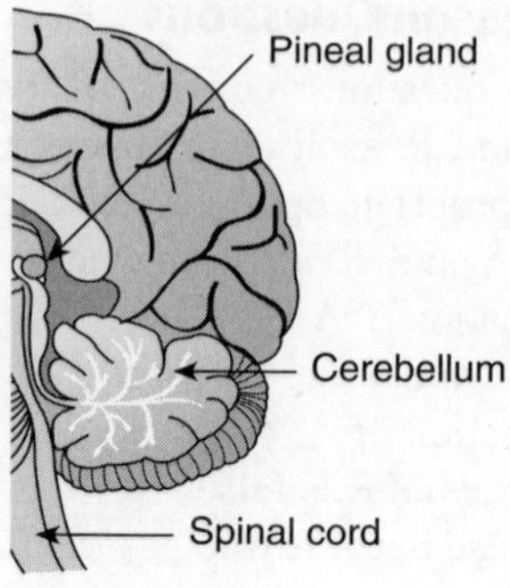

(a) Insulin producing (b) Third eye
(c) Master gland (d) Second eye

6. What is the term for this?

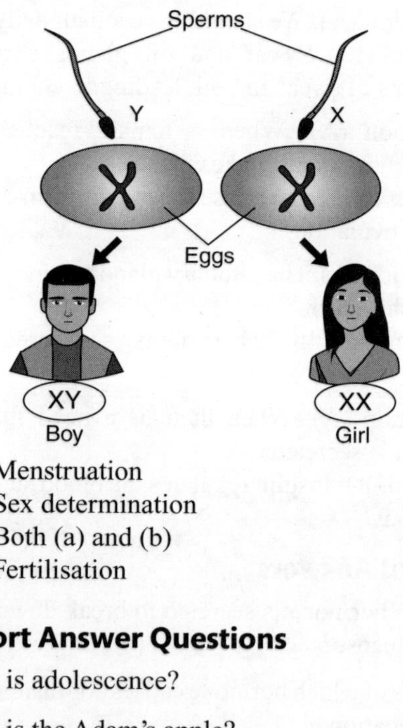

(a) Menstruation
(b) Sex determination
(c) Both (a) and (b)
(d) Fertilisation

Very Short Answer Questions

1. What is adolescence?

2. What is the Adam's apple?

3. What are the types of glands present in the human body?

4. What are the types of sex-determinant chromosomes?

5. Why does vaginal bleeding occur?

6. Which organ releases insulin?

7. Name the developed larynx in boys in adolescence.

8. Name the organism that undergoes metamorphosis.
9. Name different changes that occur in males during puberty.

Short Answer Questions

1. Write the features of hormones.
2. Write a short note on the adrenal glands.
3. What are the functions of the sweat and sebaceous glands?
4. What are the secondary sexual characters in girls?
5. Explain the significance of the pineal and pituitary glands.
6. Explain menstruation.
7. Name the two types of exocrine glands.
8. Discuss the secondary sexual characters in boys.
9. Write about the X and Y chromosomes.
10. Describe one factor leading to the well-being of adolescents.

Long Answer Questions

1. Describe the steps involved in reproductive function by hormones.
2. Describe the menstrual cycle in females.
3. What is adolescence?
4. Explain sex determination.
5. Explain the factors that lead to the well-being of adolescents.
6. Explain the endocrine glands and hormones.
7. What are the changes that humans undergo during adolescence?
8. Explain secondary sexual characters in males and females.
9. Write a short note on puberty or adolescence.

Case Study-Based Questions

1. Anita and Rakesh are expecting a child. During family discussions, the sex of the unborn child is discussed and some members claim that the mother and her eating habits are responsible for determining the sex of the child. Shyam listens to the discussion and feels the urge to argue.
   (a) What are sex chromosomes?
   (b) Which sex chromosomes are possessed by males and females?
   (c) Can you help Shyam make them understand how the sex of the child is determined?
2. Reena and Shyam went to a park to play hide and seek. While playing, Shyam encountered a snake and began to panic. Reena after noticing this went to Shyam, but Shyam started to run at twice his usual speed. After reaching home, Reena starts thinking about how this frightening situation made Shyam react differently from his usual self. Can you help Reena understand some things?
   (a) Which gland is involved in producing hormones in flight-or-fight conditions?
   (b) Which hormones are secreted during flight-or-fight conditions?
   (c) How does glucagon help in flight-or-fight conditions?
3. Bobby and Monika are slowly entering the adolescent phase of their lives along with their friends and classmates. They recently started to notice that the voice of their classmates had changed. Can you help them know how and why the voices have changed?
   (a) What are secondary sexual characters?
   (b) What is the larynx?
   (c) How do voices change during adolescence?
4. Ram has recently entered the adolescent phase and has noticed some changes in his body. At first he was taken aback, but slowly he realised they were a part of his growth and development. Can you help Ram understand more about the process?
   (a) What are the secondary sexual characters in males?
   (b) Why is maintaining hygiene important after developing secondary sexual characters?
   (c) What are the hormones responsible for development?

NCERT Corner

1. What is the term used for secretions of endocrine glands responsible for changes taking place in the body?
2. Define adolescence.
3. What is menstruation? Explain.
4. List the changes that take place in the body at puberty.
5. Prepare a table having two columns depicting the names of endocrine glands and the hormones secreted by them.
6. What are sex hormones? Why are they so named? State their function.
7. Choose the correct option:
   (a) Adolescents should be careful about what they eat, because:
      (i) Proper diet develops their brains.
      (ii) Proper diet is needed for the rapid growth taking place in their body.
      (iii) Adolescents feel hungry all the time.
      (iv) Taste buds are well developed in teenagers.
   (b) Reproductive age in women starts when their:
      (i) Menstruation starts.
      (ii) Breasts start developing.
      (iii) Body weight increases.
      (iv) Height increases.
   (c) A balanced meal for adolescents consists of:
      (i) Chips, noodles, coke.
      (ii) Chapati, dal, vegetables.
      (iii) Rice, noodles and burger.
      (iv) Vegetable cutlets, chips and lemon drink.
8. Write notes on:
   (a) Adam's apple
   (b) Secondary sexual characters
   (c) Sex determination in the unborn baby
9. Word game: Use the clues to work out the words.

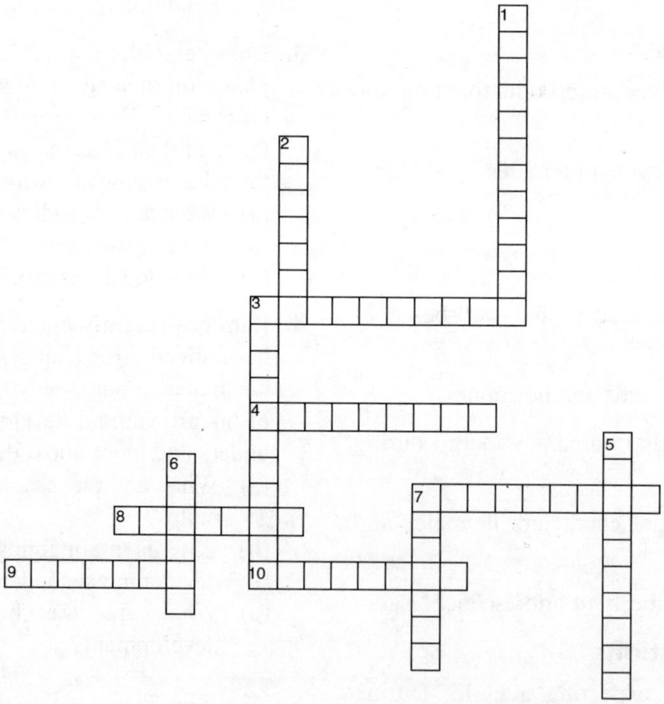

Across:
3. Protruding voice box in boys
4. Glands without ducts
7. Endocrine gland attached to brain
8. Secretion of endocrine glands
9. Pancreatic hormone
10. Female hormone

Down:
1. Male hormone
2. Secretes thyroxine
3. Another term for teenage
5. Hormone reaches here through the bloodstream
6. Voice box
7. Term for changes at adolescence

10. The table below shows the data of likely heights of boys and girls as they grow in age. Draw graphs showing height and age for both boys and girls on the same graph paper. What conclusions can be drawn from these graphs?

Age (Years)	Height (cm)	
	Boys	Girls
0	53	53
4	96	92
8	114	110
12	129	133
16	150	150
20	173	165

COMPETITION WINDOW (OBJECTIVE TYPE)

1. Hormones of the menstrual cycle are produced by:
(a) Ovaries (b) Uterus
(c) Fallopian tube (d) Both (a) and (b)

2. In females, meiosis-II is not completed until:
(a) Puberty (b) Implantation
(c) Birth (d) Fertilisation

3. The mother is completely responsible for the sex of the child. This is incorrect because:
(a) Child develops in the body of the mother
(b) Child gets one sex chromosome from the mother and another from the father
(c) Child gets sex chromosome only from mother
(d) Child gets sex chromosome only from father

4. The speciality of the endocrine gland is:
(a) Ductless gland
(b) Seen anywhere in the body
(c) Large in size
(d) No secretion of hormone

5. The hormone that increase metabolism and glucose intake in the body is:
(a) Insulin
(b) Thyroxine
(c) Adrenaline
(d) Both (a) and (b)

6. The hormones responsible for fight-or-flight response are:
(a) Melatonin and thyroxine
(b) Glucagon and insulin
(c) Epinephrine and norepinephrine
(d) Oestrogen and progestrone

7. At which stage of life is oogenesis initiated?
(a) Birth
(b) Adult
(c) Puberty
(d) Embryonic development

Answer Keys

Multiple Choice Questions

1. (a) 2. (b) 3. (d) 4. (b) 5. (a)
6. (a) 7. (c) 8. (b) 9. (a) 10. (b)

Match the Following

1. i (b), ii (a), iii (d), iv (c)
2. i (c), ii (d), iii (a), iv (b)
3. i (d), ii (c), iii (b), iv (a)

True or False

1. F 2. T 3. T 4. F 5. T
6. F 7. T 8. F 9. T 10. T
11. F 12. F 13. T 14. T 15. T

Fill in the Blanks

1. Puberty
2. Adam's apple
3. Sebaceous glands
4. Secondary
5. Menstruation
6. X
7. Metamorphosis
8. Acquired Immunodeficiency Disease
9. Adrenal
10. Ductless glands

Assertion–Reason Questions

1. (a) 2. (a) 3. (b) 4. (a) 5. (d)

One-word Answers

1. Glucagon
2. Serotonin
3. Adrenaline
4. Female (XX)
5. 14th day
6. Somatotropin
7. Thyroid hormone
8. Testosterone
9. Sebaceous gland

Define the Term

1. Secondary sexual characters are those that develop in males and females at the time of attaining sexual maturity. These characters help to distinguish a male from a female. In girls, breasts begin to develop and boys grow facial hair and hair on the chest. In both boys and girls, hair grows under the arms and in the pubic region.
2. Hormones have certain properties or characteristics such as:
   - They are secreted by the endocrine glands
   - They are responsible for controlling the behaviour of the target cells.
   - They cause long-term effects in the body like changes in behavioir, changes in growth, development of organs, and puberty
3. The circadian rhythm or sleep/wake cycle of our body is controlled by melatonin. Sunlight prevents melatonin production; it is secreted by the pineal gland. More melatonin is secreted as it gets dark at night and it increases the urge to sleep. Phones, televisions and computers reduce the amount of melatonin produced; it is advisable to avoid using devices before bedtime or to use blue light-blocking glasses at night.
4. Insulin is necessary for cells to utilise glucose in the bloodstream. Insulin regulates the blood sugar level in our body and reduces it by converting it into glycogen. Thus, in diabetes, the body is unable to use insulin or there is very less or no insulin secreted by our body.

Groups of Terms

1. Exocrine
2. Testosterone
3. Hyperthyroidism
4. XY
5. Juvenile hormone/ecdysone

Identify the Odd One Out

1. Sebaceous
2. XO
3. Sweat
4. Frog
5. Grey hair

Image-Based Questions

1. (d) 2. (b) 3. (c) 4. (c) 5. (b)
6. (b)

Very Short Answer Questions

1. The age group between 10 and 19 years is defined as adolescence.
2. At puberty, the voice box or larynx begins to grow and as boys develop larger voice boxes, it is seen as a protruding part of the throat called the Adam's apple.
3. Exocrine: sweat and sebaceous glands; endocrine: pituitary, pineal, testes, ovary, thyroid, etc.
4. X, Y
5. If there is no fertilisation of gametes, the thickened lining of the uterine wall breaks down and leaves the body along with blood. This is called menstruation.
6. Pancreas
7. Adam's apple
8. Frog
9. Change in physical appearance, secondary sexual characters

Short Answer Questions

1. Hormones have certain properties or characteristics such as:
   - They are secreted by the endocrine glands.
   - They are responsible for controlling the behaviour of the target cells.
   - They cause long-term effects such as changes in behaviour, growth, development of organs and puberty.
2. This gland produces the hormones that control the sex drive and stress hormone. The gland is located above the kidneys and secretes adrenaline which helps in managing stress and salt balance in the blood.
3. The sweat glands secrete sweat. The sebaceous glands secrete oil and many young girls and boys develop acne and pimples due to this.
4. The secondary sexual characters that occurs in girls during puberty are:
   - Breasts develop and enlarge
   - Ovaries start to release eggs; ovaries, oviducts, uterus and vagina enlarge
   - Menstruation begins
   - Hips broaden; extra fat is deposited on the hips and thighs
   - Hair grows in the armpits and in the pubic regions between the thighs
5. The pineal gland produces serotonin derivatives of melatonin, which affect sleep patterns in the human body. The pituitary helps in controlling other glands and develops hormones that trigger growth and development.
6. If there is no gamete fertilisation, the uterine wall's thickened lining breaks down and leaves the body along with blood. This is called menstruation.
7. Sweat and oil glands
8. Hair grows on the face in the form of a moustache and beard, and on other parts of the body like the chest and armpits. The testes start to make sperm. The testes and penis become larger. The chest and shoulders broaden and the body becomes muscular. Hair grows in armpits and in the pubic regions. Sexual drive associated with adulthood begins to develop.
9. Of the 23 pairs of chromosomes, 22 are similar in all respects and are known as autosomes. Only one pair is different and known as the sex chromosome, which actively takes part in the process of sex determination. Both males and females carry two sets of sex chromosomes. A male has one X and one Y (XY) sex chromosome. A female has two X (XX) sex chromosomes.
10. Personal hygiene is how one cares for one's body. It plays an important role in maintaining the mental and physical well-being of a person. In poor hygiene conditions, the body is an ideal environment for germs and microbial infections.

Long Answer Questions

1. Hint: All the endocrine glands secrete hormones, but the reproductive function is only served by the testes and ovaries. Mention every hormone secreted by them with their functions.
2. Hint: Explain these steps in detail: maturation of eggs, release of mature eggs, thickening of the wall of the uterus, break-down of thickened wall if pregnancy does not occur.
3. Hint: The age group between 10 and 19 years is defined as adolescence. Mention the secondary sexual characters associated with it.
4. Hint: Only one pair is different, which is known as the sex chromosome; it actively takes part in the process of sex determination. Explain the role of the X and Y chromosomes in sex determination and conditions of the child to be either male or female.
5. Hint: Explain in detail about nutritional health, personal hygiene, physical exercise and staying away from drugs.
6. Hint: Explain the hormones produced along with the functions of endocrine glands, pituitary, pineal, thyroid, pancreas, adrenal, testes and ovary.

7. Hint: Explain the secondary sexual characters observed in males and females, along with changes in exocrine gland secretion, endocrine gland secretion, physiological changes and menstruation.
8. Hint: Explain the secondary sexual characters observed in males and females.
9. Hint: This will include physiological changes in females and males, development of secondary sexual characters, secretion changes of exocrine and endocrine glands, menstruation and factors that are important for a healthy adolescence.

Case Study-Based Questions

1. (a) Of the 23 chromosomes, only one pair is different; it is known as the 'sex chromosome', and it plays an active role in sex determination.
   (b) Male XY, Female XX
   (c) If a sperm carrying an X chromosome fertilises an ovum (or egg) which carries an X chromosome, the child will be female. This is because the child will have an XX combination of sex chromosomes. If a sperm carrying a Y chromosome fertilises an ovum (or egg) which carries an X chromosome, the child will be male. This is because the child will have an XY combination of sex chromosomes.
2. (a) Adrenal and pancreas
   (b) Adrenaline, epinephrine, nonepinephrine, glucagon
   (c) Glycogen is broken down into glucose to produce energy during the condition.
3. (a) Secondary sexual characters are those that develop in males and females at the time of attaining sexual maturity. These characters help to distinguish a male from a female.
   (b) Change in voice is caused by the voice box or larynx.
   (c) At puberty, the voice box begins to grow and because boys develop larger voice boxes, the muscles get enlarged due to which their voice becomes hoarse. In girls, the larynx is small and is hardly visible from the outside. Girls have a higher pitched voice than boys.
4. (a) Secondary sexual characters are those that develop in males and females at the time of attaining sexual maturity. These characters help to distinguish a male from a female. Boys grow facial hair and hair on the chest.
   (b) It is important to have a daily bath as sweat and oil glands secrete large amounts of sweat and oil into the skin which can lead to bad odour and skin infections.
   (c) T3, T4, melatonin, cortisol, insulin, glucagon, prolactin, serotonin, adrenaline, somatotropin

NCERT Corner

1. Hormones
2. The period of life when the body undergoes changes, leading to reproductive maturity, is called adolescence. The period of adolescence is normally 11 years to 19 years.
3. In females, the ova or eggs begin to mature with the onset of puberty (10 to 12 years of age). One egg becomes mature and is released by one of the ovaries once in about 28 to 30 days. During this period, the wall of the uterus becomes thick so as to receive the fertilised egg. In case it is fertilised, it begins to develop. This results in pregnancy. If the fertilisation fails to take place, the ovum or the released egg and the thickened lining of the uterus along with its blood vessels are shed. This causes bleeding in women. This is called menstruation. It occurs once in 28 to 30 days.
4. The changes take place differently in the body of boys and girls at puberty. These change may be listed as below:
   - There is sudden increase in the height of both boys and girls. However, both reach their maximum height at the age of 17 to 18 years.
   - The voice of boys become hoarse and that of girls become shrill.
   - Shoulders become broader and muscles grow more prominently in boys. In girls, the regions below the waist becomes wider.
   - Hair starts growing in different parts of boys and girls.
   - Most teenagers, due to increased secretion of sweat and sebaceous glands, get acne and pimples on their face.
   - Onset of puberty brings about changes in secondary sexual characters in both male and female.
   - Boys and girls become capable of reproduction.
   - In girls, menstruation starts.
   - Different types of hormones start to release in initiating the reproductive function.

5.

Endocrine glands	Hormones
1. Pituitary gland	(i) Growth hormones
2. Ovaries	(ii) Oestrogen
3. Testes	(iii) Testosterone
4. Thyroid	(iv) Thyroxine
5. Pancreas	(v) Insulin
6. Adrenal glands	(vi) Adrenaline

6. The hormones which help and control the formation of secondary sexual characters are called sex hormones. They are named so because they control the sexual activities and are secreted by males and females separately. The functions of sex hormones are given below separately:

Male sex hormone: It is also called testosterone. It is secreted by testes and causes changes in secondary sexual character of boys like growth of facial hair like moustache, beard, etc. It also stimulates spermatogenesis.

Female sex hormone: It is also called oestrogen. It is secreted by ovaries and controls the secondary sexual characters in females, appearance of mammary glands, etc. It also maintains pregnancy.

7. (a) (ii)
(b) (i)
(c) (ii)

8. (a) At puberty, the voice box or larynx begins to grow and because boys develop larger voice boxes, it is seen protruding in the throat, called Adam's apple.

(b) Secondary sexual characters are the characters that develop in males and females at the time of attaining sexual maturity. These characters help to distinguish a male from a female.

(c) In the unborn baby, only one pair of chromosomes is different and known as the 'sex chromosomes'. These play a major role in the process of sex determination.

9.

10. From the above graph, we conclude that initially boys grow faster than girls from 0–8 years, and girls grow faster than boys from 8–16 years. But both reach their maximum height at the age of 20. Here, boys show a greater increase in height than girls.

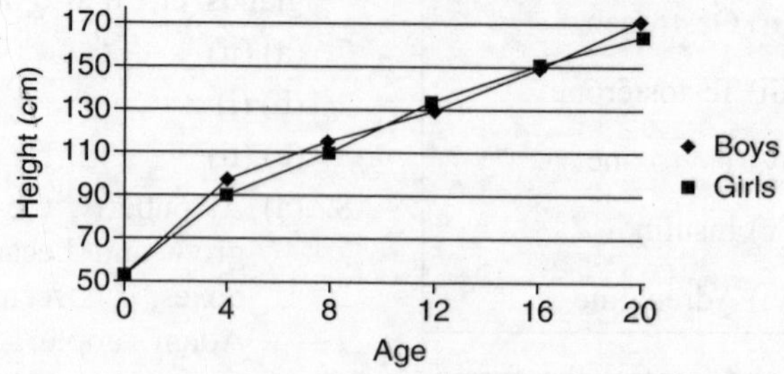

Competition Window

1. (d) 2. (d) 3. (b) 4. (a) 5. (a) 6. (c) 7. (d)

CHAPTER

3 Reproduction in Animals

PREREQUISITES

You must have knowledge of:

- Basic human anatomy
- Cells
- Glands
- Tissues
- Organ-level organisation

LEARNING OBJECTIVES

At the end of this chapter, you will be able to:

- Explain the modes of reproduction
- List the types of reproduction
- Describe sexual reproduction
- Describe the male reproductive organs
- Describe the female reproductive organs
- Explain fertilisation
- Explain the process of embryo development
- Differentiate between viviparous and oviparous animals
- Describe asexual reproduction

INTRODUCTION

Reproduction can be defined as the formation of new individuals by sexual or asexual means. Different organisms reproduce in different ways. For example, yeast and bacteria reproduce by budding or fission. Some plants have the ability to reproduce in a vegetative non-sexual way using various processes. In contrast, humans reproduce only sexually.

Fun Fact!

A natural defence system against infections in the female reproductive tract is formed by the microorganisms present in it!

MODES OF REPRODUCTION

Like in plants, on the basis of the number of parents involved, there are two modes of animal reproduction: asexual and sexual (Fig. 3.1). Asexual reproduction results in a large number of offspring that are genetically identical to the parent and which are produced at a very fast rate. On the other hand, in sexual reproduction, the genetic material of both the parents is combined to produce genetically diverse offspring that differ from their parents, which are of better fitness and which can thus maintain the two broad categories of male and female. Moreover, in sexual reproduction, only half of the population – the female half – can produce offspring. This results in fewer number of offspring than in asexual reproduction.

Let us now discuss sexual and asexual reproduction in detail.

Sexual Reproduction

In plants, sexual reproduction involves the participation of male and female reproductive parts. In animals too, male and female reproductive organs are present. They produce gametes or sex cells, known as the egg in females and sperm in males, which are microscopic in nature in humans. The gametes fuse to form a zygote that develops into a new individual. The fusion of male and female gametes to form an offspring is known as sexual

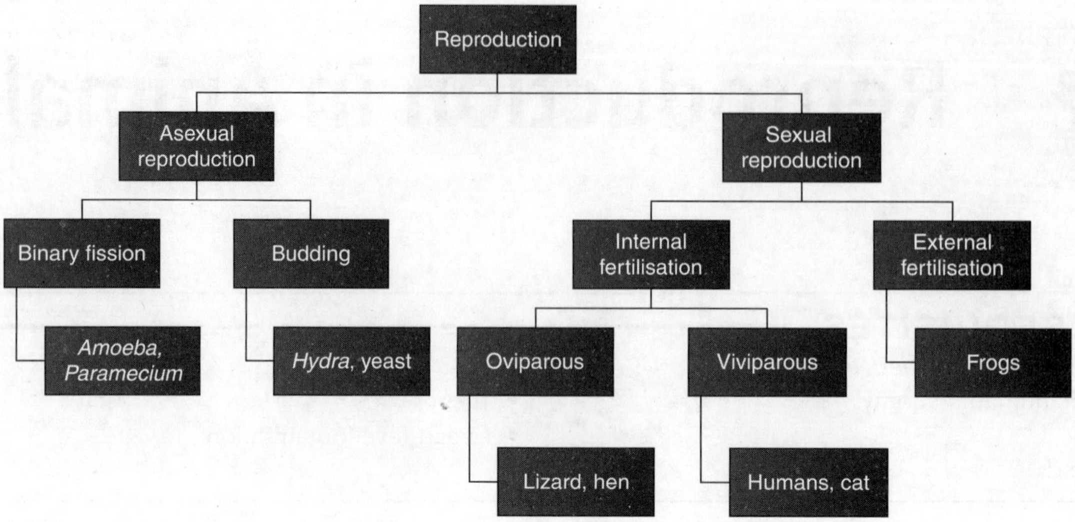

Fig. 3.1 Modes of reproduction

SCIENCE IN DAILY LIFE

Go to a nearby farm and count the young ones of the animals living there along with the animals seen on the road. Make a list of them with pictures in the following table:

Animal	Number of young ones	Picture
Cat		
Dog		
Butterfly		
Hen		
Cow		
Frog		
Pig		
Human		
Goat		

reproduction. Gametes are produced differently in males and females.

Animals have different secondary sexual characters that serve the purpose of attracting or identifying partners. These could be pheromones (odour or body smell), skin colouration, different calling sounds, behavioural patterns or body contour and they are known as cues for mating.

In humans, these secondary sexual characters are as follows:

- Skeletal structure
- Voice difference: deeper voice in males and higher pitched voice in females
- Breast development in females
- Different patterns of body hair growth: beard and moustache in males
- Muscularity: strongly built muscles in males
- Different behaviour and psyche

Reproduction in Humans

The form of sexual reproduction that results in human fertilisation is called human reproduction. In this process, the female's egg is fertilised by the sperm of the male and produces a zygote. The zygote has 46 chromosomes as the gametes have 23 chromosomes each, thereby restoring the genetic combination of humans. The organs that are involved in the reproductive system are divided into two groups:

- **Primary reproductive parts:** These include the main gonads. The testes in males and ovaries in females are the gonads of humans. Gonads produce gametes or sex cells, that is, eggs and sperms.
- **Accessory reproductive parts:** These comprise the structures involved in the transfer and meeting of gametes that leads to fertilisation and the development of the egg up to the birth of the baby. They do not participate directly in the process of sexual reproduction.

Male Reproductive System: The main components of the male reproductive system are (Fig. 3.2):

- A pair of testes that produce sperm
- Sperm duct (vas deferens) that discharges sperm in the form of semen
- Accessory glands (prostate, bulbo-urethral and seminal vesicle) that contribute to seminal fluid
- A penis that transfers the sperms into the female
- A urethra located in the penis that conveys the sperm from the sperm duct

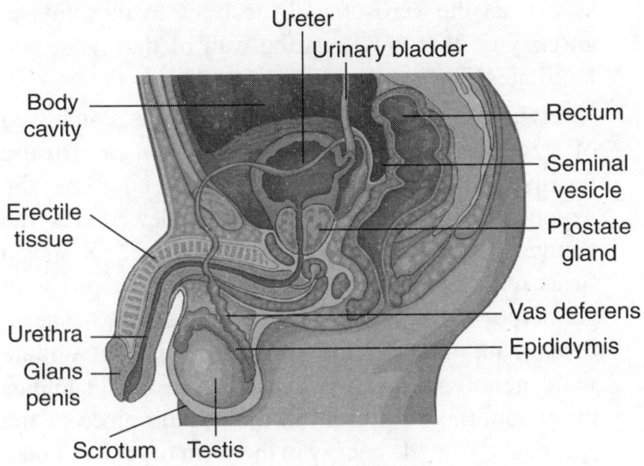

Fig. 3.2 Male reproductive system

The testes and penis are located outside the abdominal cavity. Let us take a brief look at the parts of the male reproductive system:

1. **Testes:** The testes are oval organs present as a pair and contained in the scrotum, which is a thin-walled sac of skin (also known as the scrotal sac). The testes descend into the scrotum only after birth as in the embryonic stage they are held within the abdomen. Sometimes, a condition may occur where they do not descend into the scrotum. This makes the individual sterile or incapable of producing sperm. The testes are located outside the abdominal cavity as the high temperature of the body does not allow for maturity of the sperm. The temperature required for survival of the sperm is 2–3°C less than normal body temperature (37°C). Therefore, locating the testes outside the body leads to an approximately two-fold reduction in the mutation of male germinal tissues caused by heat induction. If the testes are too close to the body, formation of the male gamete (sperm) becomes harder and this makes conception more difficult. The testes in the scrotum sac adapt according to changes in temperature. When the temperature outside is higher, the scrotum loosens and hangs away from the body. On the other hand, if it is cold, the skin of the scrotum contracts in a folded manner and the testes move closer to the body for warmth. Each testis is partitioned into 15–20 segments or lobules encased in a capsule. Each lobule consists of the following:
   - **Seminiferous tubules:** Sperms are produced and multiply (spermatogenesis) in the seminiferous tubules. The sperms that attain maturity are passed into a small network of tubes from which they travel to 12–14 ducts known as **efferent ducts**. These efferent ducts are joined by a small tubular knot known as the **epididymis**. This fits like a cap on the upper end of the testis. The **sperm duct (vas deferens)** is a distinct tube that arises where the epididymis ends, from the sides to the back of the testis. A single, 6-metre coiled tube is present inside the epididymis that continues into the sperm duct. Sperms are stored for some days in the epididymis for maturation and attaining motility.
   - **Interstitial cells:** Between the coils of the seminiferous tubules, packing tissues known as interstitial cells are present. These cells are also known as Leydig cells and they produce testosterone, which is responsible for the secondary sexual characters in males.
2. **Sperm Duct:** The vas deferens or sperm duct travels upwards into the abdomen and passes through an inguinal canal from each testis. The inguinal canal allows the testes along with their blood vessels, ducts and nerves to descend. The condition where the intestine bulges into the scrotum due to pressure in the abdomen through the inguinal canal is known as **hernia.** Over the ureters, two sperm ducts loop and come together to join the median duct or **urethra.**
3. **Accessory Glands:** In males, three accessory glands are present:
   - **Prostate gland:** Close to the origin of the bladder is a bilobed structure called the prostate gland that surrounds the urethra and which pours an alkaline secretion into the semen. This alkaline secretion neutralizes the acidity of the female vagina.
   - **Seminal vesicles:** Between the urinary bladder's posterior surface and the rectum lie a pair of lobulated glands that give rise to a duct that opens into the seminal vesicle just before it joins the urethra. Seminal vesicles produces the liquid used to transport sperms. A milky fluid known as semen, which is a mixture of different components and sperms, helps sperms to become motile.

- **Bulbo-urethral glands:** These are two ovoid glands that open into the urethra just before it enters the penis; they secrete lubricants.

4. **Penis:** This is a cylindrical organ positioned in front of the scrotum and which helps to pass urine and semen. This organ is highly vascular and consists of erectile tissues along with vascular spaces. During sexual stimulation, blood flows in large quantities to the penis, entering the sinuses or spaces, which results in rigidness and erectness of the penis. Expansion of the vascular spaces leads to compression of the veins and thus the blood drains out. This whole process is known as penile erection.
5. **Semen:** This is a mixture of sperm and secretions from various glands such as prostate, Cowper's, seminal vesicles and urethral. It is a milky fluid; a single ejaculate consists of an average amount of 2–3 ml containing around 2000,000,000–400,000,000 sperms.
6. **Sperm:** The male reproductive gamete or cell is known as the sperm (Fig. 3.3). The term comes from the Greek '*sperma*' meaning 'seed'. Motile sperm are divided into three main parts: head, middle body and tail. The tail of the sperm is known as flagellum and sperms that are mature and have a tail are known as spermatozoa. Animals produce spermatozoa while fungi and red algae have non-motile sperms known as spermatia. During spermatogenesis, the spermatogonium differentiates into spermatocytes. These spermatocytes undergo meiosis to form spermatids with 23 chromosomes, that is, half the number of chromosomes as compared to normal cells. Spermatids mature to develop a tail and become motile. The maturation of the sperm takes about 3 months from the initial phase. Sperms have a limited lifespan and cannot divide further as they are already haploid. A sperm fuses with the egg cell to form a diploid zygote. Sperms contain mitochondria which provide the energy for swimming and motility.

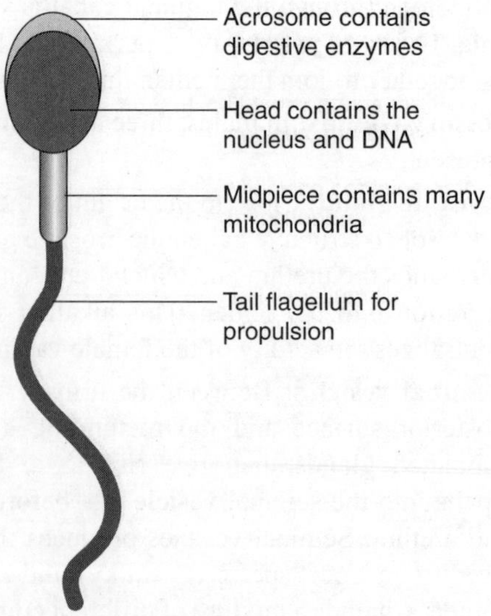

Fig. 3.3 Sperm anatomy

The main parts of the sperm involved in fertilisation are:

- **Acrosome:** The top of the head of the sperm is known as the acrosome. It secretes hyaluronidase, an enzyme that dissolves the wall of the ovum and facilitates entry of the sperm into it.
- **Nucleus:** The nucleus of the sperm is composed of 22+X or 22+Y sets of chromosomes. During fertilisation, when the egg and sperm combine, the genetic composition of the sperm determines the gender of the new individual. When a 22+X sperm fuses with the egg, the offspring will be a girl with genotype 44+XX. Similarly, when a 22+Y sperm fuses with an egg, the offspring will be a male with genotype 44+XY. Apart from the nucleus, mitochondria are present in the middle piece of the sperm and provide energy in the form of ATP to make the sperm motile.
- **Tail:** The lashing movement of the tail is used for propulsion. Sperms can stay alive in the female genital system for a few days but their power of fertilisation is limited to 2–3 days.

Fun Fact!

The secretions of seminal vesicles contain fructose, which is a type of sugar commonly found in fruits.

Female Reproductive System: The female reproductive system consists of the following organs or parts (Fig. 3.4):

Fun Fact!

Did you know that eggs are released every month by alternating the ovaries? For instance, in one month, if the left ovary releases an egg, then in the next month, the right ovary will release it.

1. **Vulva:** The external part of the female reproductive system is known as the vulva. The two fleshy and small folds present on the sides of the vulva are known as labia majora and labia minora, respectively.

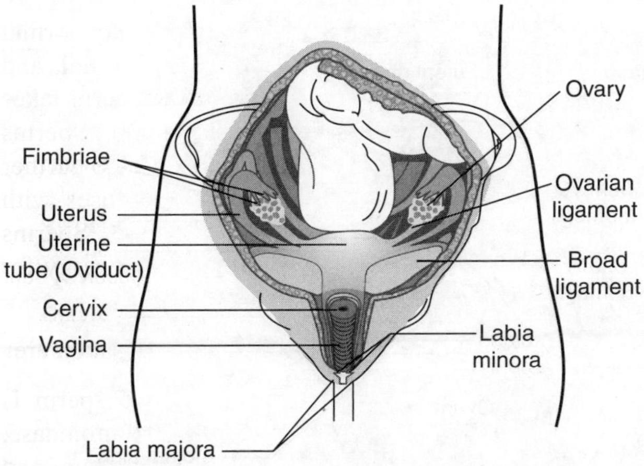

Fig. 3.4 Female reproductive anatomy

The labia minora or lesser lips are hidden by large folds of hair while the labia majora are known as greater lips. Both the structures are equivalent to the scrotum of the male reproductive system. The vulva contains openings into the urethra and vagina but both are independent of each other. The clitoris is a small erectile structure located in front of the urethral opening. It is highly sensitive and may be compared to the male penis.

2. **Vagina:** This is a 10–15-cm-long muscular tube that starts from the lower uterus and leads to the outside. It receives the male penis during copulation. It consists of elastic muscles as the elasticity of its wall allows for passage of the baby during parturition. A thin membrane known as the hymen is found at the opening of the vagina and partially covers it. It is known as the vaginal knot. The hymen is ruptured during copulation or by carrying out strenuous physical exercise such as horse riding and cycling or by a disease.
3. **Uterus:** This is a pear-shaped, hollow muscular organ with dimensions of 7 x 5 cm. It is located between the rectum and urinary bladder. The upper portion that receives the ends of the fallopian tubes is wider and the lower portion is constricted and known as the neck or cervix.
4. **Oviducts or Fallopian Tubes:** The fallopian tubes, oviducts or uterine tubes are present as a pair near the corresponding ovary. This 12-cm-long tube has a funnel-shaped opening, known as the oviducal funnel. The ovum is released into the oviduct by the ovaries; the inner walls of the oviducts are lined with cilia. A wave of muscular contractions or peristalsis of the oviduct wall is how the egg is pushed into the uterus.
5. **Ovaries:** Two small and ovoid bodies known as ovaries produce eggs. In the ovaries, the peripheral parts are responsible for egg production. A maturing egg in the cellular sac is known as a follicle. The follicle enlarges as the egg grows and is eventually filled with fluid; it is then called a Graafian follicle. When the egg is fully mature, it bulges on the surface of the ovary. The process from production of ova to attaining maturity is known as oogenesis. The process of rupture of the follicle that releases the egg is called ovulation. Fimbrae or cilia present in the oviduct pick up the released egg and the rest of the follicle, which after some time is converted into a yellow mass termed the corpus luteum, is responsible for the secretion of two hormones:
   - **Progesterone:** This hormone prepares the uterus to receive the embryo.
   - **Oestrogen:** This hormone is secreted for ovulation by the follicle and is also responsible for the secondary sexual characters.

Menstrual Cycle: From the age of 13 to 45–50 years, a continuous process associated with reproduction occurs in human females, called the menstrual cycle (Fig. 3.5). This cycle lasts 28 days or approximately a month with small variations, with a menstrual flow. The flow is stopped temporarily by pregnancy. The onset of menstruation is known as **menarche** while permanent cessation of the menstruation cycle is known as **menopause.**

The 28 days can be divided into four main phases:

- **Menstrual phase:** The first phase, which lasts 3–5 days, is when blood is discharged. The uterine lining of blood cushions, developed to protect and shield the developing zygote, is shed by rupture of the blood vessels. Onset of menstrual flow takes place and a new egg is formed in the follicle.
- **Follicular phase:** Certain hormones are produced as the follicle grows which leads to reactivation of the uterine wall to thicken and be supplied with the blood vessels. This change prepares the uterus to receive the fertilised egg and develop the future embryo.
- **Ovulatory phase:** Ovulation occurs around days 13–14 of the cycle and the egg is released and travels down the fallopian tube.
- **Luteal phase:** This phase takes place during days 15–28, when the uterus wall lining is thickened and the empty follicle, also known as corpus luteum, produces hormones for the next menstruation cycle.

At the end of the luteal phase, if the egg is fertilised then an embryo develops, but if the egg remains unfertilised, it

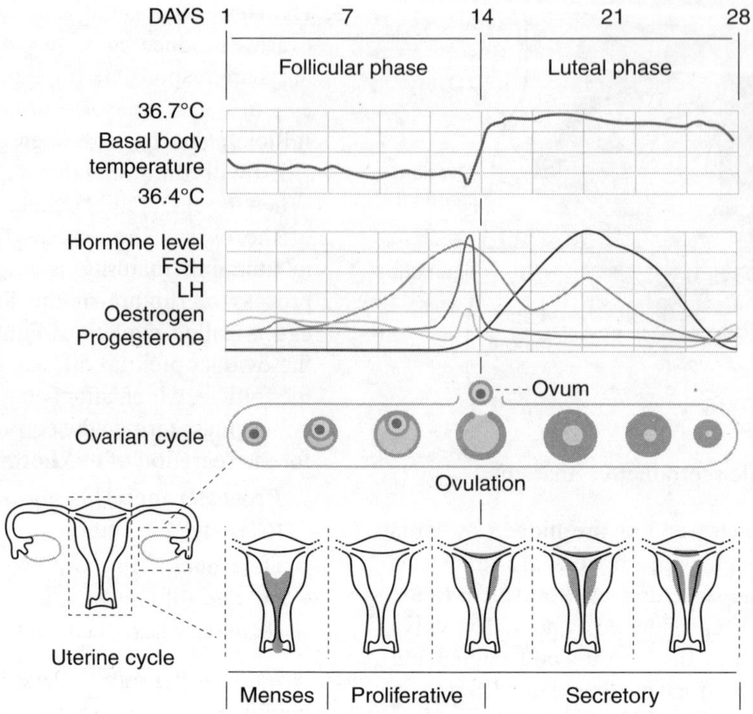

Fig. 3.5 The menstrual cycle

disintegrates. This is followed by shedding of the uterine wall lining on the 28th day. If the egg is fertilised, it is implanted into the uterus and there would be no menstrual discharge due to high levels of progesterone, the hormone released by the corpus luteum that prevents maturation of another follicle.

From Fertlisation to Birth: The process of reproduction starts from fusion of the female gamete, the ovum, and the male gamete, the sperm. During copulation, sperms are released into the vagina, near the cervix area of the uterus. Millions of sperms are released in a single ejaculation and, with the help of their tails, they start swimming into the uterus. The sperms cross the vaginal area and enter the fallopian tube area at a speed of 1.5 mm per minute. If the egg is released and present in the fallopian tube, fertilisation occurs. During fertilisation, the nuclei of the egg and sperm fuse to form a single nucleus. The chromosome number of the fused nucleus is restored to diploid (2n) with 46 chromosomes. The fertilised egg is known as the **zygote.** As the zygote is formed by fusion of gametes from both parents, the characters from both individuals are inherited. A few days after ovulation, that is, the 12th or 13th days of the cycle, are the optimum days for conception.

SCIENCE IN DAILY LIFE

Sit with your family and make a list of the characteristics that you and your siblings have inherited from your mother and father. Find out the relatable and common characteristics.

Fertilisation is of two types:

- **Internal:** Here, fertilisation takes place inside the body of the female. It occurs in many animals like humans, dog, hen and cow.
- **External:** In spring or in the rainy season, frogs and toads move to slow-flowing water bodies like ponds where the female lays hundreds of eggs. The eggs of frogs are devoid of shells and are very delicate. The only protection provided to them is the jelly coating that holds the eggs together. As the eggs are laid, the male frog deposits sperms over the eggs. Now, each sperm with a long tail swims in the water and comes in contact with these eggs. Fertilisation occurs outside the body of the female. This is very common in aquatic animals such as starfish, fish and frog.

Internal fertilisation is the preferred method for sexual reproduction. The animals that use external fertilisation lay hundreds of eggs and release millions of sperms, but not all the eggs are fertilised and grow into new individuals. This is because the sperms are exposed to the movement of water, rainfall and even wind or become the feed of the other aquatic animals. The probability of fertilisation of an egg or the success rate remains low in external fertilisation.

During internal fertilisation, only a few out of millions of sperms that are released in the vagina swim to the upper parts of the fallopian tube while the other sperms die or are absorbed. When the sperms surround the ovum and try to enter it, their tails get left behind and only the nucleus containing the head enters the egg. The nucleus of the sperm then fuses with the nucleus of the egg. As the sperm penetrates the egg, a chemical barrier is created around the egg that prevents the entry of other sperm.

Implantation and Pregnancy: After it is formed, the zygote starts dividing or cleaving into different-celled stages such as 2, 4, 8 and 16 cells (Fig. 3.6). The process of division occurs in the fallopian tubes and by the time it reaches the uterus, a small and hollow ball of various cells known as the **blastocyst** is formed. This stage is embedded in the uterus in a week to form a pitin endometrium wall.

Implantation is the process where the embryo attaches itself to the wall of the endometrium and the state of pregnancy is achieved. The cells of the embryo start to form groups that develop into tissues and organs, leading to organ systems. The embryo gradually develops body parts such as legs, head, eyes, hands and ears. After five weeks, the embryo is quite advanced as the heart and circulation are also developed. The **foetus** is the stage of the embryo where all the body parts can be identified.

Human Vs Hen: The fertilised egg in humans gives rise to the birth of a baby whereas in hens, fertilisation leads to laying of eggs. Though both animals use internal fertilisation, the development of young ones differs. When the zygote in hens divides repeatedly and travels down the fallopian tube, many layers of protection are formed around it. The hard layer in a hen's egg is a protective layer and only when this layer has been formed does the hen lay eggs. The egg takes about three weeks to develop into chicks and the hens sit on the eggs to give them sufficient warmth. After complete development of the chick, the shell cracks open and the chick comes out.

Placenta: At around eight weeks, the foetus needs to take in oxygen and food and excrete nitrogenous waste and carbon dioxide. The embryo needs to be continuously fed and excretion performed. The placenta serves this purpose. The blood vessels that connects the foetus with the placenta are rich in blood and this is known as the **umblical cord.** There are two sets of finger-like projections called **villi**: one set is seen in the uterine wall and the other extends from the embryo. The vessels of the villi are almost interlocked and the blood vessels of the embryo grow close to the vessels to be unlocked. The placenta transfers oxygen and nutrients as well as antibodies from the mother to the foetus and receives carbon dioxide and urea from the foetus.

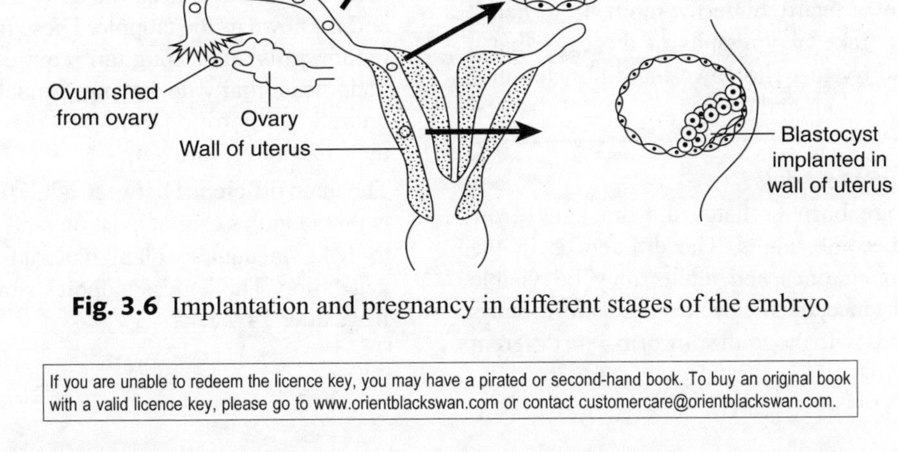

Fig. 3.6 Implantation and pregnancy in different stages of the embryo

Fun Fact!

The placenta secretes hormones such as oestrogen and progesterone. Progesterone is used to detect pregnancy.

Amniotic Fluid: The sac that developes around the embryo is known as the amnion. The fluid that fills the space between the embryo and the amnion is known as **amniotic fluid.** The functions of the amniotic fluid are:

- Prevents the foetus from sticking to the amnion
- Checks the pressure around the embryo
- Allows restricted movement to the foetus
- Protects the foetus from damage by mechanical shock or jerk

Parturition: Gestation is the term used for the development of an embryo in the uterus. It varies in different animals. For example, in humans it lasts for about 280 days and at the time of birth, powerful contractions in the muscles of the uterus lead the mother to push the baby out of her body. The head emerges first and after the baby comes out, in a few minutes, the blood vessels in the umblical cord shrink and can be cut after tying. The placenta breaks from the uterus after 15 minutes and the uterus contracts and returns to normal after a few days. When the child is born, she or he lets out a sharp cry that enables clearing of the respiratory passage.

Viviparous Vs Oviparous Animals

The differences between the hen and human egg give an idea of how animals differ in the method of giving birth. The animals that give birth to young ones are called **viviparous animals** (like humans), while those that lay eggs are known as **oviparous animals** (like hens).

SCIENCE IN DAILY LIFE

Go out in nature or observe at home the different eggs of animals such as lizard, butterfly, moth, frog, hen and other birds. Take photographs of the eggs that you observe and try to write down the difference in their size, shape, colour and texture.

Journey from Young to Adult

Individuals that are born or hatched from eggs grow and develop to become adults. The differences in the appearance of young ones and adults may be visible sometimes. For instance, the life cycle of a silkworm starts with an egg and ends with the adult silkworm with different phases in between like larvae and pupa. Similarly, the life cycle of a frog also has different stages with differences in morphology. Different organs develop in the young and in adults in order to improve their survival. In frogs, tadpoles transform into adults capable of swimming and jumping. These changes or the transformation of larvae to adult is known as **metamorphosis.**

Life Cycle of a Frog: The life cycle of frog is divided into three stages (Fig. 3.7):

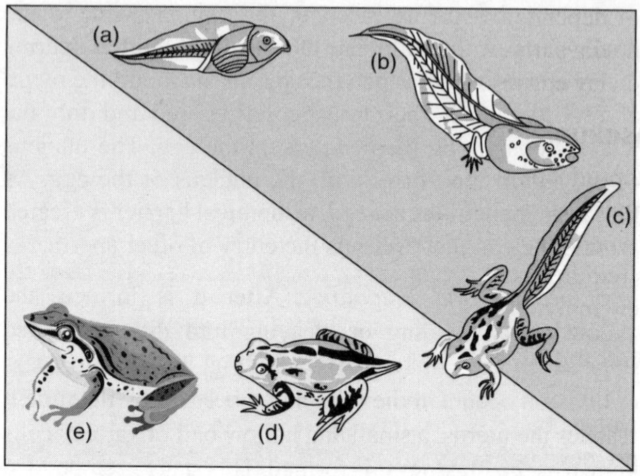

Fig. 3.7 Life cycle of frog: (a) tadpole without limbs, (b) with one pair of limbs, (c) with two pairs of limbs and tail, (d) young frog with tail and (e) without tail

- **Stage I: Eggs:** A frog lays hundreds of tiny eggs that clump together to form frogspawn. The favourable weather for this is spring, when it gets a little warmer. Frogspawn can be sighted in shallow and still water and is usually laid just below the water surface to provide protection. Tiny black dots can be observed in the eggs surrounded by a jelly-like substance.
- **Stage II: Tiny tadpoles:** After consuming the yolk of their egg in 1–3 weeks, the baby frog hatches opens and is known as the tadpole. These tiny tadpoles have mouth, gills and a long tail required for swimming. Tadpoles initially do not move much as they depend on yolk for nutrition, but once they become bigger, they start to venture out.

The main difference between adult frogs and tadpoles is that tadpoles cannot exist on land and so they have to feed on aquatic plant materials and chunks of vegetation. The legless tadpoles slowly develop into frogs after 14 weeks.

The hind legs are grown first followed by the front legs and slowly their body starts changing its shape and consuming insects. Tadpoles shrink and skin grows over their gills to develop into lungs and eardrums necessary for them to survive on land. After development, tadpoles have a crucial 24 hours to metamorphosise completely and small baby frogs are formed.

- **Stage III: Full-grown frog:** The grown-up frog is now ready to face the world and is much less dependent on water as long as it lives in the shade and does not dry out its skin. The females then mature to lay eggs and the whole process starts again.

Asexual Reproduction

Reproduction in higher animals take place by sexual mode. However, lower organisms such as *Hydra* and *Amoeba* reproduce in asexual ways. Here, only one parent is required and no genetic recombination can occur as the new individuals are clones of the parent.

Budding in Hydra

This is a form of asexual reproduction in *Hydra* that involves bulging in the parent body. The bulges develops into new individuals and are termed buds (hence, the process is called budding). This type of reproduction is also observed in yeast. Only one parent is required to give rise to new individuals. A small bulb-like projection or bud arises due to cell division at a specific site. The newly formed organism is a clone of the parent organism with no mutation. These individuals when fully mature detach from the parent body and become individuals. Budding is of two types: internal budding which is observed in parasites and external budding which observed in *Hydra* (Fig. 3.8).

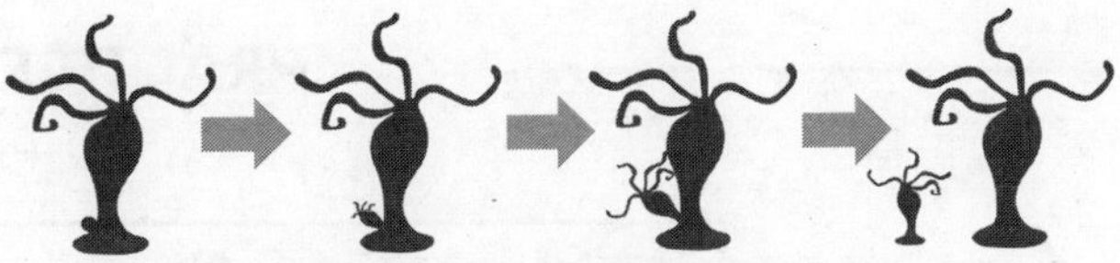

Fig. 3.8 Budding in *Hydra*

SCIENCE IN DAILY LIFE

Observe permanent slides of *Hydra* under a microscope and mark any bulges you observe in the parent body. Count the bulges and note down their size. Draw a diagram of *Hydra* and compare it with the one given in this book.

Fission in Amoeba

In single-celled *Amoeba,* the nucleus divides into two nuclei (Fig. 3.9). This is known as karyokinesis, and it leads to division of the body into two parts, known as cytokinesis. Two amoeba are formed from one parent. This type of reproduction is called binary fission as it involves division of an individual into two individuals. Binary fission is typically observed in prokaryotes and single-celled organisms. The daughter cells receive one copy of their parent's genetic material. The daughter cells are mirror copies of the parent and, in the process, the identity of the parent is completely lost.

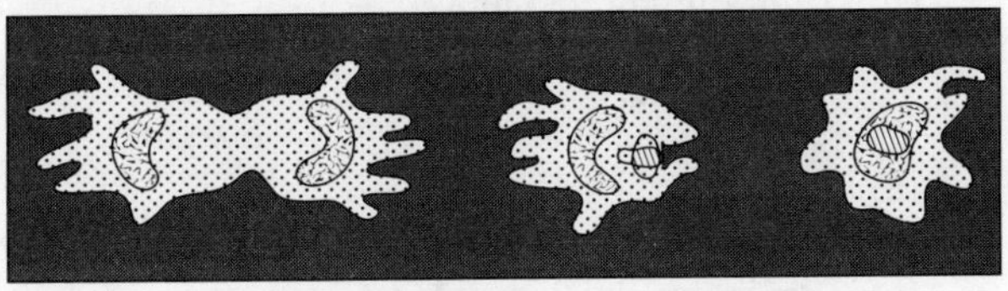

Fig. 3.9 Fission in *Amoeba*

SUMMARY

- There are two modes of reproduction: asexual and sexual.
- In animals, sexual reproduction is the mode of reproduction used to give rise to offspring.
- Sexual reproduction requires fusion of two gametes: one from the male and the other from the female. Fertilisation is of two types: internal (like in humans) and external (like in frogs).
- The human male reproductive system produces sperm, the male gamete. This involves many glands and organs. The female reproductive system produces ova, the female gametes, and also involves many organs.
- The gestation period is humans is 9 months and then the child is born. If no fertilisation occurs, the menstruation cycle resumes.
- Apart from the human reproductive system, external fertilisation enables a frog to undergo metamorphosis from an egg to a tadpole to a frog, with changes in organs and features.
- Asexual reproduction in *Hydra* and *Amoeba* gives rise to individuals that are clones of their parents.

PRACTICE QUESTIONS

Analyse Your Concepts (School Exam Based)

Multiple Choice Questions

1. The gametes fuse to form a ______ that develops into a new individual.
(a) Embryo (b) Foetus
(c) Zygote (d) Blastula

2. The zygote has ____ chromosomes as the gametes have ____ chromosomes each.
(a) 46, 23 (b) 23, 46
(c) 22, 45 (d) 45, 22

3. Spermatogenesis takes place in the:
(a) Vas deferens
(b) Epididymis
(c) Seminiferous tubules
(d) Efferent ducts

4. These are two ovoid glands that open into the urethra before entering the penis; they secrete a lubricant:
(a) Vas deferens
(b) Prostate gland
(c) Seminal vesicle
(d) Bulbo-urethral gland

5. The top of the head of the sperm is known as the _____ acrosome. It secretes ______ enzyme.
(a) Acrosome, lysosome
(b) Microsome, hyaline
(c) Acrosome, hyaluronidase
(d) Microsome, hyaluronidase

6. A thin membrane known as the ______ is found at the opening of the vagina.
(a) Hymen (b) Labia majora
(c) Hymen (d) Labia majora

7. ________ hormone prepares the uterus to receive the embryo.
(a) Progesterone
(b) Oestrogen
(c) Follicle-stimulating hormone
(d) Testosterone

8. The onset of menstruation is known as ______while permanent cessation of the menstruation cycle is known as _______.
(a) Menopause, menarche
(b) Menarche, menopause
(c) Menostart, menopause
(d) Menopause, menostart

9. The blood vessels that connect the foetus with the placenta are rich in blood and this is known as the ______.
(a) Blastocyst (b) Placenta
(c) Umbilical cord (d) None of the above

10. ______ is a form of asexual reproduction in *Hydra.*
(a) Fission (b) Fusion
(c) Egg laying (d) Budding

Match the Following

Each question contains statements given in two columns which have to be matched. Statements in column I have to be matched with statements in column II.

1. Choose the correct option:

Column I	Column II
i. Humans	a. Egg
ii. Frogs	b. Baby
iii. Hen	c. Puppy
iv. Dog	d. Tadpole

2. Choose the correct option:

Column I	Column II
i. Egg	a. Thin membrane
ii. Prostate	b. Female gamete
iii. Sperm	c. Accessory gland
iv. Hymen	d. Male gamete

3. Choose the best match:

Column I	Column II
i. Luteal phase	a. Phase III
ii. Ovulatory phase	b. Phase IV
iii. Menstrual phase	c. Phase II
iv. Follicular phase	d. Phase I

4. Choose the best match:

Column I	Column II
i. Egg	a. Stage II
ii. Tadpole	b. Stage I
iii. Tiny frog	c. Stage IV
iv. Full grown frog	d. Stage III

5. Choose the correct option:

Column I	Column II
i. Viviparous	a. Hen
ii. External fertilisation	b. Human
iii. Oviparous	c. Frog

True or False

Read the following statements and write your answer as true or false.

1. Progesterone is also produced by the placenta.
2. The voice difference in the sexes is an example of primary sexual characters.
3. Blood from the mother flows to the foetus through the placenta.
4. The male gamete is known as the sperm whereas the female gamete is known as the ovum.
5. Only two accessory glands contribute to seminal fluid.
6. The temperature required for the survival of the sperm is normally 10–12°C less than normal body temperature.
7. The prostate gland pours acidic secretions into semen.
8. The clitoris is equivalent to the male penis.
9. A long tube that has a funnel-shaped opening is known as the oviducal funnel that corresponds to the uterus.
10. The menstrual phase of the menstruation cycle is the phase where blood is discharged.
11. Frogs are an example of internal fertilisation.
12. At the blastocyst stage, the embryo is fix to the endometrium wall.
13. Amniotic fluid does not allow any kind of foetal movement.
14. Animals who lay eggs are known as oviparous animals.
15. The changes or transformation of larvae to pupa is known as metamorphosis.

Fill in the Blanks

Complete the following statements with an appropriate word/term to be filled in the blank spaces.

1. ______ is a form of asexual reproduction in *Hydra* that involves bulging in the parent body.
2. Tadpoles shrink and the skin grows over the ________.
3. The animals that give birth to young ones are called ________ like ________.
4. Placenta transfers ________ and ________ to the baby.
5. ________ is a stage of the embryo where all the body parts can be identified.
6. The only protection provided to the eggs of frogs is the layer of ________.
7. At the speed of ________, the sperms cross the vaginal area.
8. The onset of menstruation in females at about 13 years of age is known as ________.
9. The remainder of the follicle that is converted into a yellow mass is termed ______.
10. The top of the head of the sperm is known as ________.

Assertion–Reason Questions

The following questions consist of two statements: Assertion (A) and Reason (R). Answer these questions selecting the appropriate option given below:

(a) Both A and R are true and R is the correct explanation of A
(b) Both A and R are true but R is not the correct explanation of A
(c) A is true but R is false
(d) A is false but R is true

1. Assertion (A): The sac that develops around the embryo is known as the amnion.
Reason (R): The fluid that fills the space between the embryo and the amnion is known as amniotic fluid.
2. Assertion (A): The menstrual phase lasts 3–5 days when blood is discharged.
Reason (R): The uterine lining of blood is shed by rupture of the blood vessels.
3. Assertion (A): The acrosome secretes hyaluronidase.
Reason (R): It dissolves the wall of the hymen and facilitates entry of the sperm into it.

4. Assertion (A): Testosterone is responsible for the secondary sexual characters in females.
   Reason (R): Leydig cells produce testosterone in males.
5. Assertion (A): The fluid that fills the space between the embryo and the amnion is known as amniotic fluid.
   Reason (R): It protects the foetus from damage by mechanical shock or jerks.

One-Word Answers

1. Which part of the male reproductive system discharges sperm in the form of semen?
2. Which part of the seminiferous tubule fits as a cap on the upper end of the testes?
3. Which cells produce testosterone in the male reproductive cells?
4. What is the condition where the intestine bulges into the scrotum due to pressure in the abdomen through the inguinal canal?
5. Which organelle in sperms provides the energy for swimming and motility?
6. What is the follicle that enlarges with the growing egg and is filled with fluid?
7. Which hormone is secreted for ovulation by the follicle?
8. In which phase is the egg released and then travels down the fallopian tube?
9. What is the sac that develops around the embryo known as?
10. What are the changes or the transformation of larvae to adult called?

Define the Term

1. Primary reproductive parts
2. Interstitial cells
3. Semen
4. Follicular phase
5. External fertilisation

Groups of Terms

Given below are groups of terms. In each group the first pair indicates the relationship between the two terms. Rewrite and complete the second pair on a similar basis.

1. Male: sperm :: Female:______
2. Sperm: sperm duct :: Egg: ____
3. Male: testosterone :: Female:____
4. *Amoeba*: fission :: *Hydra*: _____
5. Zygote: 46 chromosomes :: Egg: _____

Identify the Odd One Out

1. Luteal, Follicular, Ovulatory, Tadpole
2. *Hydra*, *Amoeba*, Yeast, Humans
3. Progesterone, Oestrogen, Follicle stimulating hormone, Testosterone
4. Frogs, Insects, Humans
5. Testes, Seminiferous tubule, Fallopian tube, Prostate gland

Image-Based Questions

1. What is the term used to describe the male gamete?

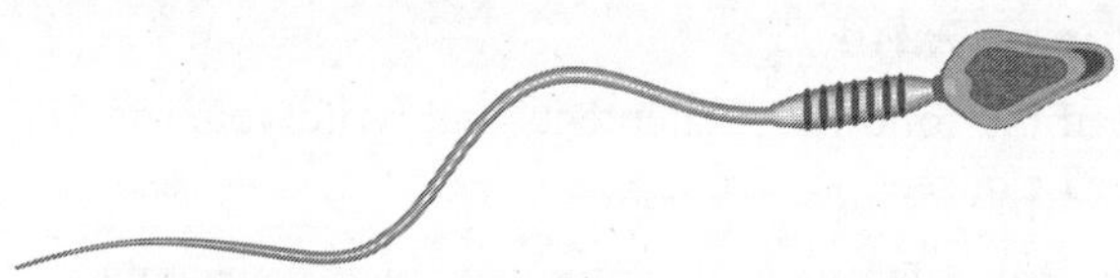

(a) Spermatia
(b) Spermatozoa
(c) Sperma
(d) Spermatogonia

2. To which category does this animal belong?

(a) Viviparous (b) Oviparous
(c) Asexual (d) None of the above

3. What is the mode of fertilisation adopted by this animal?

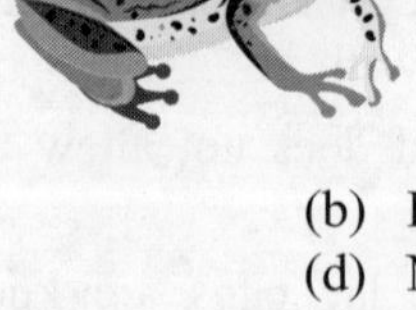

(a) External (b) Internal
(c) Both (d) None

4. What is this stage of metamorphosis?

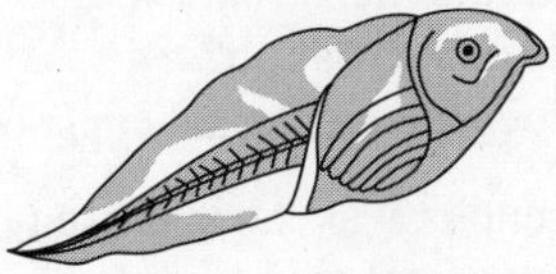

(a) Tiny frog (b) Sperm
(c) Tadpole (d) Grown up frog

5. Choose the correct term:

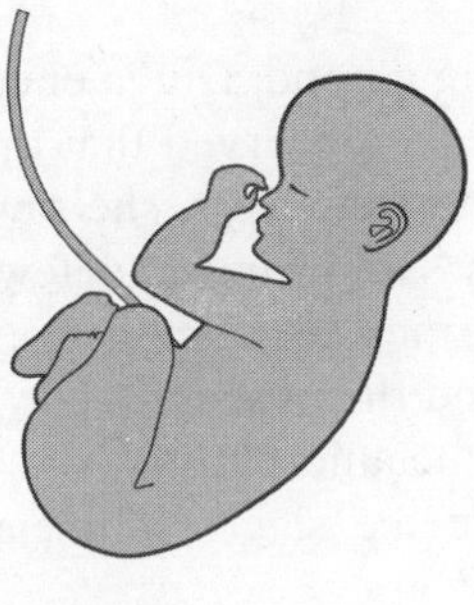

(a) Umbilical cord
(b) Cord
(c) Oviduct
(d) Vas deferens

6. What is the method of reproduction that takes place in this organism?

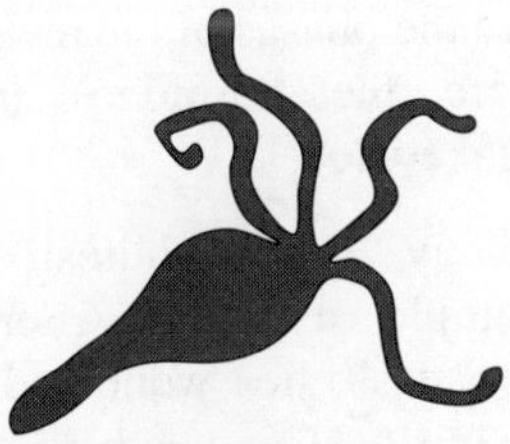

(a) Fission
(b) Budding
(c) Binary fission
(d) Vegetative reproduction

7. Choose the correct term:

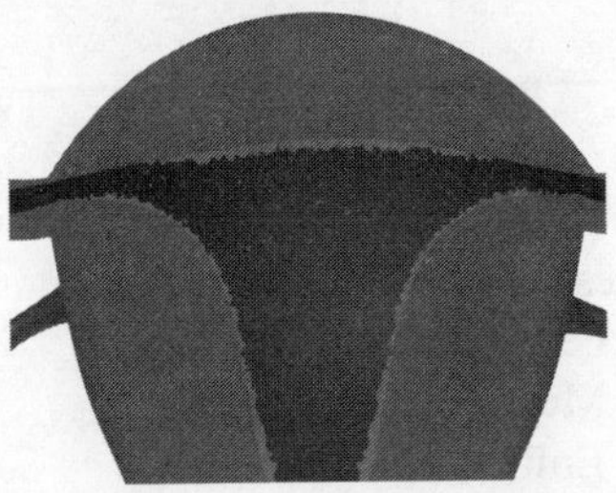

(a) Cervix (b) Hymen
(c) Oviduct (d) Uterus

8. Choose the correct option:

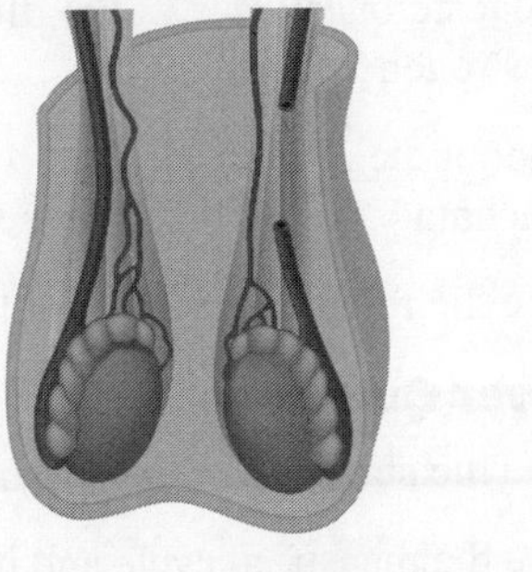

(a) Penis (b) Epididymis
(c) Sperm duct (d) Testes

Very Short Answer Questions

1. What are the two modes of reproduction?
2. What is the difference between primary and secondary sexual characters?
3. Name the primary reproductive parts of males and females.
4. Write about the role of:
(a) Seminiferous tubules (b) Interstitial cells
5. What is the acrosome?
6. What is semen?
7. Name the three accessory glands in the male reproductive system.
8. How is the sex of the baby determined?
9. What is the purpose of the tail in the sperm?
10. What is the purpose of hymen?

Short Answer Questions

1. Name the main divisions in:
(a) Male reproductive system
(b) Female reproductive system
2. Why are the testes located outside the human male body?
3. What is metamorphosis?
4. What is the difference between Graafian follicle and corpus luteum.
5. What is the difference between oviparous and viviparous animals? Explain with examples.
6. Write briefly about:
(a) Bulbo-urethral gland (b) Prostate gland
(c) Seminal vesicle

7. What is the role of:
   (a) Progesterone (b) Testosterone
   (c) Oestrogen
8. Write about the role of:
   (a) Placenta (b) Ovaries
9. Describe the role of amniotic fluid.

Long Answer Questions

1. Describe the structure of the sperm in detail.
2. Describe the menstrual cycle and its phases in detail.
3. Briefly describe the process that takes place from fertilisation to birth.
4. Explain the parts of the male reproductive system.
5. Explain the divisions of the female reproductive system.
6. Differentiate between asexual and sexual reproduction with examples.
7. Describe the life cycle of frogs.
8. Differentiate between internal and external fertilisation in detail.
9. Explain the role of:
   (a) Villi in oviduct
   (b) Acrosome in sperm
   (c) Ovulation phase in menstrual cycle
   (d) Meiosis in gamete production
10. Explain the process of fission in *Amoeba* and budding in *Hydra*.

Case Study-Based Questions

1. Reema and her classmates in school were playing and suddenly Reema experienced menarche. Her friend took her to the clinic thinking that she might be suffering from a disease. Can you help educate Reema about the menstrual cycle?
   (a) What is menstruation?
   (b) Why does the menstrual cycle occur?
2. Abhijeet's uncle was hospitalised and underwent surgery. His parents told him that his uncle suffered from hernia. Can you help Abhijeet understand what a hernia is?
   (a) What are the male reproductive parts?
   (b) What is a hernia?
3. While visiting a pond, Divya encountered a group of frogs and she observed that there were eggs in the pond. After few days, she saw small creatures with no legs swimming and few days later they were transformed into small frogs. Can you help her understand the process?
   (a) What is metamorphosis?
   (b) What are the stages of metamorphosis frogs undergo?
   (c) What is difference between external fertilisation and internal fertilisation?
4. While reading a book, Ahmad saw the graph of the phases of the menstrual cycle. Can you help him understand the role of hormones and phases involved in the cycle?
   (a) What are the phases of the menstrual cycle?
   (b) What are the hormones involved in the menstrual cycle?
5. In the biology lab, Ashutosh encountered a preserved sample of *Hydra*. After learning about its characteristics, he wondered what asexual reproduction means. Can you help him understand?
   (a) How does *Hydra* reproduce?
   (b) What is asexual reproduction?
   (c) How is it different from sexual reproduction?

NCERT Corner

1. Select the correct option:
   (a) Internal fertilisation occurs:
      (i) In the female body
      (ii) Outside the female body
      (iii) In the male body
      (iv) Outside the male body
   (b) A tadpole develops into an adult frog by the process of:
      (i) Fertilisation
      (ii) Metamorphosis
      (iii) Embedding
      (iv) Budding

(c) The number of nuclei present in a zygote is:
(i) 0 (ii) 1
(iii) 2 (iv)

2. Indicate whether the following statements are True (T) or False (F):
(a) Oviparous animals give birth to young ones.
(b) Each sperm is a single cell.
(c) External fertilisation takes place in the frog.
(d) A new human individual develops from a cell called gamete.
(e) The egg laid after fertilisation is made up of a single cell.
(f) *Amoeba* reproduces by budding.
(g) Fertilisation is necessary even in asexual reproduction.
(h) Binary fission is a method of asexual reproduction.
(i) A zygote is formed as a result of fertilisation.
(j) An embryo is made up of a single cell.

3. Give two differences between a zygote and a foetus.

4. Define asexual reproduction. Describe two methods of asexual reproduction in animals.

5. In which female reproductive organ does the embryo get embedded?

6. What is metamorphosis? Give examples.

7. Differentiate between internal fertilisation and external fertilisation.

8. Describe the process of fertilisation in human beings.

COMPETITION WINDOW (OBJECTIVE TYPE)

1. Secondary sexual characters develop in females due to:
(a) Oestrogen
(b) Absence of androgen
(c) Androgen
(d) Testosterone

2. Which extra-embryonic membrane in humans prevents the desiccation of the embryo inside the uterus?
(a) Amnion (b) Yolk sac
(c) Chorion (d) None of the above

3. Hysterectomy is the surgical removal of the:
(a) Uterus (b) Mammary gland
(c) Prostate (d) Vas deferens

4. The Leydig cells found in the human body are the secretory source of:
(a) Progesterone (b) Glucagon
(c) Androgen (d) Intestinal mucus

5. The secretory phase in the human menstrual cycle is also called:
(a) Luteal phase that lasts for 6 days
(b) Follicular phase that lasts for 6 days
(c) Luteal phase that last for 13 days
(d) Follicular phase that lasts for 13 days

6. Seminal plasma in human males is rich in:
(a) Glucose and calcium
(b) Fructose and calcium
(c) DNA and testosterone
(d) Ribosome and potassium

7. Secretions from which one of the following are rich in fructose, calcium and some enzymes?
(a) Male accessory glands
(b) Liver
(c) Pancreas
(d) Salivary gland

8. Seminal plasma in humans is rich in:
(a) Fructose and calcium and no enzymes
(b) Glucose and enzymes but no calcium
(c) Fructose and calcium but no enzymes
(d) Fructose, calcium and enzymes

9. Which of these hormones is secreted by the placenta?
(a) HCG (b) Oestrogen
(c) Progesterone (d) Prolactin

10. The middle piece of the sperm contains:
(a) Protein (b) Centriole
(c) Mitochondria (d) Nucleus

11. After ovulation, the Graafian follicle regresses into:
(a) Corpus callosum (b) Corpus albicans
(c) Corpus luteum (d) Corpus artesia

12. The mammalian corpus luteum produces:
(a) Luteotrophic hormone
(b) Luteinising hormone
(c) Oestrogen
(d) Progesterone

13. Male hormone is produced in the testes by the cells of:
(a) Sertoli (b) Epithelial
(c) Spermatocyte (d) Leydig

14. Freshly released human egg has:
(a) One X chromosome
(b) One Y chromosome
(c) Two X chromosome
(d) One X chromosome and one Y chromosome

15. Middle piece of mammalian sperm possesses:
(a) Mitochondria only
(b) Centriole only
(c) Mitochondria and centriole
(d) Nucleus and mitochondria

16. Which of the following is NOT a characteristic feature of sexual reproduction?
(a) It involves the fusion of two reproductive cells
(b) It is common in most organisms
(c) Fertilisation may take place inside or outside the body of the female
(d) The offspring receives its characteristics from one parent only

17. Which of the following is/are paired structure in human reproductive system?
(a) Ovary
(b) Testes
(c) Fallopian tube
(d) All of these

18. How are *Paramecium* and *Amoeba* similar?
(a) Unicellular (b) Devoid of nucleus
(c) Binary fission (d) (a) and (c)

19. In which of the following parts of the female reproductive system does fertilisation occur?
(a) Ovary (b) Vagina
(c) Uterus (d) Fallopian tube

20. The substances that are transported through umbilical cord are:
(a) Urea (b) Antibodies
(c) Carbon dioxide (d) All of these

Answer Keys

Multiple Choice Questions

1. (c)	2. (a)	3. (c)	4. (d)	5. (c)
6. (c)	7. (a)	8. (b)	9. (c)	10. (d)

Match the Following

1. i (b), ii (d), iii (a), iv (c)
2. i (b), ii (c), iii (d), iv (a)
3. i (b), ii (a), iii (d), iv (c)
4. i (b), ii (a), iii (d), iv (c)
5. i (b), ii (c), iii (a)

True or False

1. T	2. F	3. T	4. T	5. F
6. F	7. F	8. T	9. F	10. T
11. F	12. T	13. F	14. T	15. F

Fill in the Blanks

1. Budding
2. Gills
3. Viviparous, humans
4. Oxygen, nutrition or antibodies
5. Foetus
6. Jelly coating
7. 1.5 mm per minute
8. Menarche
9. Corpus luteum
10. Acrosome

Assertion–Reason Questions

1. (b)	2. (a)	3. (d)	4. (c)	5. (a)

One-word Answers

1. Seminiferous tubule
2. Epididymis
3. Leydig/Interstitial cells
4. Hernia
5. Mitochondria
6. Graafian follicle
7. Oestrogen
8. Ovulatory phase
9. Amnion
10. Metamorphosis

Define the Term

1. The primary reproductive parts are the main gonads. Testes in males and ovaries in females are the gonads of humans. Gonads produce gametes or sex cells: eggs and sperms.
2. Between the coils of seminiferous tubules, there are packing tissues known as interstitial cells. These cells are also known as Leydig cells and they produce testosterone, the male hormone. It is responsible for secondary sexual characters in males.
3. This is a mixture of sperm and secretions from various glands like prostate, Cowper's, seminal vesicles and urethral. It is a milky fluid that is ejaculated in an average amount of 2–3 mL each time and which contains around 2000,000,000 to 400,000,000 sperms.
4. Certain hormones are produced as the follicle grows and that leads to reactivation of the uterine wall to be supplied with blood vessels and thicken. This change occurs to prepare the uterus to receive the fertilised egg and develop the future embryo.
5. Frogs and toads in the rainy or spring season move to slow-flowing streams like ponds where the female lays hundreds of eggs. The eggs of frogs are devoid of shells and are very delicate. The only protection is the layer of jelly that holds the eggs together. After the eggs are laid, the male frog deposits sperms over them. Now, each sperm with its long tail swims in water and comes in contact with the eggs. Fertilisation occurs outside the body of the female and is one of the most common practices in aquatic animals like starfish, fish, frogs, etc.

Groups of Terms

1. Ovule
2. Fallopian tube/Oviduct
3. Oestrogen
4. Budding
5. 23 chromosomes

Identify the Odd One Out

1. Tadpole
2. Humans
3. Testosterone
4. Humans
5. Fallopian tube

Image-Based Questions

1. (b) 2. (b) 3. (a) 4. (c) 5. (a)
6. (b) 7. (d) 8. (d)

Very Short Answer Questions

1. Asexual and Sexual
2. Primary sexual characters are responsible for the internal maturitaion of the sexual organs. There are secondary sexual characteristics that are developed in both sexes that show phenotypically differentiating features.
3. Testes in males and ovaries in females are the gonads of humans.
4. (a) This tubule is where spermatogenesis occurs.
   (b) They produce testosterone, the male hormone.
5. The top of the head of the sperm is known as the acrosome and it secretes hyaluronidase, an enzyme that dissolves the wall of the ovum and facilitates the entry of sperm into it.
6. A milky fluid known as semen, which is a mixture of different components, helps sperms to transform from sluggish to active.
7. Prostate, bulbo-urethral and seminal vesicle
8. By sex chromosomes: X, Y
9. The movement of the tail is used for propulsion by lashing movements.
10. A thin membrane known as the hymen at the opening of the vagina makes it partially closed.

Short Answer Questions

1. The female reproductive system consists of the following organs or parts:
   Vulva (the outermost part)
   Vagina, Uterus (a sac-like organ to develop the embryo)
   A pair of fallopian tubes to transfer the egg from ovaries to uterus
   A pair of ovaries
   The following are main divisions in the male reproductive system:
   A pair of testes that produces sperm
   Sperm duct (vas deferens) that discharges sperm in the form of semen
   Accessory glands (prostate, bulbo-urethral and seminal vesicle) that contribute to seminal fluid
   Penis that transfers the sperms into the female
   Urethra that is located in the penis and conveys the sperm from the sperm duct
2. The temperature required for the survival of the sperm is normally 2–3°C less than the normal body temperature of the humans (37°C). In the normal conditions, the scrotum sac carries the testes in the suspended condition outside the body in order to escape the body heat. The testes undergoes adjustments in the scrotum sac according to temperature changes.
3. Different organs develop in the young and in the adult in order to improve their survival rate and transform into adults. In frogs, the tadpoles transform into adults capable of swimming and jumping. These changes or transformation of larvae to adult is known as metamorphosis.
4. The follicle enlarges as the egg grows larger and eventually gets filled with fluid; this is called the Graafian follicle. Fimbrae or cilia present in the oviduct pick up the released egg and the remainder of the follicle which after some time is converted into a yellow mass termed corpus luteum; it is responsible for the secretion of two hormones: progesterone and oestrogen.
5. The animals that give birth to young ones are called viviparous animals, like humans. The animals that lay eggs are known as oviparous animals, like hens.
6. Prostate gland: Close to the bladder origin, this is a bilobed structure that surrounds the urethra that pours the alkaline secretion into the semen. This alkaline secretion neutralises the acidity of the female's vagina.
   Seminal vesicles: Between the urinary bladder's posterior surface and the rectum lie a pair of lobulated glands that give rise to the duct that opens into the seminal vesicle just before it joins the urethra. Seminal vesicles produces the liquid which is important for the transportation of sperms. A milky fluid known as semen, which is a mixture of different components, helps sperms to transform from sluggish to active.
   Bulbo-urethral gland: Just before entering the penis, these are two ovoid glands that open into the urethra and which secrete lubricants.
7. (a) Responsible for preparing the uterus to receive the embryo
   (b) Testosterone is responsible for the secondary sexual characters in males.
   (c) The hormone that is secreted for ovulation by the follicle and which is also responsible for the secondary sexual characters.

8. (a) Placenta transfers oxygen, nutrients and antibodies to the baby and receives carbon dioxide and urea from the foetus.
   (b) Ovary produces gametes or sex cells, called eggs in females.
9. A sac which develops around the embryo is known as amnion. The fluid that fills the space between the embryo and amnion is known as amniotic fluid. Functions of amniotic fluid: Prevents the foetus from sticking to the amnion; checks the pressure around the embryo; allows restricted movement to the foetus; protects the foetus from damage by mechanical shocks or jerks.

Long Answer Questions

1. Hint: Describe the parts of the sperm with a labelled diagram, as mentioned in the chapter.
2. Hint: Explain the menstrual cycle and its phases.
3. Hint: Explain all the stages of fertilisation till parturition.
4. Hint: Write about the different parts of the male reproductive system and their roles.
5. Hint: Write about the different parts of the female reproductive system and their roles.
6. Hint: Explain the reproduction of humans and *Hydra* or yeast or *Amoeba*.
7. Hint: Define metamorphosis, the stages of the life cycle of frogs.
8. Hint: Internal fertilisation is observed in humans while external fertilisation is seen in frogs. Elaborate on this.
9. Hint: Explain all the parts and their roles, as mentioned in the chapter.
10. Hint: Explain fission and budding, as mentioned in the chapter, with a diagram.

Case Study-Based Questions

1. (a) The sexual reproduction of humans leads to a gestation period of 9 months and then the child is born, but if no fertilisation occurs, then the female body goes through the menstrual cycle.
   (b) From the age of 13 to 45–50 years, a continuous reproductive period occurs in human females. There is a characteristic event that is repeated every 28 days or approximately a month with small variations in the dates with a menstrual flow. This menstrual flow is stopped temporarily by pregnancy. The onset of menstruation in females at about 13 years of age is known as menarche while the permanent stoppage of menstruation at the age of 45 is known as menopause.
2. (a) The following are the main divisions in the male reproductive system: A pair of testes (singular: testis) that produces sperm, sperm duct (vas deferens) that discharges sperm in the form of semen, accessory glands (prostate, bulbo-urethral and seminal vesicles) that contribute to seminal fluid, penis that transfers the sperms into the female, and a urethra that is located in the penis and which conveys the sperm from the sperm duct.
   (b) The condition where the intestine bulges into the scrotum due to pressure in the abdomen through the inguinal canal is known as hernia.
3. (a) In frogs, tadpoles transform into adults capable of swimming and jumping. These changes or transformation of larvae to adult is known as metamorphosis.
   (b) Eggs $\rightarrow$ Tiny tadpole $\rightarrow$ Grown-up frog
   (c) Internal Fertilisation: The fertilisation that takes place inside the body of the female is known as internal fertilisation. It occurs in many animals like humans, dog, hen and cow.

   External Fertilisation: Frogs and toads in the rainy or spring season move to slow flowing streams like ponds where the female lays hundreds of eggs. The eggs of frogs are devoid of shells and are very delicate. The only protection provided to the egg is the layer of jelly that holds them together. As the eggs are laid, the male frog deposits sperms over them. Now, each sperm with a long tail swims in water and comes in contact with these eggs. Fertilisation occurs outside the body of the female and is one of the most common practices in aquatic animals like starfish, fish, frogs, etc.
4. (a) Menstrual phase, Follicular phase, Ovulatory phase, Luteal phase
   (b) Progesterone, Oestrogen LSH, FH
5. (a) A form of asexual reproduction in *Hydra* that involves bulging in the parent body. The bulge develops into new individuals termed buds. In this type of reproduction, only one parent is required to give rise to new individuals. The individuals develop from the buds of *Hydra*; so it is known as budding. A small bulb-like projection that is known as the bud arises due to cell division at a specific site. The newly formed organism is a clone of the parent organism with no mutation. These individuals, when fully mature, detach

from the parent body and become individuals independent from their parent.

(b) Microscopic organisms like *Amoeba* reproduce in different ways. In asexual reproduction, only a single parent is required and no genetic recombination will occur as the new individuals are the clones of the parent.

(c) No fusion of gametes will occur and the offspring will be identical to the parent.

NCERT Corner

1. a. (i) b. (ii) c. (iii)

2. a. F f. F
b. T g. F
c. T h. T
d. F i. T
e. F j. F

3. Hint: Explain the differences between zygote and foetus, as mentioned in this chapter.

4. Hint: Explain asexual reproduction, binary fission and budding.

5. Uterus

6. Hint: Explain the process, from egg to tadpole to tiny frog to full grown frog

7. Internal Fertilisation: The fertilisation that takes place inside the body of the female is known as internal fertilisation. It occurs in many animals like humans, dog, hen and cow.

External Fertilisation: Frogs and toads in the rainy or spring season move to slow flowing streams like ponds where the female lays hundreds of eggs. The eggs of frogs are devoid of shells and are very delicate. The only protection provided to the egg is the layer of jelly that holds them together. As the eggs are laid, the male frog deposits sperms over them Now, each sperm with a long tail swims in water and comes in contact with these eggs. Fertilisation occurs outside the body of the female and is one of the most common practices in aquatic animals like starfish, fish, frogs, etc.

8. Hint: Explain all the steps in human reproduction, as mentioned in this chapter.

Competition Window

1. (a)	**2.** (a)	**3.** (a)	**4.** (c)	**5.** (c)
6. (b)	**7.** (a)	**8.** (d)	**9.** (a)	**10.** (c)
11. (c)	**12.** (d)	**13.** (d)	**14.** (a)	**15.** (a)
16. (d)	**17.** (d)	**18.** (d)	**19.** (c)	**20.** (d)

CHAPTER 4

The Circulatory System

PREREQUISITES

You must have knowledge of:

- Human body parts
- Basic human physiology
- Cells, tissues and other chemical components of the body

LEARNING OBJECTIVES

At the end of this chapter, you will be able to:

- Explain the composition of blood
- List the functions of blood
- Describe the blood group system and transfusion
- Explain the process of blood clotting
- Describe the functions and structure of the heart
- Explain blood circulation
- Describe the lymphatic system

INTRODUCTION

Living cells require oxygen, nutrients and other essential substances along with removal of harmful substances or waste for the proper functioning of tissues. Therefore, it is important to have efficient mechanisms for the movement of these substances. Different methods of transport have evolved in different groups of animals.

In animals such as coelenterates and sponges, water circulates from their surroundings through their body cavities in order to carry out exchange of gases. As the complexity of animals increases, special fluids are used within the body to transport materials. Blood is one of the most common body fluids used by higher organisms like humans. Lymph is also used to transport certain substances.

In this chapter, you will learn more about the circulatory system.

CIRCULATORY FLUIDS OF THE BODY

The three principle fluids of the body are:

- Blood, which is present in the blood vessels and heart.
- Lymph, which is present in the lymph vessels and lymphatic organs.
- Tissue fluid, which occupies the empty spaces between cells in organs.

There is a connection between these fluids as blood circulates within tubular blood vessels and forms a closed circulatory system in humans, whereas insects have an open circulatory system.

Blood

Animals that possess a circulatory system have blood. It is one of the most important components of life. According to the theory of evolution, blood arose from a single type of cell that was known to provide nutrition and to carry out phagocytosis. Later, the circulatory system and blood evolved to become more complex.

Blood is a type of loose connective tissue that is circulated throughout the body to deliver nutrients and oxygen to various organs (cells and tissues). An average human adult has around 5–6 litres of blood and it makes up about 8% of the total body weight. It consists of a fluid matrix called plasma and formed elements. Blood is thick and dynamic in nature. It has a pH of 7.3–7.4, which makes it slightly alkaline and taste salty.

Fun Fact!

Did you know that oxygen-rich blood is bright red in colour while carbon dioxide-rich blood is dark red in colour?

Plasma

This is a viscous, straw-coloured fluid that constitutes approximately 55% of blood. Plasma is composed of water and proteins that are important for health. The water makes up about 90–92% while the proteins contribute about 6–8%. Apart from these constituents, it also contains salts, enzymes and minerals such as Mg^{++}, Cl^-, HCO_3, Na^+ and Ca^{++}. Blood plasma transfusions are carried out in patients with life-threatening injuries and liver failure.

The main proteins present in plasma are:

- **Fibrinogen**: This protein is required for the coagulation of blood. It is a glycoprotein complex that is circulated in the blood of vertebrates and is produced by the liver. In vascular or tissue injury, it is converted by thrombin to fibrin to enable clotting of blood.
- **Globulin**: This is involved in the defence mechanisms of the body. The different types of globulin are $\alpha 1$, $\alpha 2$, β and γ. They are responsible for exerting oncotic pressure (osmotic pressure exerted by proteins) and their deficiency results in oedema due to increased osmotic pressure, loss of carrier function and increased susceptibility to infections.
- **Albumin**: This is involved in maintaining the osmotic balance. It belongs to the globular protein family, is water soluble and can denature (destroy the properties of proteins) at high heat.

The plasma that is devoid of clotting factors is known as **serum**.

Formed Elements

The formed elements that make up 45% of blood usually comprise erythrocytes (RBCs), leucocytes (WBCs) and platelets (Fig. 4.1).

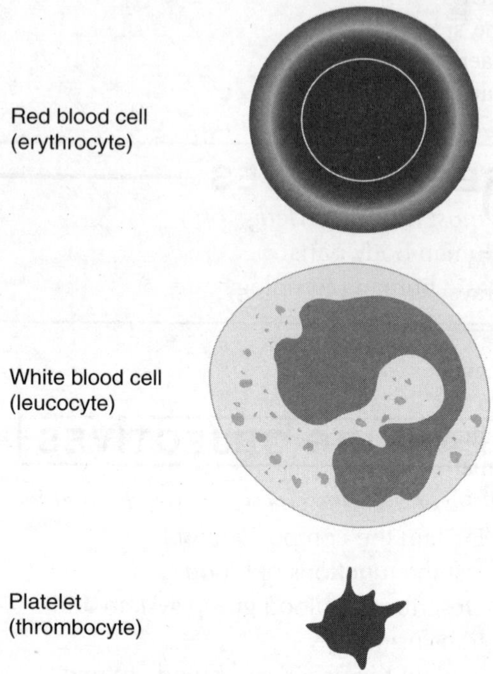

Fig. 4.1 Types of blood cells

- **Erythrocytes**: These are the most abundant cells in blood and are formed in the red bone marrow of the long bones of adults, such as the ribs, ilium and breastbone. However, in the embryo, it is formed in the spleen and liver. A healthy human has 5–5.5 million RBCs or red blood cells per mm^3 of blood. They are concave in shape. Initially, they have a nucleus but it disappears when they become mature. Mature RBCs do not have endoplasmic reticulum, mitochondria and a nucleus, making them more efficient. This is because they have a bigger surface area and they only transport oxygen and do not consume it. They also transport glucose and are more flexible as well. RBCs are flat and thin in the centre and rounded and thick at the edges. This unique structure provides a larger surface area for oxygen transport and can also travel through small capillaries easily. The average diameter of an erythrocyte is 7 microns. It has a complex protein containing iron, known as haemoglobin. This is responsible for its red colour and is the respiratory pigment. It consists of two parts: haem (iron containing) and globin (protein). Haemoglobin combines with oxygen to make an unstable compound called oxyhaemoglobin. This is readily released in the tissues that require oxygen. Similarly, it combines with carbon dioxide to form carbaminohaemoglobin. The lifespan of RBCs is 120 days, after which they are destroyed in the spleen, bone marrow and liver. The iron part of haemoglobin is retained in the liver while the protein part is excreted as bilirubin. In adult humans, around 1% of RBCs are destroyed every day!

Fun Fact!

Did you know that the spleen is known as the graveyard of RBCs?

- **Leucocytes:** White blood cells (WBCs) or leucocytes do not have haemoglobin and so are colourless in nature. They possess a nucleus and are present in smaller amounts, that is, around 6000–8000 per mm^3 of blood. They are usually amoeboid in shape and possess pseudopodia which helps them squeeze through the walls of narrow capillaries. This is known as diapedesis. They are produced in the lymph nodes, bone marrow and sometimes in the spleen and liver. The average lifespan is about two weeks and older WBCs are destroyed in the same way as RBCs. On the basis of their structure and other characteristics, they are categorised into two main groups:
  - **Granulocytes:** These are immune cells with small particles or granules with enzymes present inside them that are released during allergic reactions, asthma and infection. These are further divided into:
    1. **Neutrophils:** These are the most abundant leucocytes (60–65%). They are phagocytic in nature and have the ability to destroy foreign substances that enter the body. They are produced in the bone marrow. Their nucleus has 3–4 lobes and the cytoplasm is granulated. Neutral dyes are used to stain them.
    2. **Basophils:** These are the least abundant WBCs (0.5–1.0%). They are produced in the bone marrow. Basophils secrete serotonin, histamine and heparin. They are involved in inflammation as they dilate the blood vessels (histamine). Basophils are indistinctly lobed with a large nucleus and granulated cytoplasm. They can be stained by basic dyes like methylene blue.
    3. **Eosinophils:** These cells can resist infections and are associated with allergic reactions. They comprise 2–3% of all WBCs. They have a bilobed nucleus with granulated cytoplasm that is bigger than in the others. They can be stained with acidic dyes like eosin. They are produced in the bone marrow and perform phagocytosis as well as secretion of antitoxins.
  - **Agranulocytes:** These are white blood cells that are devoid of granules. They are further divided into:
    1. **Monocytes:** These are phagocytic in nature and have the ability to destroy foreign substances. They have a large, kidney-shaped nucleus and are produced in the bone marrow. They transform into macrophages at the site of infection.
    2. **Lymphocytes:** Around 20–25% of WBCs are lymphocytes. They are of two types: B and T. They are responsible for the immune response of the body. They are the smallest in size with a single large nucleus and no granules in the cytoplasm. B and T cells are produced in the bone marrow; B cells mature there but T cells mature in the thyroid glands. The antibodies produced by lymphocytes act against microbes, toxins and foreign particles and eliminate them. However, memory of these foreign substances remains; this forms the principle of vaccination. The toxins or weakened microbes are introduced in the body and this induces the production of antibodies. These antibodies are recalled later when the microbes enter the body again, and they are easily eliminated.

Inflammation is characterised by swelling, pain, redness and heat at the site of entry of microbes or allergens or at the site of infection. Leucocytes like monocytes and neutrophils reach the site by diapedesis (Fig. 4.2) and destroy the damaged cells. Pus (contains dead WBCs and tissue cells) sometimes forms at the site.

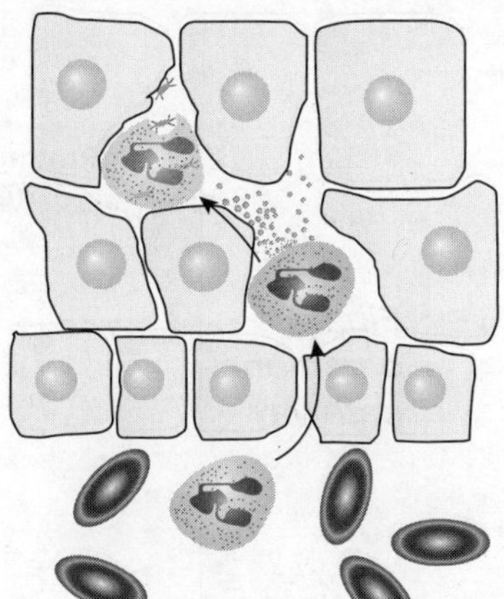

Fig. 4.2 Diapedesis

- **Platelets:** They are known as thrombocytes and are cell fragments produced by special cells in the bone marrow, known as megakaryocytes. Blood contains around 1,500,00–3,500,00 platelets per mm^3. Platelets release multiple substances that are involved in the clotting of blood. Decrease in the number of platelets leads to clotting disorders which can result in excessive blood loss. Platelets are non-nucleated with an oval shape. The bone marrow produces megakaryocytes, which branch off platelets to form a structure encased in a membrane. They have a short lifespan of 3–5 days and old platelets are destroyed in the spleen. Platelets play an important role at the site of injury as they release thrombokinase, which initiates the clotting process.
  - **Clotting:** Coagulation of blood or clotting is the process by which a clot is formed at the site of injury; if it is not initiated, bleeding will not stop (Fig. 4.3).

Platelets present at the site of injury disintegrate to release thrombokinase or thromboplastin (also known as Factor X or Stuart factor). Prothrombin is an inactive form which is converted into its active form, thrombin, by calcium ions present in thrombokinase and plasma. The fibrinogen present reacts with thrombin in the presence of calcium to give rise to insoluble fibrin. These fibrins form microscopic threads that create a mesh at the site of injury. Blood cells are trapped in this mesh and when it shrinks, the plasma is squeezed out in the form of serum. The water component is removed and a clot or thrombus is left. This in turn prevents bleeding and so platelets are important for initiation of clotting. If the platelet count decreases, the coagulation process will occur very slowly and could lead to haemorrhage. For example, in dengue, the platelet number becomes very low.

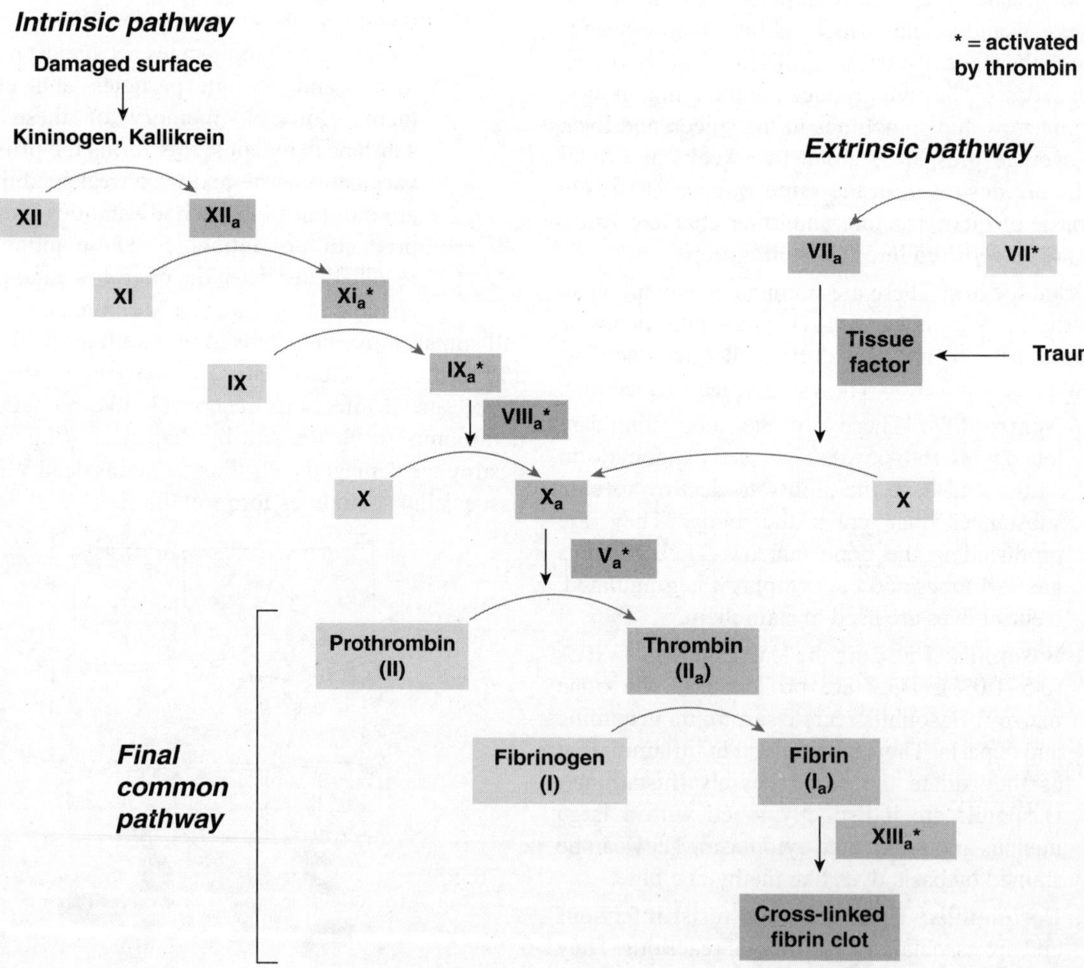

Fig. 4.3 Blood clotting cascade

Blood Group System and Transfusion

The surface of an RBC consists of antigens or protein molecules. Different kinds of antigens are present on the RBCs of different persons. Antibodies that are complementary to these antigens are also present in blood plasma. Based on the antigens present on the RBCs, blood can be divided into different groups. This concept was discovered by Karl Landsteiner. In humans, two main types of antigens are seen: antigen A and antigen B. Depending on the absence or presence of these antigens, the four types of blood groups are A, B, AB and O. This system is also known as the **ABO blood grouping system**. Other systems are also available but the most important ones are the ABO and Rh systems. The basics of the ABO system are as follows:

- Group A has antigen A and antibody B
- Group B has antigen B and antibody A
- Group AB has antigens A and B and no antibodies
- Group O has no antigen and A and B antibodies

Blood group	Antibody	Antigen	Donor's group
O	Anti-A, Anti-B	-	All
A	Anti-B	A	A, O
B	Anti-A	B	B, O
AB	-	A, B	A, B, AB, O

Blood transfusion is the process by which blood is taken from a donor and transferred to a recipient during critical surgical processes where the recipients lose quite a lot of blood. Before the transfusion, blood group matching between the donor and recipient is performed and this is known as **blood group compatibility**.

Fun Fact!

Did you know that type O is known as the universal donor as people with this type can donate blood to persons with A, B, AB or O blood groups, whereas type AB is known as the universal recipient as people with this type can receive blood from persons with any blood group!

A person with blood type A can receive blood from persons with types O and A, while a person with blood type B can receive blood from persons with types O and B. In the **Rh system**, blood groups are divided on the basis of absence or presence of the D antigen. People who have the D antigen are said to be Rh-positive or Rh+ while those without it are said to be Rh-negative or Rh–. This system obtained its name from the Rhesus monkey, in which this antigen was first discovered. It is used along with ABO for compatibility determination.

If blood from an Rh+ person is transfused to an Rh– person, the recipient will start to develop antibodies against the D antigen in about two weeks; this is known as **sensitization**. This sensitization will worse if the transfusion is performed a second time without checking compatibility. It will lead to RBC destruction with the D antigen, resulting in the death of the person. This is a similar reaction to allergy.

Fun Fact!

Erythroblastosis fetalis is a condition where if a pregnant Rh– mother carries an Rh+ child, she becomes sensitive to the Rh factor. This will not affect the first child but if a second child is also Rh+, then it can lead to severe anaemia or jaundice in the baby and can sometimes result in abortion or even death. Therefore, screening is important during pregnancy.

Functions of Blood

The primary functions of blood are:

- **Protection**: When there is a cut or wound, a clot is formed at the site of injury which prevents further loss of blood and the entry of microbes into the body. WBCs engulf the bacteria and eliminate them. Anti-toxins and antibodies are produced by specific cells that nullify the effects of the toxins produced by the microbes.
- **Transport**: Nutrients from digested food are transported to the organs from the alimentary canal by blood. These nutrients are carbohydrates like glucose, amino acids, vitamins and minerals. Apart from nutrients, respiratory gases like oxygen and carbon dioxide are transported between the lungs and tissues. In RBCs, haemoglobin attaches itself to oxygen to form oxyhaemoglobin which breaks down when it reaches the tissues to deliver oxygen with the reaction:

$$\text{Hb} + \text{O} \rightleftharpoons \text{Hb.O (Oxyhaemoglobin)}$$

Similarly, the carbon dioxide molecules from tissues combine with haemoglobin to form carbaminohaemoglobin.

$$\text{Hb} + \text{CO} \rightleftharpoons \text{Hb.CO (Carbaminohaemoglobin)}$$

Other Functions

- The digested nutrients such as minerals, vitamins, proteins and glucose are absorbed by blood through the capillaries present in the villi of the small intestine.
- Hormones that are secreted by the endocrine glands are transported to different tissues and organs by blood.
- It is the fluid connective tissue of the body.
- Oxygen from the lungs is transported to different cells and carbon dioxide from them is transported to the lungs by blood.
- The internal body temperature is maintained by blood by releasing or absorbing heat.
- Blood is filtered in the kidneys to remove nitrogenous waste from plasma; toxins are removed by the liver.
- White blood cells fight against infections and also multiply rapidly during infections.

BLOOD VESSELS

The tubular structures that carry blood and which extend from the heart to different parts of the body are known as blood vessels. A network of hollow tubes is formed of these vessels to carry blood in both directions. There are three types of blood vessels: arteries, veins and capillaries (Fig. 4.4).

There are three **layers** in the veins and arteries:

- **Tunica intima**: This is the thinnest and innermost layer of arteries and veins. It has direct contact with blood.
- **Tunica media**: The middle layer is made up of smooth muscle cells.
- **Tunica externa**: This lies adjacent to the tunica media and is composed of collagen; it has elastic lamina in the arteries.

Functions

- Help to maintain homeostasis
- Play a role in gaseous exchange
- Involved in circulation of oxygenated and deoxygenated blood
- Protect against loss of blood at the time of injury
- Help in water, hormone, mineral and nutrient transport

Arteries

Blood is transported from the heart to the organs of the body through the arteries in a downward direction. These

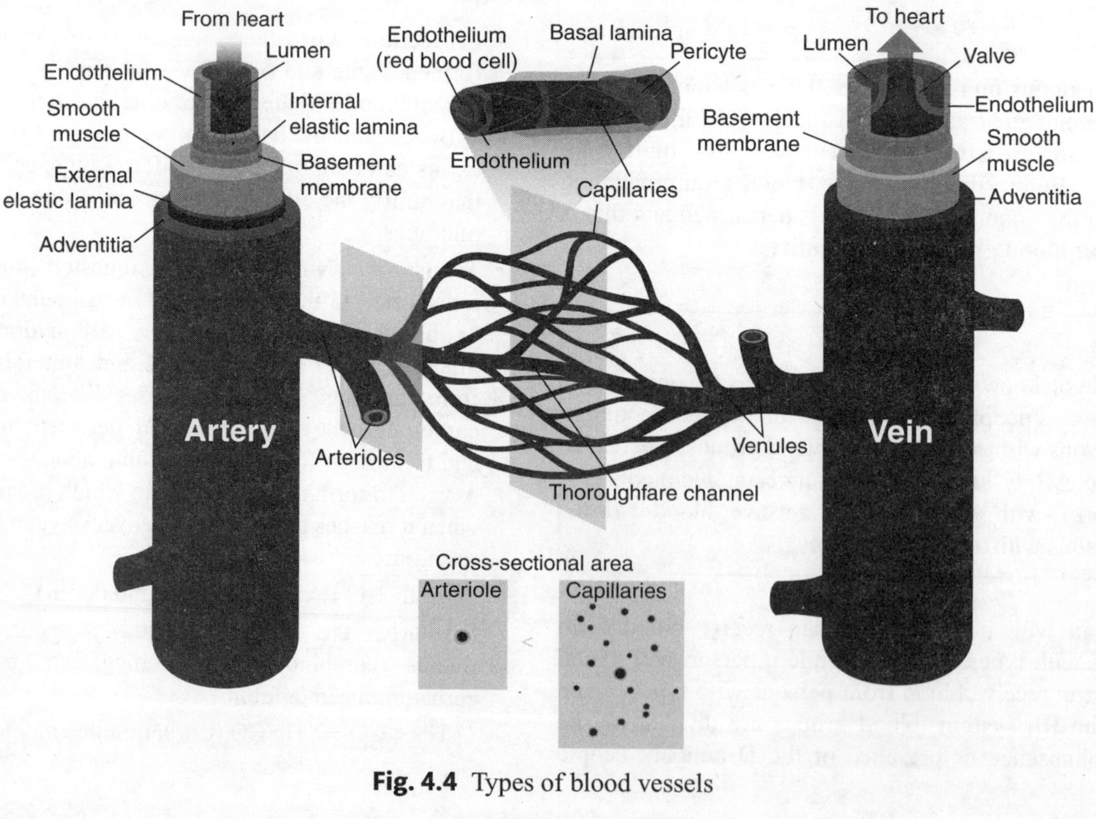

Fig. 4.4 Types of blood vessels

red-coloured blood vessels have thick and elastic walls with a narrow lumen with no valves. They are placed deeper in the body and can dilate or constrict to regulate the flow. Having narrow lumens, arteries enable blood to flow with great pressure and in spurts. The spurts correspond to the ventricle contractions. Apart from the pulmonary artery, all other arteries carry oxygen-rich or oxygenated blood. When the arteries are empty, they do not collapse. They progressively branch out, decreasing in size; the smallest arteries are called arterioles. Arterioles can change their diameter and branch to form capillaries. Arteries are strong with three distinct layers that are rigid, thick and muscular.

Veins

The blood vessels that carry deoxygenated blood from the body parts to the heart are known as veins. These translucent, blue-coloured blood vessels have a wide lumen with non-elastic, thin, muscular walls with thin valves. The valves open in the direction of the heart to avoid backflow and the movement of blood is at low pressure. They are placed superficially, near the surface of the skin and the blood flows uniformly. Apart from the pulmonary vein, all other veins carry deoxygenated blood. Veins do not constrict and collapse when empty. Their size increases as the veins progress, the smallest veins arise from unification of venules. The capillaries combine or reunite to form venules.

Capillaries

These tiny and short blood vessels are found in tissues. They bring about exchange of substances between tissues and blood. They connect the venous system and the arterial system in order to help in exchanging substances. Diffusion of oxygen into intercellular fluid from blood along with diffusion of carbon dioxide to blood from intercellular fluid is enabled by capillaries. Along with respiratory gases, diffusion of substances like urea, amino acids, hormones and glucose also takes place in the capillaries. Leucocytes can squeeze out of capillaries by diapedesis. Regulation of blood supply to different parts of the body is carried out by capillaries by undergoing vasodilation or vasoconstriction. For example, the colour of skin on a very hot day darkens and becomes pink as blood flow increases whereas in harsh winters, blood flow decreases on the face and it appears blue or pale.

CIRCULATORY PATHWAYS

There are two types of circulation: open and closed. The open circulatory system is present in lower animals like molluscs and arthropods in which large vessels are passages for blood pumped from the heart or from body cavities called sinuses. Chordates and annelids have a closed circulatory system, where blood is pumped in a closed network of blood vessels. Fish have a two-chambered heart in which the heart pumps deoxygenated blood out and it returns to the heart; this is known as **single circulation**. In reptiles and amphibians with a three-chambered heart, the left atrium receives oxygenated blood from the gills or lungs or skin. The right atrium receives deoxygenated blood from other parts and these two are mixed in a single ventricle and the mixed blood is pumped out. In mammals and birds with a four-chambered heart, oxygenated and deoxygenated blood are separated into the left and right atria, respectively, and pass on to the ventricles of the same side. The ventricles do not pump mixed blood in separated circulation pathways. This is known as **double circulation**.

HUMAN CIRCULATORY SYSTEM

The circulatory system is known as the blood vascular system. Blood is circulated in a closed circulatory system with the help of the heart and blood vessels. It comprises the heart, arteries, vein and capillaries.

Heart

The heart originates in the mesoderm and is located in the thoracic cavity between the two lungs and above the diaphragm (Fig. 4.5). The heart has a narrow end that points towards the left; contraction of the heart is most powerful at this end. The size of the heart is similar to that of a clenched fist. The outer surface of the heart is covered by a **pericardium** (double-walled membrane). The walls are filled with **pericardial fluid** which is a lubricating fluid. It reduces friction and protects the heart from mechanical injury. The human heart is divided into four chambers:

- The two upper chambers are known as **atria or auricles**
- The two lower chambers are known as **ventricles**

Auricles have thin walls as compared to ventricles as their function is to receive blood and pump it into ventricles, whereas the walls of ventricles are thick and muscular as they pump blood away from the heart. There is a thin, muscular wall that separates the left and right atria; it is known as the **interatrial septum.** A thick-walled septum separates the right and left ventricles and is known as the **interventricular septum**. The ventricle and atrium present on the same side are separated by an **atrio-ventricular septum.** These septa have openings by which the two chambers of the same side are connected. The

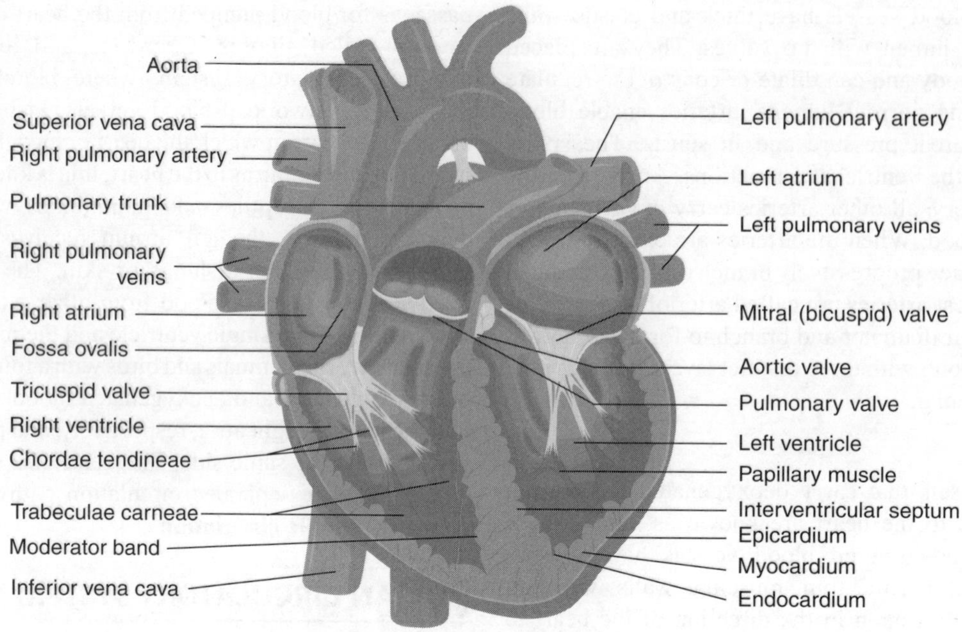

Fig. 4.5 Structure of the heart

flow of blood is regulated by **valves**. Valves are flap-like structures that help to maintain the direction of flow of blood:

- **Right atrio-ventricular valve/tricuspid valve**: This is located between the left atrium and the right ventricle and has triangular leaf-life cusps or flaps. The papillary muscles are muscular projections on the ventricular walls that give rise to **chordae tendinae** which hold the tips of the flaps together.
- **Left atrio-ventricular valve/bicuspid valve/mitral valve/pulmonary semilunar valves**: These are located between the left atrium and left ventricle. They are pocket shaped and seen at the opening of the right ventricle into the pulmonary artery.
- **Aortic semilunar valves**: These are three pocket-shaped valves present at the opening of the left ventricle into the aorta.

Fun Fact!

Did you know that the anterior vena cava, precaval and superior vena cava are the largest veins that bring deoxygenated blood to the right auricle of the heart? The pulmonary vein carries oxygenated blood.

Arteries, Veins and Capillaries

There are four types of pulmonary veins. The pulmonary artery carries deoxygenated blood to the lungs whereas the aorta carries oxygenated blood from the left ventricle of the heart. Two coronary arteries arise from the aorta and supply blood to the heart. The blood vessels involved in the circulatory system are as follows:

- **Aorta:** This is the main artery that carries blood to the body from the heart through the aortic valve. Blood travels through the aorta which makes a cane-shaped curve, allowing other arteries to deliver oxygen-rich blood to the muscles, brain and different cells. The aorta is made up of three layers: intima, media and adventitia, and is more than an inch wide at some places. It is the largest blood vessel in the body and it begins at the left ventricle of the heart and extends upwards into the chest to form an arch leading down into the abdomen, which then eventually branches into the iliac arteries.
- **Pulmonary trunk:** This short artery is responsible for transferring deoxygenated blood from the heart to the lungs; it is also known as the pulmonary artery. This trunk arises from the right ventricle base of the heart. When it passes behind the ascending aorta, the pulmonary trunk splits into the right and left pulmonary arteries to provide blood to be oxygenated in the lungs. This trunk is about 3 cm in diameter and

5 cm in length and starts at the conus arteriosus with direct continuation of the outflow tract of the right ventricular flow.

- **Coronary arteries:** These are the main blood vessels in the human body that supply blood to the heart. The right coronary artery (RCA) and left coronary artery (LCA) are the primary components. Each coronary artery contains smaller branches that go deep inside the heart muscles. The coronary arteries branch off the aorta and begin at the root of aorta; they are considered as the first of the many branches of the aorta. RCA supplies blood to the right atrium and right ventricle. It also delivers blood to one-third of the interventricular septum. LCA supplies blood to the left atrium and left ventricle and to two-thirds of the interventricular septum.
- **Vena cava:** This large vein plays a crucial role in carrying deoxygenated blood from the middle and lower body to the right atrium; it is formed by the common iliac veins of the left and right. There are two venae cavae: the inferior vena cava carries deoxygenated blood from the lower body to the heart and the superior vena cava carries blood from the upper half of the body to the heart. This large vein lies posterior to the abdominal cavity and enters the right auricle low at the back of the heart.

Heart attack or **myocardial infarction** is the condition where a particular area of heart muscle stops or dies. This is caused by blockage in the coronary artery or any of its branches. Due to this, there is insufficient supply of blood to the heart that leads to chest pain called **angina pectoris.**

The heart is made of cardiac muscles and the walls of the atria are thinner than those of the ventricles. The specialised cardiac muscle that is distributed in the heart is known as **nodal tissue**. A patch of tissue present in the right upper corner of the right atrium is known as the **sino-atrial node (SAN)** while the tissue seen in the lower left corner of the right atrium is known as the **atrio-ventricular node (AVN).**

AV bundle (atrio-ventricular bundle) is a bundle of nodal fibres that starts from the AVN, passes through the AV septa and ends at the top of the interventricular septum. This bundle divides into right and left. It results in generation of minute fibres throughout the ventricle, known as **Purkinje fibres**. Action potentials are generated by nodal muscles without any external stimuli, yet many potentials could be generated at different parts of the nodal system in a minute. The maximum number of potentials generated by SAN is 70–75 per minute; this is responsible for initiation and maintenance of the rhythmic contractile activity of the heart and is called the **pacemaker**.

Circulation and the Cardiac Cycle

Initially, all the chambers of the heart are in a relaxed state known as **joint diastole**. When SAN generates the action potential stimulating both atria, blood from the venae cavae is brought to the right auricle and from the pulmonary vein to the left auricle. Circulation starts, as after receiving blood, the auricles contract; this known as **atrial systole**. This leads to increase in blood flow by 30%. Simultaneously the ventricles go into a relaxed state and are empty. Blood passes easily from the auricles to the ventricles. The auricles then move into a relaxed state **(diastole)** and AVN generates the action potential along with the AV bundle from where the **bundle of His** transmits it through the ventricles and they contract; this is known as **ventricular systole**. To prevent backflow of blood to the auricles, the bicuspid valves tighten and remain closed with the help of chordae tendinae. Blood flows from the left ventricle to the aorta and from the right ventricle to the pulmonary artery without any hindrance from valves. When the ventricles are relaxed **(ventricular diastole),** the valves close between the passages to prevent backflow of blood. The pressure inside the ventricles declines whereas in the atria it increases, resulting in opening of the tricuspid and bicuspid valves, which leads to emptying of blood into the veins. Blood moves once again to the ventricles. The relaxed state of both the ventricles and the atria is called **joint diastole**. SAN again generates a new action potential and the cycle continues.

One heart or **cardiac cycle** is the contraction and relaxation of the auricles and ventricles. Contraction is known as systole and relaxation as diastole. One heart beat lasts 0.85 seconds. During each cycle, around 70 mL of blood is pumped by each ventricle; this is known as **stroke volume**. When stroke volume is multiplied by heart rate, it gives the **cardiac output**. It can be defined as the volume of blood pumped by each ventricle in one minute. The average blood in a healthy individual is 5 litres. The stroke volume can be altered by the body as well as the heart rate that affects the cardiac output. The cardiac output of an athlete is higher than in an average person.

The heart goes through different phases during circulation and the time taken for one cycle is indirectly counted as a pulse. Ventricle contraction and blood flow through the arteries leads to expansion of the arterial wall. This expansion can be felt on a superficial artery like the radial artery in the wrist. This alternating expansion and contraction of the arterial wall is known as a **pulse.** The **stethoscope** is an instrument that is used to hear heartbeats in the chest area. The sounds that can be heard are:

- **Lubb:** Closing of the atrio-ventricular valves at the end of a ventricular systole
- **Dub:** Closing of the semi-lunar valves at the beginning of a ventricular systole

This is followed by a gap. The heart rate differs between species and individuals of the same species on the basis of size; the smaller the body, the faster the heart rate. The rate of growth as well as metabolism is high in the growing stages of an individual and this is accompanied by a faster heart rate that helps in removing metabolic substances along with balancing the supply of metabolites. A small body loses heat at a faster rate and so a faster heart rate helps in faster heat distribution.

ECG (Electrocardiograph)

The machine that gives a graphical representation of the electrical activity in a cardiac cycle of the heart is called the electrocardiograph or ECG (Fig. 4.6). The patient is connected to the machine with three electrical leads (one on each wrist and one on the left ankle) to monitor the activity of the heart. For a more detailed evaluation, the chest region is attached with multiple leads. This is important as any abnormality in this graph could give an idea of a disease or any other health condition. Each peak in an ECG is identified from P to T and is related to specific electrical activities of the heart:

- **P wave**: Electrical excitation or depolarisation of the atria that leads to contraction of the atria
- **QRS complex**: Depolarisation of the ventricles; ventricular contraction initiated where contraction starts after Q
- **T wave**: Return of the ventricles to normal state from excited state or repolarisation; also marks the end of systole

The rate of heartbeats can be determined by counting the number of QRS complexes that occur in a time period.

Double Circulation

Blood flows through the blood vessels when the right ventricle pumps blood into the pulmonary artery (which divides into two branches) and the left ventricle into the aorta. Deoxygenated blood is passed to the lungs by the pulmonary artery whereas oxygenated blood is passed to the left atrium by the pulmonary vein. This pathway is called **pulmonary circulation**. It involves blood circulation between the lungs and the heart. The network of arterioles, arteries and capillaries carries oxygenated blood to the aorta; deoxygenated blood is collected by the veins, venules and vena cava. This is known as **systemic circulation**. This type of circulation provides oxygen,

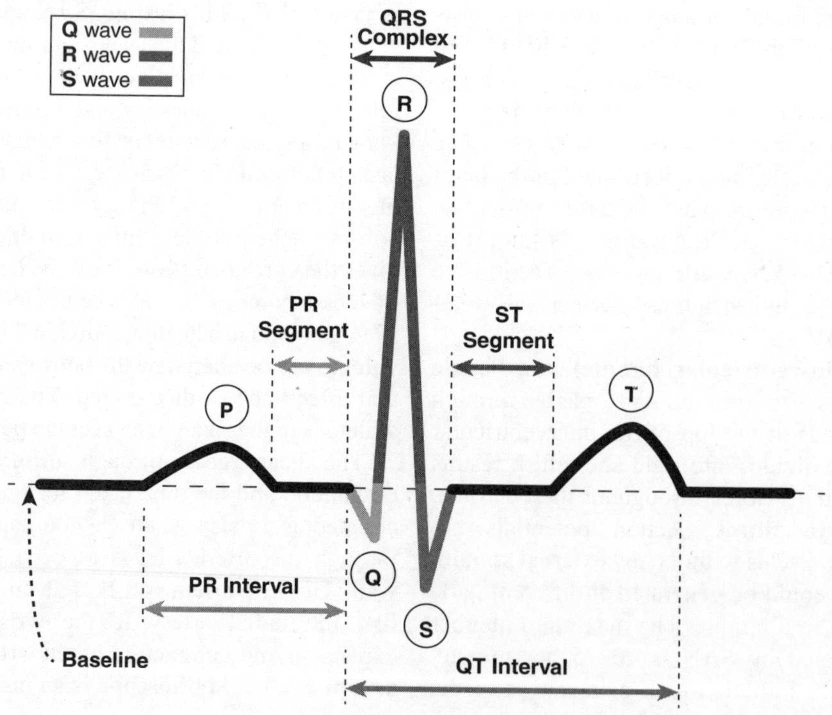

Fig. 4.6 ECG

nutrients and other essential substances to tissues while eliminating carbon dioxide and other substances (Fig. 4.7).

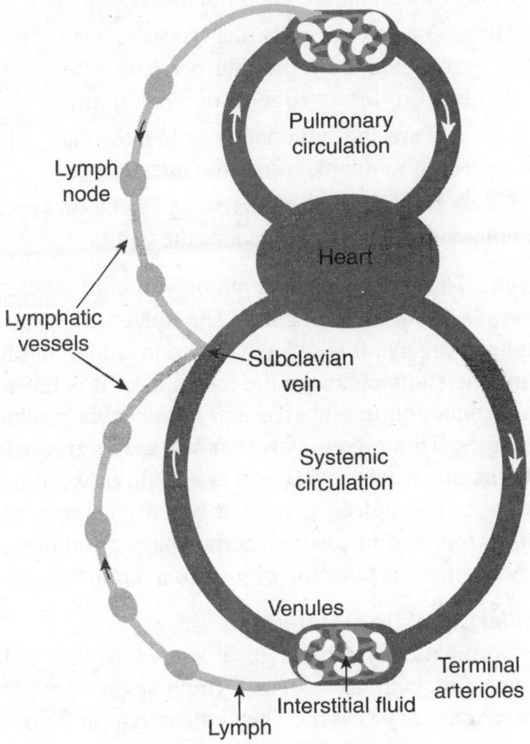

Fig. 4.7 Double circulation

A vascular connection between the digestive tract and liver is known as the **hepatic portal system.** This system starts and ends in capillaries. The capillaries in which blood is carried from the stomach cells to the intestines combine to form veins. These veins combine to form the **hepatic portal vein** that enters the liver, where it branches out into capillaries. The hepatic portal vein carries blood to the liver from the intestine before it gets delivered into systemic circulation. This system is important for nutrient transport from the digestive tract to the liver along with detoxification of nutrients by the liver. For circulation of blood from the cardiac muscles, a special coronary system of blood vessels is present.

Cardiac Activity Regulation

The activities of the heart are regulated by nodal tissue intrinsically; hence the term **myogenic.** The medulla oblongata has a special neural centre that moderates cardiac function through the autonomic nervous system (ANS). The neural signals generated from the ANS from the sympathetic nerves increase the heart rate and strength of contraction, which increases the cardiac output. In the parasympathetic system, neural signals decrease the heart rate and conduction of action potential speed is also reduced, which decreases the cardiac output. Cardiac output could also be increased by adrenal medullary hormones.

Pacemaker: The heart undergoes contractions and relaxations at a particular speed and these contractions are triggered by electrical impulses generated by specialised cells. These impulses are propagated by SAN (sino-atrial node) which helps in regulating the heart rate. It is known as the pacemaker of the heart. SAN is located on the right auricle wall. A secondary pacemaker known as AVN (atrio-ventricular node) is present near the interauricular septum. These nodes and Purkinje fibres together form an electrical conduction system. An artificial pacemaker is placed in the body of a person whose heart rate becomes inconsistent.

Disorders

- **High blood pressure or hypertension:** Blood that flows through arteries exerts pressure on the walls of the arteries; this is known as blood pressure. Normal blood pressure is 120/80 mmHg. There are two limits to this pressure: systolic pressure or pumping pressure is 120 mmHg and diastolic or resting pressure is 80 mmHg. Hypertension is a condition where a person's blood pressure is 140/90 mmHg or higher. Hypertension leads to heart disease with severe effects on the brain and kidneys.
- **Coronary artery disease or atherosclerosis:** This disease affects the vessels which supply blood to the heart muscles. The causative agents are fat, cholesterol, calcium and fibrous tissue which are deposited in the vessels, making them narrower.
- **Angina pectoris:** This is acute chest pain that appears with deficiency of oxygen in the heart muscles. It can occur at any age but is common among elderly and middle-aged people.
- **Heart failure:** This is a state where the heart is not pumping enough blood to meet the body's needs. It is also known as **congestive heart failure** as congestion of the lungs is one of the primary symptoms. It is different from cardiac arrest or a heart attack which occurs when the heart stops beating or when heart muscles suddenly get damaged due to insufficient blood supply.
- **Palpitations:** The feeling of a fast-beating or pounding heart could be triggered by medication, exercise or stress, or a medical condition. These palpitations are harmless but worrisome as they could be considered a symptom of a serious heart condition like arrhythmia. It can be felt in the neck or throat along with the chest.

- **Cardiac arrest:** The abrupt loss of function of the heart, lungs or brain by loss in breathing and consciousness is known as sudden cardiac arrest. It usually results from a problem in the heart's electrical system which induces disruption in the action of the heart and ceases blood flow in body. This could lead to death if not treated immediately as survival is only possible with fast as well as appropriate medical care like CPR or giving chest compressions. The symptoms could range from discomfort in the chest, shortness of breath, palpitations and weakness to sudden collapse, no pulse, no breathing and loss of consciousness.

LYMPH AND THE LYMPHATIC SYSTEM

The tissue fluid that is not reabsorbed in the blood vessels but which flows through another type of vessel is known as the **lymphatics or lymph.** This tissue fluid is made up of water (94%) and solids (6%) which include proteins, fats, enzymes, antibodies, carbohydrates, etc. The flow of lymph is enabled by muscle contraction of the channels and it is carried and drained by the lymph nodes. From these nodes, another lymph channel arises that carries the lymph to the anterior veins which are close to the entry of lymph in the right auricle. This is how lymph is circulated (Fig. 4.8).

Functions of Lymph

- Lymph supplies nutrients and oxygen to the parts where blood cannot reach like interior areas.
- The major defence cells like lymphocytes are present in lymph and they provide protection to cells and tissues from foreign agents or harmful microbes.
- Lacteals are the lymphatics near intestinal villi that absorb fat molecules from the intestine.
- Tissue fluid that is not reabsorbed in blood vessels is drained as lymph and reaches the heart.

Spleen: The organs that contain lymphoid tissues are known as **lymphatic organs**. The spleen is one of the lymphatic organs. It is reddish brown in colour and found behind the stomach above the left kidney. It is the site of RBC production in embryos and lymphocyte production in all ages. This organ is also known as the graveyard of RBCs as this is where RBCs that are old or worn out are destroyed. The spleen acts as a blood reservoir. Stored blood is released in case of haemorrhage or emotional or physical stress or poisoning by carbon monoxide.

Parallel Circulatory System

The lymphatic system is also known as the parallel system of the venous system. Lymph vessels are parallel to the circulatory vessels. They collect lymph fluids from tissues as part of an open-ended system and transfer them into the venous system (Fig. 4.9).

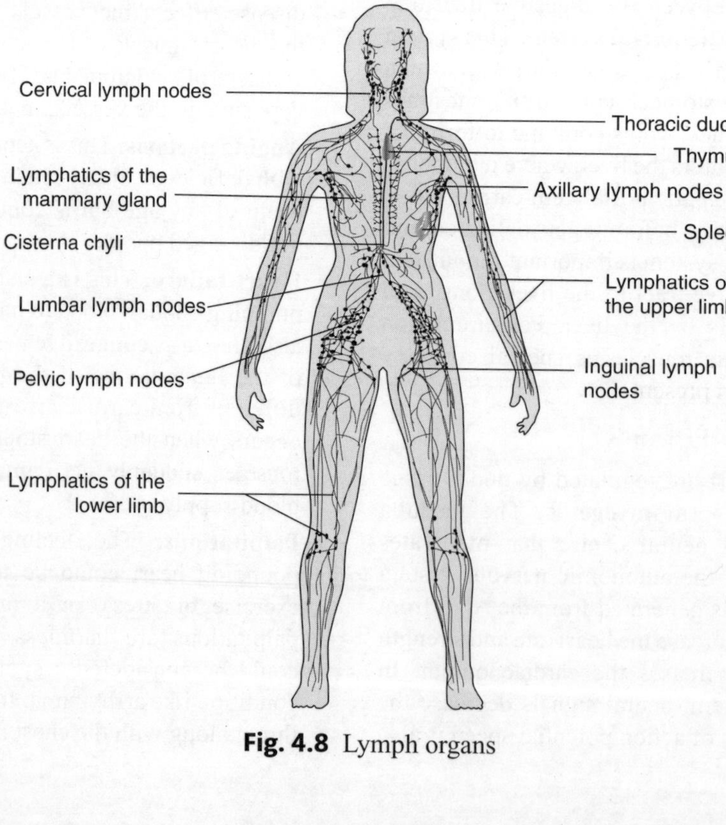

Fig. 4.8 Lymph organs

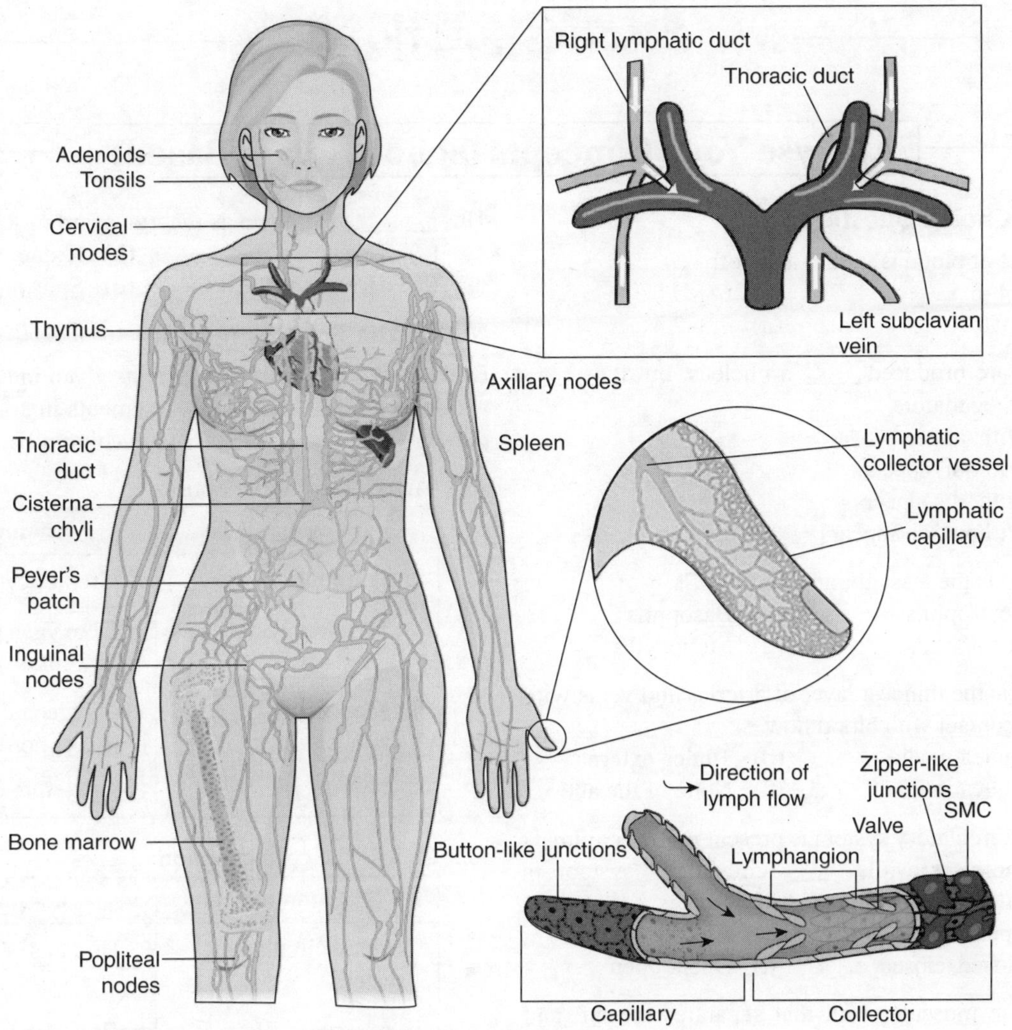

Fig. 4.9 Lymphatic system

SUMMARY

- Circulation in vertebrates is performed by a fluid connective tissue called blood which is important for essential substances to be transported to the cells and carry waste as well.
- Lymph is another tissue fluid which is used to transport substances.
- Blood is made up of formed elements and plasma, where the formed elements contain RBCs, WBCs and platelets.
- There are four blood groups in the ABO system: A, B, AB and O. The Rh blood group system is based on the D antigen.
- There are two types of circulatory systems: closed and open.
- Blood vessels are of three types: veins, arteries and capillaries.
- Circulation takes place by SAN and AVN action potential generation and a cardiac cycle is completed with the rhythmic motion of the heart.
- Electrocardiogram is an instrument used to measure impulses to detect diseases of the heart.
- Humans have double circulation; pulmonary and hepatic circulation take place.
- The lymphatic system with lymph circulation serves many purposes such as defence.

PRACTICE QUESTIONS

Analyse Your Concepts (School Exam Based)

Multiple Choice Questions

1. The pH of blood is in the range of:
(a) 6.4–6.5 (b) 6–8
(c) 9–10 (d) 7.3–7.4

2. RBCs are produced _____a nucleus, but it ______ when they mature.
(a) With, appears
(b) Without, appears
(c) With, disappears
(d) Without, disappears

3. ______ is the least abundant in WBCs.
(a) Neutrophils (b) Basophils
(c) Eosinophils (d) Monocytes

4. _____ is the thinnest layer of arteries and veins with direct contact with blood flow.
(a) Tunica media (b) Tunica externa
(c) Tunica intima (d) None of the above

5. _____ circulatory system is present in lower animals like molluscs while ____ circulatory system is present in higher animals like chordates.
(a) Open, closed (b) Closed, open
(c) Closed, closed (d) Open, open

6. The thin muscular wall that separates the left and right atria is known as:
(a) Atrio-ventricular septum
(b) Interatrial septum
(c) Interventricular septum
(d) Valve

7. Insufficient supply of blood to the heart leads to chest pain called:
(a) Cardiac arrest (b) Palpitations
(c) Heart failure (d) Angina pectoris

8. Minute fibres throughout the ventricle are known as:
(a) Purkinje fibres (b) Bundle of His
(c) Pacemaker (d) Nodal tissue

9. The vascular connection between the digestive tract and the liver is known as:
(a) Pulmonary portal system
(b) Hepatic portal system
(c) Nephrotic portal system
(d) None of the above

10. _______ is known as the graveyard of RBCs.
(a) Liver (b) Kidney
(c) Heart (d) Spleen

Match the Following

Each question contains statements given in two columns which have to be matched. Statements in column I have to be matched with statements in column II.

1. Choose the correct option:

Column I	Column II
i. Pulmonary artery	a. Granulocyte
ii. Eosinophil	b. Deoxygenated blood (exception)
iii. Monocytes	c. Oxygenated blood (exception)
iv. Pulmonary vein	d. Agranulocyte

2. Choose the correct option:

Column I	Column II
i. Open circulatory system	a. Humans
ii. Artery	b. Blue colour
iii. Closed circulatory system	c. Molluscs
iv. Veins	d. Red colour

3. Choose the best match:

Column I	Column II
i. AVN	a. Repolarisation
ii. P wave	b. Right atrium upper right corner
iii. SAN	c. Depolarisation of atria
iv. T wave	d. Right atrium lower left corner

4. Choose the best match:

Column I	Column II
i. Pericardium	a. Lymphoid organ
ii. Spleen	b. Outermost layer of blood vessel
iii. Tunica externa	c. Heart wall

5. Choose the correct option:

Column I	Column II
i. Lubb	a. Insufficient supply of blood to heart
ii. Chordae tendineae	b. Closing of semi-lunar valves
iii. Dubb	c. Closing of atrio-ventricular valves
iv. Angina pectoris	d. Hold the tips of flaps together

True or False

Read the following statements and write your answer as true or false.

1. Blood is an epithelial tissue.
2. Plasma makes up around 55% of blood.
3. Globulins are involved in the defence of the body.
4. Neutrophils and eosinophils are agranulocytes.
5. Plasma is also known as thrombocyte.
6. Thromboplastin is also known as Factor VIII.
7. A person with blood type A can receive blood from O and A blood types.
8. Except the pulmonary artery, all other arteries carry oxygen-rich or oxygenated blood.
9. The heart is covered by a pericardium.
10. A patch of tissue is present in the upper right corner of the atrium, known as atrio-ventricular node (AVN).
11. When stroke volume is multiplied by heart rate, it gives the stroke output.
12. The each peak in ECG is identified from P to T that is related to specific electrical activities of the heart.
13. The vascular connection between the digestive tract and the liver is known as the pulmonary system.
14. Lacteals are the lymphatics near intestinal villi that absorb fat molecules from the intestine.
15. Spleen is one of the lymphatic organs.

Fill in the Blanks

Complete the following statements with an appropriate word/term to be filled in the blank spaces.

1. ______ is the glycoprotein complex required for the coagulation of blood.
2. Haemoglobin combines with oxygen to make an unstable compound called ______.
3. ________ is the process by which WBCs squeeze through the walls of narrow capillaries.
4. ______ are the bilobed WBCs involved in allergic reactions.
5. The concept of blood group system was discovered by ________.
6. There are three layers in veins and arteries: _____, ________ and ________.
7. The ventricle and atrium present on the same side are separated by _____.
8. ________ is the condition when the function of a particular area of heart muscle stops or deadens.
9. During each cycle, around 70 mL of blood is pumped by each ventricle, known as _____.
10. The alternating expansion and contraction of the arterial wall is known as ______.

Assertion–Reason Questions

The following questions consist of two statements: Assertion (A) and Reason (R). Answer these questions selecting the appropriate option given below:

(a) Both A and R are true and R is the correct explanation of A
(b) Both A and R are true but R is not the correct explanation of A
(c) A is true but R is false
(d) A is false but R is true

1. Assertion (A): Haemoglobin combines with oxygen.
   Reason (R): It makes a stable compound called oxyhaemoglobin.

2. Assertion (A): WBCs are usually amoeboid in shape and possess pseudopodia.
Reason (R): This helps them squeeze through the walls of narrow capillaries.

3. Assertion (A): Increase in the number of platelets leads to clotting disorders.
Reason (R): Decrease in the number of platelets can result in excessive blood loss.

4. Assertion (A): The walls of the heart are filled with pericardial fluid.
Reason (R): It reduces friction and protects the heart from mechanical injury.

5. Assertion (A): The spleen is a lymphatic organ which is reddish brown in colour.
Reason (R): The spleen is known as the graveyard of RBCs.

One-word Answers

1. The causative agents of which disease are deposition of fat and cholesterol in the vessels which make them narrower?
2. What is known as the pacemaker of the heart?
3. Which wave of ECG is known to cause electrical excitation or depolarisation of atria?
4. What is the relaxed state of both ventricles and atria called?
5. What covers the outer surface of the heart?

Define the Term

1. Leucocytes
2. Lymphocytes
3. Pacemaker
4. Hypertension
5. Cardiac cycle

Groups of Terms

Given below are groups of terms. In each group the first pair indicates the relationship between the two terms. Rewrite and complete the second pair on a similar basis.

1. Oxygenated: arteries :: Deoxygenated: _____
2. Oxygen: oxyhaemoglobin :: Carbon dioxide: _____
3. Group A: Antibody B :: Group B: ______
4. Phagocytic: neutrophils :: Histamine: ____
5. Outermost : tunica externa :: Innermost: _____

Identify the Odd One Out

1. Mg^{++}, Cl^-, HCO_3, K
2. Alpha 1 globulin, kappa globulin, alpha 2, beta globulin
3. Erythrocytes, leucocyte, hepatocytes, platelets
4. Lymph, lymphatic vessels, spleen, lymphocytes
5. Asthma, atherosclerosis, hypertension, angina pectoris,

Image-Based Questions

1. What is the term used to describe this cell?

(a) WBC (b) RBC
(c) Platelets (d) Fibrinogen

2. What is the wall of this organ called?

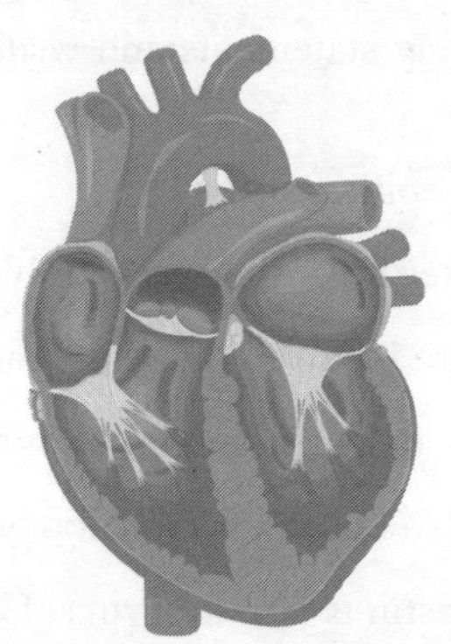

(a) Perimetrium (b) Pericardium
(c) Peritoneal (d) Peripheral

3. What does this graph represent?

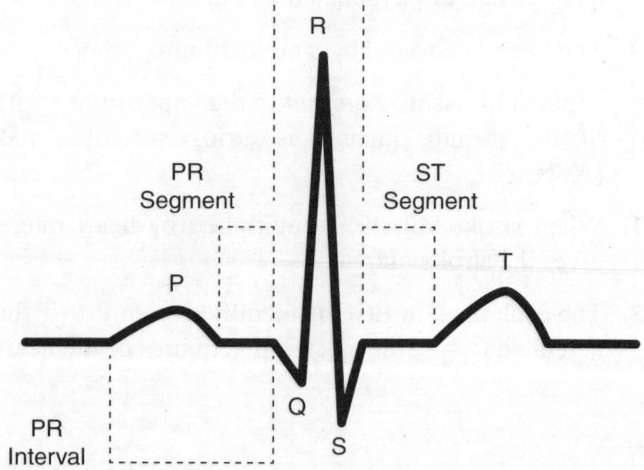

(a) BCG (b) ECG
(c) PQR (d) TRS

4. What is the name of the muscle in this organ?

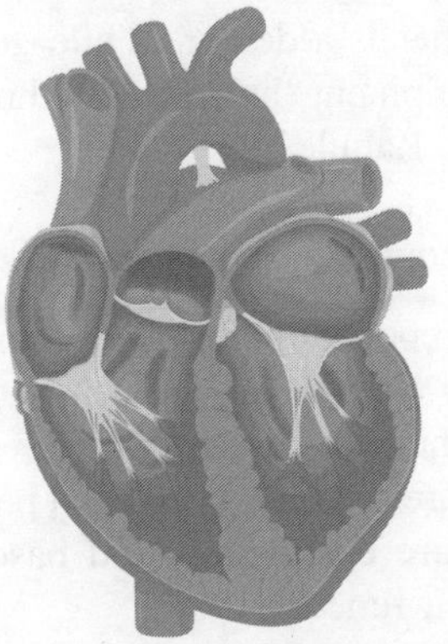

(a) Skeletal (b) Myoskeletal
(c) Smooth (d) Cardiac

5. What is the name of the connective tissue circulated by this organ?

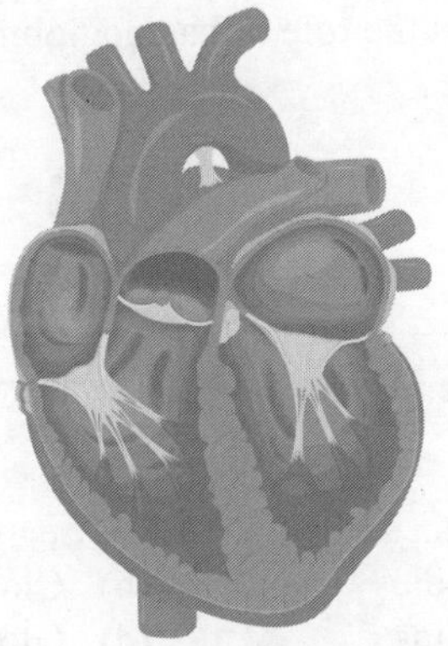

(a) Lymph (b) Blood
(c) Water (d) Areolar

6. Choose the name of this circulatory system:

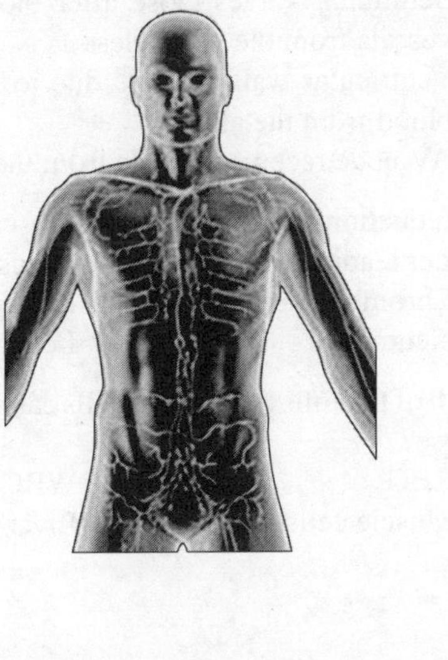

(a) Blood (b) Lymphatic
(c) Both (a) and (b) (d) None of the above

Very Short Answer Questions

1. What is the instrument used to hear heartbeats?
2. Name three principal fluids of the body.
3. What are the types of formed elements found in blood?
4. What are the types of blood vessels and what are their layers?
5. Name the valves in the heart.
6. Who discovered the blood group system? What are the different types of blood groups?
7. What are the different diseases or defects of the heart?
8. Name the two sounds of heartbeat with their origin.
9. Name all the waves and the meaning of ECG.
10. Name the different antigens found on the surface of RBCs, along with their antibodies in a blood group system.

Short Answer Questions

1. Write about the defects of the human heart. Explain any two.
2. Write a short note on blood.
3. What are the functions of plasma? Explain briefly.
4. What are the functions of blood vessels?
5. Explain the significance of ECG.
6. Explain the ABO blood group system in brief.
7. Explain double circulation.
8. Discuss the three types of cells found in plasma.
9. Write about the cardiac cycle.
10. Describe the pacemaker briefly.
11. Explain the pulmonary trunk and vena cava.

Long Answer Questions

1. Describe the components of blood in detail.
2. Describe the structure of the heart.
3. What is circulation and double circulation? Explain in detail.
4. Explain circulation and the cardiac cycle in detail.

5. What are the types of body fluids involved in circulation? Explain each in detail.
6. Explain the defects of the heart and the significance of the pacemaker.
7. Explain the lymphatic system in detail.
8. Explain ECG and the significance of each wave.
9. Write a short note on the pulmonary and hepatic circulatory systems.
10. What are blood vessels? What is their structure and function? Explain each type.

Case Study-Based Questions

1. Aditya went to visit his grandmother in hospital and heard that she needs a blood transfusion. He offered to donate his blood. But, Aditya's father said that he could not due to differing blood groups. This puzzled Aditya. Can you help Aditya understand why?
   (a) What is a blood group?
   (b) What are the antigens and antibodies present in different blood groups?
2. Neha and Ritu were playing when they saw an ECG report with an analysis of a healthy heart.They tried to understand the report but were unable to do so. Can you help them?
   (a) What is ECG and why is it used?
   (b) What do the P, Q, R, S and T waves signify?
3. Rahul's father is a doctor. While going through his father's equipment, he saw a stethoscope. Can you explain it to Rahul?
   (a) What is a stethoscope?
   (b) What sound does the heart make?
4. In a blood report of her mother, Zainab encountered several new terms. Can you help her understand the terms and their functions?
   (a) What are red blood cells and platelets?
   (b) What are eosinophils and basophils and what are their functions?
5. Paula and her family went on a trip to the mountains, and after a few days, she noticed that their cheeks became pink, like the other residents. Can you solve her query?
   (a) Why did their cheeks become pink?
   (b) What is the role of haemoglobin?

COMPETITION WINDOW (OBJECTIVE TYPE)

1. RBCs are enucleated. Which statement is most appropriate for this feature?
   I. Somatic cells
   II. Do not reproduce
   III. All space is available for oxygen transport
   IV. Do not metabolise
   A. I, II, IV B. III, IV
   C. III D. I
2. AB blood group is a universal recipient because it possesses:
   (a) No antibodies
   (b) A and B antibodies
   (c) No antigens and no antibodies
   (d) Both antigens and no antibodies
3. Pulmonary artery has blood pressure that is:
   (a) More than in the carotid
   (b) Same as in the aorta
   (c) Less than in the vena cava
   (d) More than that of the pulmonary vein
4. Serum lacks:
   (a) Antibodies (b) Clotting factors
   (c) Albumins (d) Globulin
5. Doctors use a stethoscope to hear the sounds of the cardiac cycle. When does the second sound occur?
   (a) AV valves open up
   (b) Semilunar valves close after blood flows into vessels from the ventricles
   (c) Ventricular walls vibrate due to gushing in of blood from the atria
   (d) AV node receives signals from the SA node
6. The reduction in which blood cells causes clotting disorder leading to excessive blood loss?
   (a) Thrombocyte (b) Erythrocyte
   (c) Neutrophil (d) Leucocyte
7. Which of the following blood cells cannot metabolise glucose?
   (a) RBC (b) WBC
   (c) Muscle cells (d) Liver cells

8. Which valves guard the right atrium and right ventricle?
   (a) Bicuspid (b) Tricuspid
   (c) Mitral (d) Semilunar

9. Superior vena cavae receive blood from:
   (a) Head, neck, chest
   (b) Digestive
   (c) Lower limbs
   (d) Diaphragm and legs

10. What is the function of RBCs?
    (a) Transport oxygen (b) Transport water
    (c) Clotting (d) Defence

11. What is the composition of blood?
    (a) Blood vessels, platelets, RBC
    (b) RBC, plasma, WBC, platelets
    (c) RBC, WBC, platelets
    (d) Plasma, WBC, RBC

12. Which of the following carries blood from the lungs to the heart?
    (a) Pulmonary artery (b) Pulmonary vein
    (c) Coronary artery (d) Renal artery

13. Blood from the left side of the heart is bright red due to:
    (a) It contains RBC
    (b) Rich in haemoglobin
    (c) Rich in oxyhaemoglobin
    (d) Closest to the lungs

14. The function performed by the right side of the heart:
    (a) Pumps blood to the rest of the body
    (b) Receives blood from the body
    (c) Receives blood from the lungs
    (d) Pumps blood to the brain

Answer Keys

Multiple Choice Questions

1. (d) 2. (c) 3. (b) 4. (c) 5. (a)
6. (b) 7. (d) 8. (a) 9. (b) 10. (d)

Match the Following

1. i (b), ii (a), iii (d), iv (c)
2. i (c), ii (d), iii (a), iv (b)
3. i (d), ii (c), iii (b), iv (a)
4. i (c), ii (a), iii (b)
5. i (c), ii (d), iii (b), iv (a)

True or False

1. F 2. T 3. T 4. F 5. F
6. F 7. T 8. T 9. T 10. F
11. F 12. T 13. F 14. T 15. T

Fill in the Blanks

1. Fibrinogen
2. Oxyhaemoglobin
3. Diapedesis
4. Eosinophils
5. Karl Landsteiner
6. Tunica intima, Tunica media, Tunica externa
7. Atrio-ventricular septum
8. Myocardial infarction
9. Stroke volume
10. Pulse

Assertion–Reason Questions

1. (c) 2. (a) 3. (d) 4. (a) 5. (b)

One-word Answers

1. Atherosclerosis
2. SAN
3. P wave
4. Joint diastole
5. Pericardium

Define the Term

1. White blood cells (WBCs) or leucocytes do not have haemoglobin and so are colourless in nature. They possess a nucleus and are present in smaller amounts, around 6000–8000 mm^3 of blood. They are usually amoeboid in shape and possess pseudopodia which helps them to squeeze through the walls of narrow capillaries, through diapedesis. They are produced in the lymph nodes, bone marrow and sometimes in the spleen and liver. The average lifespan of WBCs is about two weeks and old ones are destroyed like RBCs.
2. Around 20–25% of WBCs are lymphocytes, which are of two types: B and T. They are responsible for the immune responses of the body. They are very small in size, with a single large nucleus and no granules in the cytoplasm. B and T cells are produced in the bone marrow; while the former mature there, the latter mature in the thyroid gland. The antibodies

produced by the lymphocytes act against microbes, toxins and foreign particles.

3. The heart undergoes contractions and relaxations at a particular speed and these contractions are triggered by electrical impulses generated by specialised cells. These impulses are propagated by SAN (sino-atrial node) which helps in regulating the heart rate; it is known as the pacemaker of the heart. SAN is located on the right auricle walls. A secondary pacemaker known as AVN (atrio-ventricular node) is present near the interauricular septum. Nodes and Purkinje fibres together form an electrical conduction system.
4. Blood exerts pressure on the walls of the arteries, known as blood pressure. Normal blood pressure is 120/80 mmHg. There are two limits to this pressure: systolic or pumping pressure is 120 mm Hg while diastolic or resting pressure is 80 mmHg. Hypertension is a condition where a person's blood pressure is 140/90 mmHg or higher. Hypertension leads to heart disease with severe effects on the brain and kidneys.
5. One heart or cardiac cycle is the contraction and relaxation of the auricles and ventricles. Contraction is known as systole and relaxation is diastole. One heartbeat lasts about 0.85 seconds. During each cycle, around 70 mL of blood is pumped by each ventricle; this is known as stroke volume. When the stroke volume is multiplied by the heart rate, it gives the cardiac output. It could be defined as the volume of blood pumped by each ventricle in one minute.

Groups of Terms

1. Veins
2. Carboxyhaemoglobin
3. Antibody A
4. Basophils
5. Tunica intima

Identify the Odd One Out

1. K
2. Kappa globulin
3. Hepatocytes
4. Lymphocytes
5. Asthma

Image-Based Questions

1. (d) 2. (c) 3. (b) 4. (d) 5. (b)
6. (b)

Very Short Answer Questions

1. Stethoscope is an instrument that is used to hear heartbeats over the chest area.
2. Blood, lymph and tissue fluids are the three principal fluids of the body.
3. Erythrocytes, leucocytes and platelets are the three types of formed elements found in blood.
4. The three types of blood vessels are artery, vein and çapillary. The three layers of blood vessels are tunica intima, tunica media and tunica externa.
5. Right atrioventricular valve/Tricuspid valve, left atrioventricular valve/bicuspid valve/mitral valve/ pulmonary semilunar valves, aortic semilunar valves
6. The concept of blood group system was discovered by Karl Landsteiner. The different blood groups in humans are A, B, AB and O.
7. Atherosclerosis, hypertension, angina pectoris, heart failure, palpitations and cardiac arrest are the defects of the heart.
8. The sounds that can be heard are lubb or the closing of the atrioventricular valves at the end of ventricular systole and dub or the closing of the semilunar valves at the beginning of ventricular systole.
9. Each peak in ECG is identified from P to T that is related to a specific electrical activity of the heart. P wave: electrical excitation or depolarisation of atria that leads to contraction of atria; QRS complex: depolarisation of ventricles with ventricular contraction initiated where contraction starts after Q; T wave: the return of the ventricle to normal state from excited state or repolarisation. This also marks the end of systole.
10. The antigens present in the different blood groups are:

    Group A: Antigen A and antibody B

    Group B: Antigen B and antibody A

    Group AB: Antigens A and B and no antibodies

    Group O: No antigens but A and B antibodies

Short Answer Questions

1. Atherosclerosis, hypertension, angina pectoris, heart failure, palpitations and cardiac arrest are the defects of the heart.

   Coronary Artery Disease or Atherosclerosis: This affects the vessels which supply blood to heart muscles. The causative agents are fat, cholesterol, calcium and fibrous tissue which are deposited in vessels and make them narrower.

Angina Pectoris: Acute chest pain caused by deficiency of oxygen in reaching the heart muscle. It can occur in persons of any age but is commonly seen in elderly and middle-aged people.

2. Animals that possess a circulatory system have blood. According to the theory of evolution, blood arose from a single type of cell that was known to enable phagocytosis and provide nutrition. Blood is a type of loose connective tissue that is circulated throughout the body to deliver nutrients and oxygen to various organs (cells and tissues). An average human adult has around 5–6 litres of blood, which makes up around 8% of total body weight. It consists of plasma and formed elements. Blood is thick and dynamic in nature. It has a pH of 7.3–7.4 which makes it slightly alkaline; it tastes salty.

3. This viscous, straw-coloured fluid constitutes approximately 55% of blood. Plasma is composed of water (90–92%) and proteins (8%). It also has salts, enzymes and minerals like Mg^{++}, Cl^-, HCO_3, Na^+ and Ca^{++}. Blood plasma transfusions are carried out in patients with life-threatening injuries and failure of the liver. The main proteins present in plasma are fibrinogen, globulin and albumins. Fibrinogen is required for coagulation of blood. It is a glycoprotein complex which is circulated in vertebrate blood and is produced by the liver. During vascular or tissue injury, it is converted by thrombin to fibrin. Globulins are involved in the defence mechanisms of the body. The types of globulin are alpha 1, alpha 2, beta and gamma. They exert oncotic pressure and their deficiency results in oedema due to increased osmotic pressure; loss of carrier functions and susceptibility to infections is also decreased. Albumin is involved in maintaining osmotic balance. It belongs to the globular protein family and is water soluble and can denature in high heat.

4. The tubular structures that carry blood and which extend from the heart to different parts of the body are known as blood vessels. There are three types of blood vessels: artery, veins and capillaries. The functions of blood vessels are:
   - Help to maintain homeostasis
   - Play a role in gaseous exchange
   - Involved in the circulation of oxygenated and deoxygenated blood
   - Protect against loss of blood at the time of injury
   - Help in water, hormone, mineral and nutrient transport

5. The ECG machine gives a graphical representation of the electrical activity during a cardiac cycle. The patient is connected to a machine with three electrical leads (two wrist and one left ankle) to monitor the activity of the heart. For a more detailed evaluation, the chest region is attached to multiple leads. This is important as any abnormality in this graph could give an idea about a disease or other health condition. Each peak in ECG is identified from P to T that is related to a specific electrical activity of the heart.

   P wave: Electrical excitation or depolarisation of atria that leads to contraction of atria

   QRS complex: Depolarisation of ventricles with ventricular contraction initiated where contraction starts after Q

   T wave: The return of the ventricle to a normal state from an excited state or repolarisation. This also marks the end of systole.

   The rate of heartbeat can be recognised by counting the number of QRS complexes that occur in a time period.

6. In humans, there are two main types of antigens on the RBC surface: Antigen A and Antigen B. Depending on the absence or presence of these antigens, there are four types of blood groups: A, B, AB and O. This system is also known as ABO blood grouping system. There are other systems available, but ABO and Rh are the important systems. The basics of the ABO blood group system is:

   Group A: Antigen A and antibody B

   Group B: Antigen B and antibody A

   Group AB: Antigen A and B and no antibodies

   Group O: No antigen, but A and B antibodies

7. The right ventricle pumps blood into the pulmonary artery, which divides into two branches; the left ventricle pumps blood into the aorta. Deoxygenated blood is passed to the lungs by the pulmonary artery while oxygenated blood is sent to the left atrium by the pulmonary vein. This pathway involves blood circulation between the lungs and heart. The network of arterioles, arteries and capillaries carries oxygenated blood to the aorta from where deoxygenated blood is collected by the veins, venules and vena cava. This is known as systemic circulation. This type of circulation provides oxygen, nutrients and other essential substances to tissues while eliminating carbon dioxide and other substances. A vascular connection between the digestive tract and the liver is known as the hepatic portal system. It starts and ends in capillaries. The capillaries in which blood is carried from stomach cells to the intestines combine to form veins. These veins combine to form

a hepatic portal vein that enters the liver where it branches out into capillaries. The hepatic portal vein carries blood to the liver from the intestine before delivering it to systemic circulation. This system is important for nutrient transport from the digestive tract to the liver along with the detoxification of nutrients by the liver. For the circulation of blood from cardiac muscles, a special coronary system of blood vessels is present.

8. Formed elements that makes up 45% of blood usually consist of three cells:

Erythrocytes: These are the most abundant cells in blood which are formed in the red bone marrow of long bones of adults like the ribs, ilium and breastbone. They do not have a nucleus and are concave in shape. Mature RBCs do not have endoplasmic reticulum, mitochondria and nucleus, making them more efficient as they have more surface area and they only transport oxygen (and do not consume it) along with glucose. It contains haemoglobin which is the respiratory pigment. This consists of two parts: haem (iron-containing) and globin (protein).

Leucocytes: White blood cells (WBCs) do not have haemoglobin and so are colourless. They possess a nucleus and are present in smaller amounts in blood. They are usually amoeboid in shape and possess pseudopodia which helps them to squeeze through the walls of narrow capillaries. They are produced in the lymph nodes, bone marrow and spleen and liver. On the basis of their structure and other characteristics, they are categorised into two main groups:

- Granulocytes: Neutrophils, basophils and eosinophils
- Agranulocytes: Monocytes and lymphocytes

Platelets: Thrombocytes are cell fragments produced by special cells present in the bone marrow known as megakaryocytes. They release multiple substances involved in clotting of blood. Decrease in the number of platelets can lead to clotting disorders which can result in excessive blood loss. Platelets are non-nucleated with an oval shape. Platelets release thrombokinase at a site of injury; this initiates the clotting process.

9. The beginning of the cycle is marked by contraction of the auricles followed by blood flow into ventricles; the auricles then enter a relaxed state. This leads to contraction of the ventricles as in ventricular systole, followed by ventricular diastole as the ventricles move into a relaxed state. When both ventricles and auricles are in the relaxed state, it is known as joint diastole. This sequence marks one cardiac cycle. The ventricle valves prevent backflow. The phases of the heart during circulation and the time distribution for different phases of one cycle is indirectly counted as a pulse. The ventricle contraction and blood flow through arteries leads to expansion of the arterial wall. This expansion can be felt by pressing on a superficial artery like the radial artery at the wrist. The alternating expansion and contraction of the arterial walls is known as a pulse. The sounds that can be heard are:

- Lubb: Closing of the atrio-ventricular valves at the end of ventricular systole
- Dub: Closing of the semilunar valves at the beginning of ventricular systole

10. The heart undergoes contractions and relaxations at a particular speed and these contractions are triggered by electrical impulses generated by specialised cells. These impulses are propagated by SAN (sino-atrial node) which helps in regulating the heart rate; it is known as the pacemaker of the heart. SAN is located on the right auricle wall. A secondary pacemaker known as AVN (atrio-ventricular node) is present near the interauricular septum. Nodes and Purkinje fibres together form the electrical conduction system. An artificial pacemaker is placed in the heart of the person whose heart rate becomes inconsistent.

11. (a) The short artery is responsible for transferring deoxygenated blood from the heart to the lungs, known as the pulmonary artery. This trunk arises from the right ventricle base of the heart. When it passes behind the ascending aorta, the pulmonary trunk splits into the right and left pulmonary arteries to provide blood to be oxygenated in the lungs. This trunk is about 5 cm long and starts at the conus arteriosus with direct continuation of the outflow tract of right ventricular flow.

(b) The large vein that plays a crucial role in carrying deoxygenated blood from the middle and lower body into the right atrium is formed by the common iliac veins of the left and right side. There are two vena cavae, in which the inferior vena cava carries deoxygenated blood from the lower body to the heart and the superior vena cava carries blood from the upper half of the body. This large vein lies posterior to the abdominal cavity and enters the right auricle low at the back of the heart.

Long Answer Questions

1. Hint: The formed elements and fluids present in blood should be explained in detail along with a diagram, as described in this chapter.
2. Hint: All the parts of the heart with their functions and characteristics should be explained with the help of a well-labelled diagram, as described in this chapter.
3. Hint: Explain circulation and double circulation with a labelled diagram, as described in this chapter.
4. Hint: Explain circulation and relate it to the cardiac cycle by defining its features and significance, as described in this chapter
5. Hint: Explain blood and lymph along with their types and components in detail, as described in this chapter.
6. Hint: Define the reason and mechanism followed in all defects of the heart. Define the pacemaker and explain why it is important, as described in this chapter.
7. Hint: Explain lymph and the lymphatic system in detail, as described in this chapter.
8. Hint: Explain ECG, its waves and their significance along with when they are created, as described in this chapter.
9. Hint: Explain the hepatic and pulmonary circulatory systems, as described in this chapter.
10. Hint: Explain the blood vessels, their types, functions and layers along with the characteristics of each, as described in this chapter.

Case Study-Based Questions

1. (a) Different kinds of antigens are present on RBCs in different persons. The antibodies that are complementary to these antigens are also present in blood plasma. According to antigens present on the RBC surface, blood can be divided into different groups. This concept was discovered by Karl Landsteiner.

   (b) The basics of the ABO blood group system are:

   Group A: has Antigen A and antibody B

   Group B: has Antigen B and antibody A

   Group AB: has Antigen A and B and no antibodies

   Group O: has no antigen and both A and B antibodies

2. (a) The ECG machine gives a graphical representation of the electrical activity during a heart's cardiac cycle. The patient is connected to a machine with three electrical leads to monitor the activity of the heart. For a more detailed evaluation, the chest region is attached with multiple leads. This is important as any abnormality in this graph could give an idea about the disease or any other health condition.

   (b) Each peak in ECG is identified from P to T that is related to specific electrical activity of the heart.
   - P wave: Electrical excitation or depolarisation of atria that leads to contraction of atria
   - QRS complex: Depolarisation of ventricles with ventricular contraction initiated where contraction starts after Q

   T wave: Return of ventricle to normal state from excited state or repolarisation. This also marks the end of systole.

3. (a) A stethoscope is an instrument that is used to hear heartbeats over the chest area.

   (b) Sounds that can be heard are:
   - Lubb: Closing of atrioventricular valves at the end of ventricular systole
   - Dub: Closing of semi-lunar valves at the beginning of ventricular systole

4. (a) RBCs: The most abundant cells in blood are formed in the red bone marrow of the long bones of adults like the ribs, ilium and breastbone. It contains a complex protein with iron in it known as haemoglobin which is responsible for its red colour.

   Platelets: They are known as thrombocytes and are cell fragments produced by special cells present in the bone marrow known as megakaryocytes. A decrease in the number of platelets leads to clotting disorders. Platelets are non-nucleated with an oval shape. Platelets release thrombokinase at the sites of injury; this initiates the clotting process.

   (b) Basophils: Least abundant in WBCs (0.5–1%) and are produced in the bone marrow. It secretes serotonin, histamine, heparin, etc. involved in inflammation as they dilate blood vessels (histamine). This is indistinctly lobed with a large nucleus and granulated cytoplasm. They can be stained with basic dyes like methylene blue.

   Eosinophils: These cells can resist infections and are associated with allergic reactions and comprise 2–3% of all WBCs. They have a bilobed nucleus with granulated cytoplasm that is bigger than others. They can be stained with acidic dyes like eosin. They are produced in the bone marrow and perform phagocytosis as well as secretion of antitoxins.

5. (a) Regulation of blood supply to different parts of the body is performed by capillaries by undergoing vasodilation or vasoconstriction. At high altitudes, haemoglobin is increased as more oxygen needs to be transported due to low availability.

(b) Erythrocytes contain a complex protein containing iron known as haemoglobin which is responsible for the red colour and is the respiratory pigment. It consists of two parts: haem (iron-containing) and globin (protein). Haemoglobin when combined with oxygen makes an unstable compound called oxyhaemoglobin. This unstable compound releases in the tissues that require oxygen.

Competition Window

1. (c)	**2.** (d)	**3.** (d)	**4.** (b)	**5.** (b)
6. (a)	**7.** (a)	**8.** (b)	**9.** (a)	**10.** (a)
11. (b)	**12.** (b)	**13.** (c)	**14.** (b)	

CHAPTER 5

Neural and Endocrine Systems

PREREQUISITES

You must have knowledge of:

- Human physiology
- Common biology terminology

LEARNING OBJECTIVES

At the end of this chapter, you will be able to:

- Explain the role of the nervous system and endocrine system in control and coordination
- Identify the main components of the nervous system and list their functions
- Explain the structure and role of the neuron in the transmission of impulses
- Describe the role of some endocrine glands in regulating our growth and behaviour
- Explain coordination in plants and the plant hormones

INTRODUCTION

Have you ever wondered how our body is controlled to perform various functions? The body moves only when our muscles are activated. Can you explain how different organs work together to perform their functions at the right time? Are you aware of which organs are in charge of your thoughts, feelings, emotions and actions?

The neural and endocrine systems ensure that the body functions in a coordinated and controlled manner. The brain, spinal cord, sense organs and nerves make up the nervous system in humans. The endocrine system is controlled by hormones, which are created by specific glands and released directly into the bloodstream. The neural system and endocrine system collaborate to communicate, integrate and coordinate the operations of our body's many organs and systems.

Even in plants that lack muscles or a nervous system, movements are not just random and spontaneous but occur in a highly controlled and coordinated manner due to the presence of hormones secreted by the plant cells that help in the growth and development of the plant.

Touching a hot object is an urgent and dangerous situation for us. We need to detect it and respond to it. How does that happen? All the information from our environment is detected by the specialised tips of some nerve cells. These receptors are usually located in our sense organs, such as the inner ear, nose, tongue, and so on. Gustatory receptors detect taste while olfactory receptors detect smell.

Before taking a closer look at the nervous system, you should be aware of the following terms:

- **Stimulus**: A sudden change in the external environment that leads to change in organisms or body parts
- **Response**: The body's reaction to a stimulus
- **Impulse**: An electric disturbance or wave
- **Receptors**: Specialised epithelial cells that receive stimuli and send impulses to the brain or spinal cord
- **Effectors**: Glands or muscles that receive impulses from the spinal cord or brain and contract or secrete substances

THE NERVOUS SYSTEM

The nervous system is a complex regulatory system that not only coordinates actions but also transmits sensory information and signals to and from the different parts of the body. Its function is dependent on its ability to perceive and respond to stimuli in an environment.

A stimulus is an agent or a change in the environment that can cause the body to respond. There are a variety of stimuli: physical (touch, prick, pressure), auditory, chemical, radiant (light), heat or cold, or electrical.

Importance of the Nervous System

- It controls and harmonises the various voluntary activities of our body, like running
- It keep us informed about our surroundings with the help of our sense organs
- It regulates involuntary activities, like breathing
- It help us to think, reason and remember things

Structure

Let us look at how neurons (nerve cells) communicate with each another and other tissues to receive and transmit information throughout the body. Figure 5.1 depicts the generalised structure of a neuron. It is divided into three parts:

- **Dendrites**: These are cytoplasmic extensions from the cell body that are branched. The dendritic tips of nerve cells receive impulses and trigger a chemical reaction that produces an electrical impulse, which is then conveyed to the cell body.
- **Cell body**: The nucleus of the cell body is highly defined and surrounded by a cytoplasm. It has organelles. The impulse received is passed on to the axon by the cell body.
- **Axon**: This is one of the cell body's branches that is unusually long in comparison to the others. The point where two neurons meet is called the **synapse** and the area between two neurons at the synapse is termed the synaptic cleft. In most neurons, the axon has a **myelin sheath** surrounding its body. The myelin sheath is further enclosed by an outer thin sheath known as **neurolemma.** There are gaps between these sheaths along the length and they are known as **nodes of Ranvier.**

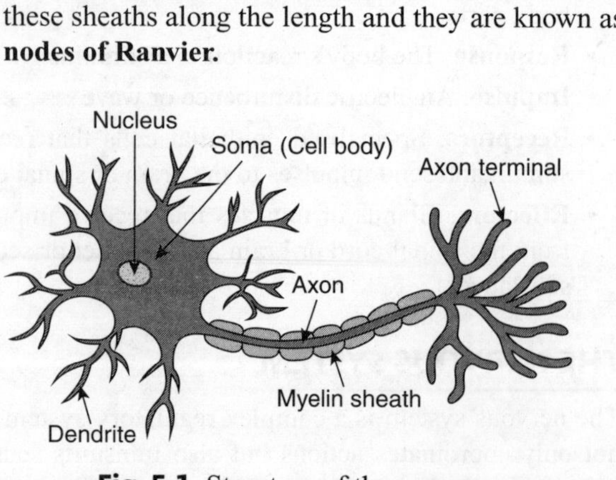

Fig. 5.1 Structure of the neuron

The nervous tissue is thus made up of an organised network of nerve cells or neurons, and is specialised for conducting information through electrical impulses.

Types of Neurons

- **Sensory neuron**: This type of neuron carries impulses from the sense organ receptor to the main nervous system, which is the brain or spinal cord.
- **Motor neuron**: This type of neuron carries impulses from the main nervous system to an effector such as a muscle or gland.
- **Association neuron**: This type of neuron is located in the brain and spinal cord and interconnected with sensory and motor neurons.

Structure of Motor/Efferent Neuron: The motor neuron (Fig. 5.2) can be divided into the following components:

- **Soma**: This is the body of the cell where the nucleus is located. It controls the cell by producing proteins and maintains its functions.
- **Axon**: This is a long extension from the soma, where transmission of the information received by dendrites takes place.
- **Dendrites**: These are branch-like structures located at the end of the neuron. They are used to send as well as receive signals or information from other neurons.
- **Myelin sheath**: These cells protect the neuron from damage; they also prevent leakage and weakening of signals.

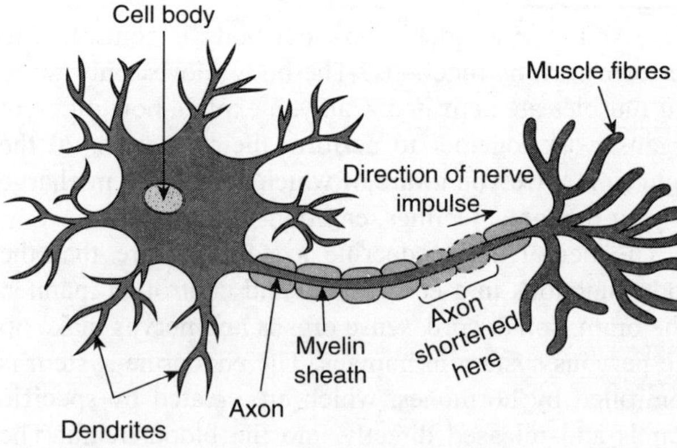

Fig. 5.2 Motor neuron

Nerves

A bundle of nerve fibres or axons of different and separate neurons that are enclosed in a tubular sheath is known as a nerve. It is a thread-like white structure that emerges from the spinal cord or brain and which branches out to all the parts of the body. The myelin sheath acts as insulation and prevents mixing of different impulses in adjacent axons. There are three types of nerves:

- **Sensory nerve**: This contains only sensory fibres that bring impulses from receptors. Example: optic nerve.
- **Motor nerve**: This contains only motor fibres that carry impulses from the spinal cord or brain to effector organs. Example: nerves from the brain to the eyeball muscles to rotate the eyes.
- **Mixed nerve**: This type of nerve contains both motor and sensory fibres. Example: spinal nerve.

Cranial Nerves

Information from the brain is carried to the other parts of the body, primarily to the neck and head, by the cranial nerves. These are paired nerves present on both sides of the body and are responsible for vision, hearing, smell and muscle movement in the head, neck and facial regions. They arise directly from the brain and exit through its foramina. There are a total of 12 cranial nerves, numbered (in roman numerals) in the order in which they emerge from the brain. They are a part of the peripheral nervous system (PNS) even though smell and vision are part of the central nervous system (CNS). The somatic system has most of the cranial nerves and is responsible for the motor and sensory neurons as well as mixed neurons. Sensory cranial nerves I and II are the olfactory and optical nerves (Table 5.1). The rest of the cranial nerves have both afferent and efferent fibres (mixed cranial nerves).

Spinal Nerves

They are 12 in number and originate from the spinal cord (Fig. 5.3). They carry messages from and to the CNS and are made up of mixed nerves.

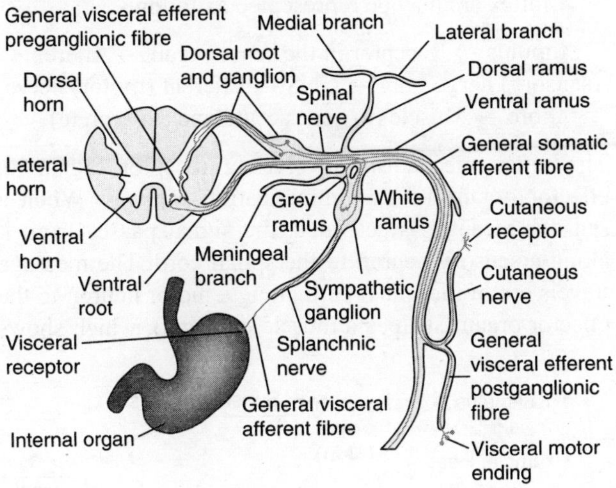

Fig. 5.3 Spinal nerves

Fun Fact!

The aggregates of nerve cells at the point where the nerve fibres enter the cells and arise from there are known as ganglia.

Table 5.1 The 12 cranial nerves, location, type and function

Cranial nerves	Location	Type	Function
Olfactory (I)	Cribriform plate	Sensory	Smell
Optic (II)	Optic foramen	Sensory	Vision
Oculomotor (III)	Superior orbital fissure	Motor	Eye movement
Trochlear (IV)	Superior orbital fissure	Motor	Eye movement
Trigeminal (V)	Superior orbital fissure	Mixed	Facial sensation
Abducens (VI)	Superior orbital fissure	Motor	Eye movement
Facial (VII)	Internal auditory canal	Mixed	Facial expression
Vestibulocochlear (VIII) (auditory vestibular nerve)	Internal auditory canal	Sensory	Hearing and balance
Glossopharyngeal (IX)	Jugular foramen	Mixed	Oral sensation and taste
Vagus (X)	Jugular foramen	Mixed	Vagus nerve
Accessory (XI)	Jugular foramen	Motor	Shoulder elevation and head-turning
Hypoglossal (XII)	Hypoglossal	Motor	Tongue movement

Reflex

The automatic response of the body to a stimulus is known as reflex action (Fig. 5.4). A reflex arc (Fig. 5.5) is the pathway taken by nerve impulses in a reflex action.

A reflex arc may be represented as follows:

Stimulus → Receptor in the sense organ → Afferent (sensory) nerve fibre → CNS → Efferent (motor) nerve fibre → Muscle (to contract)/Gland (to secrete)

The reflex mechanism involves a receptor organ, an effector organ and a communication pathway. When a sensory receptor is activated, the signal passes from it along a sensory neuron to the spinal cord. The message travels out of the spinal cord along a motor neuron to the effector organ (such as a muscle or gland), which shows

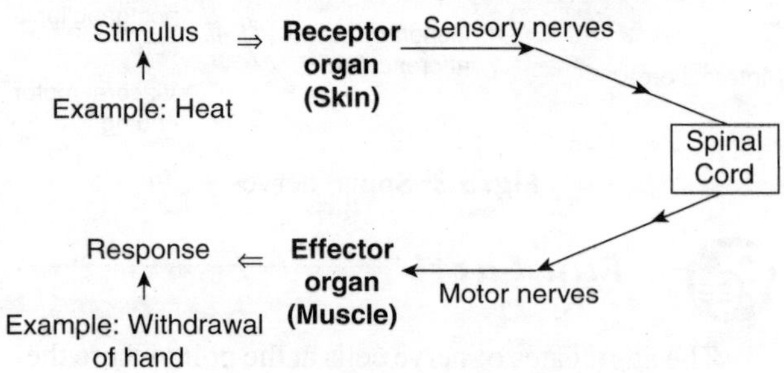

Fig. 5.4 Reflex action

a response. Such a pathway is called a reflex arc. In most cases, the basic physiological mechanism behind a reflex is far more complicated than a simple reflex arc would suggest. Additional nerve cells capable of communicating with other parts are present in reflex circuits.

Classification of the Nervous System

The nervous system is divided into two parts: the central nervous system (CNS), which contains the brain and spinal cord, and the peripheral nervous system (PNS), which comprises the nerves arising from the brain and spinal cord.

The main divisions of the nervous system are summarised in Fig. 5.6.

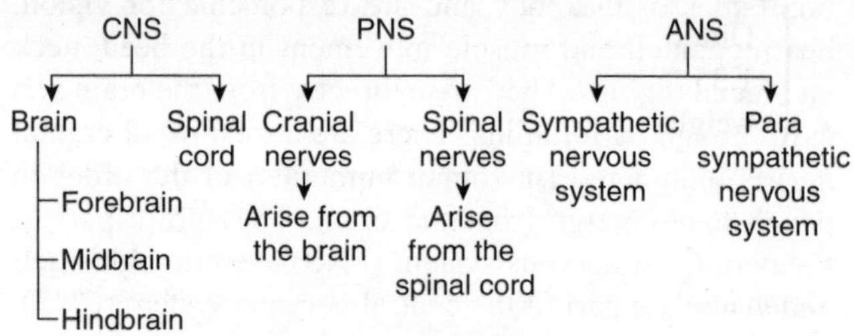

Fig. 5.6 Classification of the nervous system

The Brain

The human brain is enclosed in a cranium (brain box). It is further protected by three meninges (membranous

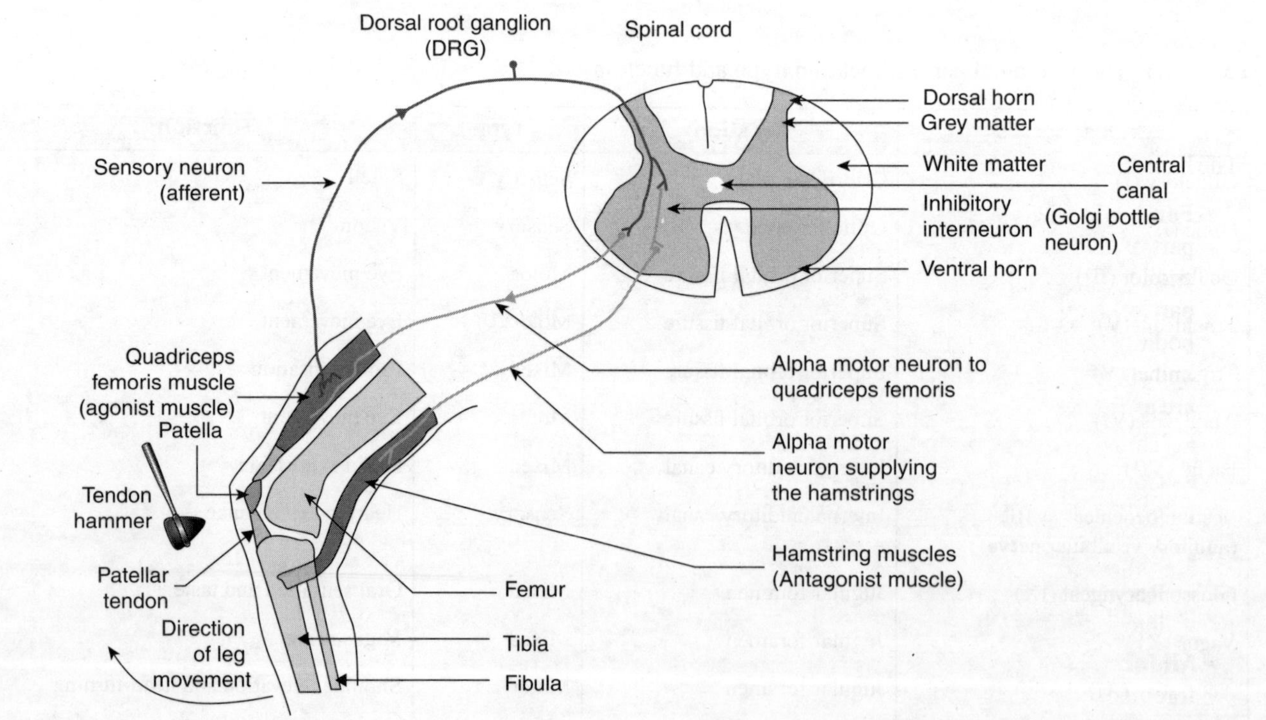

Fig. 5.5 Tendon reflex arc

coverings) that continue rearward over the spinal cord. The three meninges are:

- **Dura mater**: The outermost tough, fibrous membrane
- **Arachnoid**: A thin, delicate middle layer
- **Pia mater**: The innermost highly vascular membrane

A watery fluid called **cerebrospinal fluid** fills the area between the covering membranes, acting as a cushion to protect the brain from shocks. It is 80% water and consumes more than one-fourth of the total oxygen taken into the body.

Fun Fact!

Did you know that the adult brain weighs about 1.35 kg, which is about 2% of the total body weight?

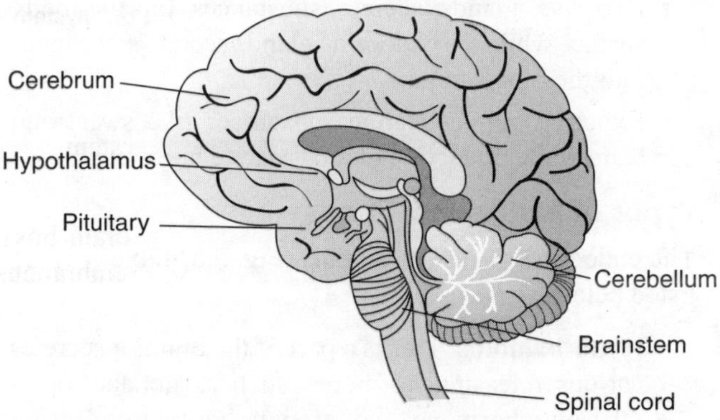

Fig. 5.7 Parts of the brain

The human brain has three main parts or regions (Fig. 5.7):

- **Forebrain (Cerebrum)**: The cerebrum is the largest part of the brain and divided vertically into the right and left cerebral hemispheres. It is the most complex part of the brain. The outermost part contains cell bodies of neurons which are greyish in colour and known as **grey matter.** The functions of the cerebrum are as follows:
  - Thinking part of the brain
  - Controls voluntary actions
  - Stores information (memory)
  - Centre associated with hunger
  - Receives sensory impulses from the body and integrates and transports it to other body parts.
- **Midbrain**: This is responsible for connecting the forebrain with the hindbrain. It is the area of the central nervous system associated with vision, hearing, motor control, sleep and wake cycles, arousal (alertness) and temperature regulation.
- **Hindbrain (Cerebellum)**: The cerebellum, which lies beneath the cerebrum, is a considerably smaller part of the brain. It does not have any convolutions, but contains furrows. It also has a grey matter outer cortex and a white matter inner cortex. The cerebrum and cerebellum function together. When you stand up and walk, for example, the cerebrum generates the wish to do so. Walking necessitates the coordinated action of numerous muscles. The cerebellum is responsible for the proper coordination and timing of muscle contraction and relaxation.

Medulla oblongata: This is the lowermost portion of brain and is located at the base of the skull. It controls the activities of internal organs such as peristalsis and heartbeat.

The Spinal Cord

The spinal cord is a bundle of nerves that runs through the backbone. It is an essential component of the central nervous system. It starts at the medulla oblongata and runs the length of the backbone and is located within the neural canal. The arrangement of white and grey matter is reversed here; the grey matter has motor cell bodies and association neurons on the inner side while the white matter is on the outer side. There is a small central canal that has cavities; it is filled with cerebrospinal fluid that acts as a shock-proof cushion and forms a medium for the exchange of gases, food materials and waste products. The spinal cord performs the following functions:

- It transmits sensory signals from the skin and muscles to the brain
- It transmits motor responses from the brain to the trunk and limb muscles

THE ENDOCRINE SYSTEM

As mentioned earlier, the nervous system and endocrine system are together used for control and coordination in animals. The endocrine system is responsible for chemical coordination. Numerous involuntary physiological activities are under the control of the endocrine system. It consists of glands which release hormones. Hormones play a vital role in various activities in the body including growth and development. They also support the nervous system. Endocrine glands are also called ductless glands.

There is another set of glands, called exocrine glands, that secrete hormones. However, these hormones do not flow directly into the bloodstream but are secreted through ducts.

Exocrine Glands

The glands that secrete molecules through ducts on the surface of the body are known as exocrine glands. For example, sweat, tears, milk, saliva and digestive juices are secreted by the exocrine glands. Glands are cells that work as a unit to create and secrete these molecules and are found in many organs of the body with different functions.

The different **types** of exocrine glands are as follows:

- **Sebaceous glands**: These are present on the skin and as hair follicles. They secrete sebum, which is an oily substance that lubricates and protects the skin and hair.
- **Sweat glands**: These glands secrete sweat. Eccrine sweat glands cover the entire body surface and produce sweat, which is a non-oily secretion that helps in temperature regulation.
- **Lacrimal glands**: Tear or lacrimal glands are located above the upper eyelids and secrete fluid every time we blink in order to keep the eyes moisturised.
- **Salivary glands**: These secrete saliva, which plays a crucial role in chewing, swallowing and digesting food. It lubricates the food and also digests starch due to the presence of ptyalin. Lysozymes in saliva help protect the body.
- **Stomach glands**: The enzymes secreted by the stomach glands break down food and absorb nutrients.
- **Mammary glands**: These glands produce milk which provides nutrients as well as immunoglobulins for the child.
- **Ceruminous glands**: The ear glands produce ear wax or cerumen to protect the ear from infection or physical damage.
- **Brunner glands**: Located in the duodenum of the small intestine, these glands produce mucus which protects the duodenum from the acid secreted by the stomach. It also helps to digest food and absorb nutrients.
- **Liver and pancreas**: These are also considered as exocrine glands as the liver secretes bile through ducts and the pancreas secretes pancreatic juices through ducts. However, they are also a part of the endocrine system as they secrete hormones directly into the bloodstream.

The **functions** of the exocrine glands are as follows:

- Digesting food
- Protecting the inner lining of organs
- Absorbing nutrients
- Controlling body temperature
- Lubricating skin and hair

There are **three ways** in which exocrine glands secrete their substances:

- **Apocrine gland**: Buds of the cell membrane break off into the duct and cause the glands to lose a part of the membrane. Example: mammary glands.
- **Merocrine gland**: These glands release their substances by exocytosis, where cells are not damaged. Example: sweat glands.
- **Holocrine gland**: In order to release the substance, the cell membrane bursts. Example: sebaceous glands.

Exocrine versus Endocrine

- Exocrine glands have ducts to secrete substances while endocrine are ductless glands.
- Exocrine glands secrete substances on the body surface while the endocrine glands secrete substances into the bloodstream.
- Exocrine glands secrete substances like sweat and tears while endocrine glands secrete hormones.

Types of Endocrine Glands

The endocrine glands in animals (Fig. 5.8, Table 5.2) are listed below:

- **Hypothalamus**: This is a part of the brain. It secretes various releasing hormones such as gonadotropin-releasing hormone and growth hormone-releasing hormone. These hormones work on the pituitary glands to stimulate other glands.

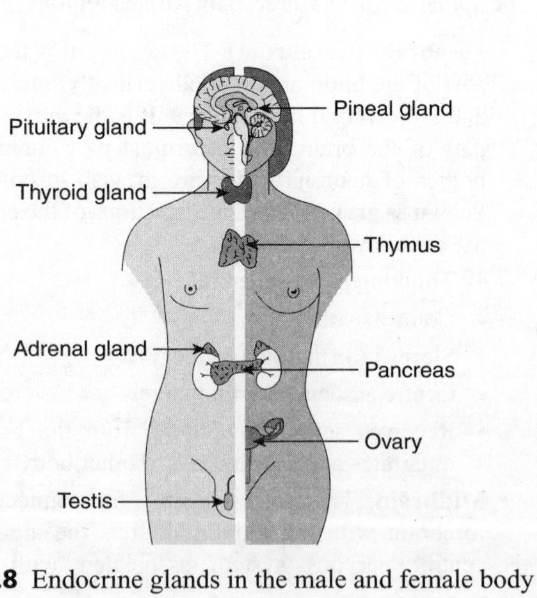

Fig. 5.8 Endocrine glands in the male and female body

Table 5.2 Endocrine glands, hormones and functions

Name of the gland	Hormones secreted	Functions
Pituitary	Several hormones, including growth hormone, thyroid stimulating hormones, etc.	(i) Produces thyroxine hormone and stimulates the thyroid gland. (ii) Controls growth. (iii) Keeps a check on the activity of other glands.
Adrenal	Adrenaline	(i) Regulates blood pressure, breathing rate, heart beat, etc. (ii) Helps the body to adjust to stress conditions.
Thyroid	Thyroxine	(i) Helps in controlling growth, respiration and metabolism. (ii) Prevents goitre.
Testes	Male hormone or testosterone	Develops secondary sexual characters and controls release of sperms.
Ovaries	Female hormone or oestrogen	Develops secondary sexual characters, maturation and release of ovum and helps in maintaining the growth of the uterus during pregnancy.
Pancreas	Insulin	Controls sugar level and prevents diabetes.

- **Pituitary glands**: Also called the master gland, it is located below the brain. It controls and regulates other glands in the body. Examples of hormones released by this gland are growth hormone and thyroid-stimulating hormone.
- **Pineal glands**: This is located in the brain. It releases melatonin, which regulates the sleep and wake cycle and helps in immunity of the body.
- **Thyroid glands**: This is a bilobed gland present below the neck, on either side of the wind pipe. It releases thyroxine (T4), which is an iodine-containing protein that regulates metabolism. Iodine is vital for thyroxine release. Thyroxine stimulates growth in young ones. Over-secretion of this hormone can cause enlargement of the glands, which increases the heartbeat, creates a feeling of restlessness and increase the metabolic rate.
- **Pancreas**: This is an endocrine as well as exocrine gland. It is situated behind the stomach. It plays a vital role in digestion and secretes hormones such as:
  - **Insulin**: This removes excess glucose from the bloodstream in three ways:
    - Induces cells to burn extra glucose to produce energy and heat for our body
    - Converts extra glucose to glycogen which is stored in muscles and the liver
    - Causes cells to convert extra sugar to fat
  - **Glucagon**: This stimulates glycogen breakdown in the liver to form glucose and raises sugar levels in blood.

Fun Fact!

The inner lining of the duodenum produces hormones to stimulate the flow of bile juice!

- **Adrenal glands**: These are made up of two regions:
  - **Cortex**: This region secretes cortisol and androgens. These hormones influence the metabolism of fat, protein and carbohydrate. The cortex is also known to regulate water and salt balance in the body in response to stress such as heat, cold and infection.
  - **Medulla**: This region secretes adrenaline and noradrenaline under conditions of stress, anger or fear. Adrenaline is known as the 'fight or flight' hormone. It provides extra energy and strength in unfavourable conditions by increasing the rate of respiration and heartbeat by raising blood pressure and increasing the supply of blood to muscles; more glucose is released into the blood by the liver.
- **Gonads**: These are reproductive glands. The male gonad is a pair of testes which releases testosterone. This is responsible for the secondary sexual characteristics in males. The female gonad consists of

a pair of ovaries. It secretes two hormones: oestrogen and progesterone.

Fun Fact!

Did you know that even our stomach acts as an endocrine gland? The wall of the stomach produces hormones to stimulate gastric juice secretion!

Disorders of the Endocrine Glands

Excess or insufficient secretion of hormones may lead to some diseases. The endocrine glands have their own feedback mechanisms to counter hormone imbalance, but sometimes, this fails and leads to some disorders or diseases.

- **Dwarfism**: The growth hormone secreted by the pituitary glands is responsible for dwarfism or shortness.
- **Acromegaly**: Also known as gigantism, this is a growth-hormone–related disorder. In contrast to dwarfism, gigantism is an outcome of excess secretion of the growth hormone by the pituitary.
- **Cretinism**: It is an endocrine disorder caused by deficiency of thyroid hormones. Cretinism is a type of hypothyroidism that occurs in infants. It leads to mental and physical growth impairment.
- **Myxoedema**: This defective state occurs when the thyroid is not functioning properly and the patient shows sluggishness of metabolism with swelling in the hands and face.
- **Goitre**: When the level of the thyroid hormone is high in blood, it shows symptoms like excessive sweating and weight loss. It leads to swelling of the thyroid gland. It is due to iodine deficiency.
- **Graves' disease**: It is known as exophthalmic goitre, a type of hyperthyroidism. It is an autoimmune disorder which is commonly seen in children. In this condition, the thyroid gland secretes excess thyroid hormones.
- **Diabetes mellitus**: The pancreas produces two types of hormones for balancing the blood glucose level: glucagon and insulin. Diabetes mellitus, also known as diabetes, is a condition in which the blood glucose level is high because of insufficient secretion of insulin. Excessive thirst, weight loss and frequent urination are symptoms of the disease. A diabetic person has a high concentration of sugar in blood and excretes high amounts of urine that is loaded with sugar and so feels thirsty. The patient also loses weight and becomes weak. There are cases where the glucose level becomes very low due to over-secretion of insulin, which is very harmful.
- **Addison's disease**: This is a condition which shows symptoms of extreme weakness, fatigue and dehydration. Addison's disease is a hormonal disorder caused by inefficiency in the adrenal cortex to produce hormones like corticosteroids.

STRESS MANAGEMENT

The World Health Organization has defined stress as "any type of change that leads to changes in emotional, physical or psychological strain". It is a response to something other than normal stimuli that requires attention and action in the near future. Everyone experiences stress to some degree and there are different ways in which an individual may respond to it, which makes a big difference. Both the brain and the body are affected by stress, and a little amount of stress helps people put in their best performance. However, too much stress acts in an antagonistic way.

During stress, the ANS starts controlling the heart rate, vision, breathing and other vital functions. The body is made ready to face the stressful situations in a fight-or-flight response; in chronic stress, continued activation of the same response results in wear and tear of the body. The physical symptoms of stress are:

- Pain
- Aches
- Chest pain
- Palpitations
- Exhaustion
- Insomnia
- Hypertension
- Dizziness/Shaking
- Jaw clenching
- Weak immune system
- Gastrointestinal problems

The mental and emotional symptoms include:

- Depression
- Anxiety
- Irritability
- Panic attacks
- Sadness

Stress is non-measurable; only the person experiencing it can determine its presence or absence. Different questionnaires prepared by healthcare providers help us to find out how stress affects our life.

Ways to Tackle Stress

Sometimes people try to manage stress by unhealthy behaviours such as:

- Drinking alcohol
- Gambling
- Overeating
- More usage of phone and electronic devices
- Drug abuse

However, there are better ways to tackle stress:

- **Yoga**: This refers to posture-based ways to maintain physical fitness, and relax and relieve stress. It mainly consists of movements or positions called asanas. There are different types of asanas in traditional yoga and modern yoga. A number of studies have shown that yoga can reduce anxiety and stress by enhancing mood and overall wellbeing.
- **Meditation**: This is a practice where the individual uses mindfulness or focus on a particular thought, object or activity to achieve mental peace and clarity. After meditation of a few minutes to an hour, a stressful mind may calm down and attain a stable state. The essentials of meditation are:
  - Awareness of the current moment
  - Ignoring and suppressing negative thoughts
  - Awareness, attention, calm, stability
  - Unwavering commitment
- **Time management**: When you start assessing your approach towards work, the importance of time management becomes clear. To asses how good you are at organising time is the first step towards management of time. When your tasks are divided into a schedule, it automatically reduces stress. Some tips to improve time management to reduce stress are:
  - Setting reminders for all tasks
  - Maintaining a daily planner
  - Maintaining a time limit for every task
  - Blocking out distractions
  - Establishing and following a routine
  - Creating a priority list of tasks
- **Sports**: Performing any form of physical activity or game to maintain physical and mental stability is known as sports. Sports distracts your mind from stress and utilises your energy in physical activity that helps you relax. For example, cricket, badminton, etc.
- **Hobby**: Any activity apart from study that makes a person happy and relaxes their mind by awakening the creative side is known as a hobby. Examples are painting, singing, dancing and travel.
- **Rational thinking**: The ability of a person to think with logic and reason and to draw sensible conclusions from facts, data and logic is known as rational thinking. Sometimes, decisions have to made from facts and not emotions. For example, to relax your mind and know how tasks should be prioritised in order to tackle stress. It is important in achieving goals and solving problems.

COORDINATION IN PLANTS

Plants lack muscles and nerves but changes and movement in living beings is one of the essential characteristics. These changes could be due to conversion or growth in the body as well as the components of the body that are connected with one another. The movements shown by plants are not random but are highly controlled and coordinated. Coordination refers to the order of working of different parts of plants that are interrelated to each other to perform activities.

Phytohormones are hormones that are responsible for the behaviour of plants and which affect their growth, which can be directional or non-directional:

- **Directional movements**: Tropic movements are those where movement occurs in the direction of the stimulus received. For example, positive phototropism is the direction movement shown by the bending of a plant's shoots towards sunlight. Similarly, negative geotropism is shown by shoots as they grow away from the ground. Negative phototropism, as well as positive geotropism, is shown by the roots of plants as they grow against the light and towards the ground, respectively. The response to chemical stimuli such as the growth of the pollen tube towards the ovule is also a directional movement known as chemotropism.
- **Non-directional movement**: The nastic response is concerned with the movement of parts of a plant. These movements are independent of growth but the cells involved change their shape in order for this to occur. It takes place in the opposite direction to the stimulus. For example, a daisy bloom closes at dusk and opens again before dawn, a phenomenon known as sleep movement.

Tropism is the movement of plant organs towards or away from a stimulus. The following are the several forms of tropic motion in plants:

- **Phototropism**: The movement of plant components towards or away from light.
- **Geotropism**: The movement of plant organs in response to gravity.
- **Thigmotropism**: The movement of plant organs in response to stimuli induced by physical contact with solid objects.
- **Hydrotropism**: The movement of plant organs in reaction to water. Roots are positively hydrotropic, meaning they grow towards a source of moisture.
- **Chemotropism**: The movement of plant organs in response to a chemical stimulus.

PLANT HORMONES

Plants lack a nervous system and instead use various hormones to coordinate their behaviour against environmental changes (Fig. 5.9). The stimulated cells release chemical compounds called hormones that diffuse around the cell. The hormones present in plants are:

- **Auxin**: Responsible for plant growth by elongating the cell at the tip of the stem.
- **Gibberellin**: Responsible for seed germination, stem growth and flowering.
- **Cytokinins**: In fruits and seeds, the area responsible for cell division contains cytokinins that help in growth and in opening of stomata.
- **Abscisic acid**: Responsible for inhibition of growth and closure of stomata.

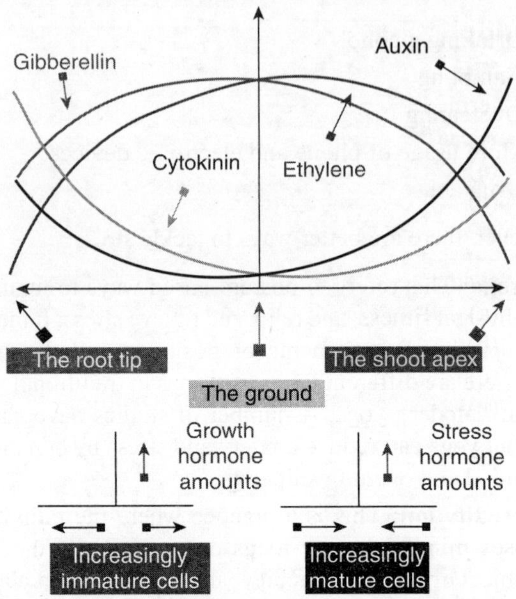

Fig. 5.9 Changes in secretion of hormones based on location and maturity of parts of plants

SUMMARY

- Control and coordination are the functions of the nervous system and hormones.
- The responses of the nervous system can be classified as reflex action, voluntary action and involuntary action.
- The nervous system uses electrical signals to transmit messages.
- The nervous system receives information from the sense organs and acts through our muscles, also called effectors.
- The human brain has three main parts or regions: forebrain, midbrain and hindbrain.
- The nervous system has two basic divisions of spinal and cranial nerves.
- The automatic response of the body to a stimulus is known as a reflex action. A reflex arc is the pathway taken by nerve impulses in a reflex action.
- Chemical coordination is seen in both plants and animals.
- Hormones are produced in one part of an organism and travel to another part to achieve the desired effect. A feedback mechanism controls the action of the hormones.
- Exocrine glands have ducts to secrete substances while endocrine glands are ductless.
- Different endocrine organs are responsible for the secretion of hormones such as the hypothalamus, pituitary gland, pineal gland and adrenal gland.
- Excess or insufficient secretion of hormones may lead to diseases like Addison's disease.
- Phytohormones are fluid secretions in plants that help them to coordinate their chemical processes. They are essential for control of growth at various stages of development.

PRACTICE QUESTIONS

Analyse Your Concepts (School Exam Based)

Multiple Choice Questions

1. The area between two neurons at the synapse is termed synaptic cleft, and the joining of two neurons is called:
   (a) Synapse (b) Nodes of Ranvier
   (c) Neurolemma (d) Myelin sheath

2. _______ carry impulses from the main nervous system to an effector such as a muscle or gland:
   (a) Sensory neurons
   (b) Association neurons
   (c) Motor neurons
   (d) None of the above

3. There are a total of _______ cranial nerves which are numbered in roman numerals in the order of their emergence from the brain:
   (a) 10 (b) 12
   (c) 9 (d) 11

4. The nervous system is divided into two parts: _________ and ________:
   (a) Central nervous system, parasympathetic nervous system
   (b) Central nervous system, sympathetic nervous system
   (c) Peripheral nervous system, axial nervous system
   (d) Central nervous system, peripheral nervous system

5. A watery fluid called ______fills the area between the covering membranes, acting as a cushion to protect the brain from shocks:
   (a) Cerebrospinal fluid (b) Perinuclear fluid
   (c) Spinal fluid (d) Perilobular fluid

6. The outermost part of it contains cell bodies of neurons which are greyish in colour and known as:
   (a) White matter (b) Black matter
   (c) Red matter (d) Grey matter

7. This inhibits growth, cause wilting of leaves:
   (a) Abscisic acid (b) Cytokine
   (c) Auxin (d) Gibberellin

8. ________are located above the upper eyelids and secrete as well as create the fluid every time the eye blinks in order to keep them moisturised:
   (a) Sweat glands (b) Lacrimal glands
   (c) Sebaceous glands (d) Salivary glands

9. ______ causes cells to convert extra sugar to fat:
   (a) Insulin (b) Glucagon
   (c) Thyroid (d) Aldosterone

10. _____ is the condition where the thyroid gland secretes too much thyroid hormones:
   (a) Diabetes (b) Goitre
   (c) Addison's disease (d) Grave's disease

Match the Following

Each question contains statements given in two columns which have to be matched. Statements in column I have to be matched with statements in column II.

1. Choose the correct option:

Column I	Column II
i. Insulin	a. Thyroid
ii. Thyroxine	b. Pancreas
iii. Oestrogen	c. Cortex
iv. Cortisone	d. Gonads

2. Choose the correct option:

Column I	Column II
i. Sensory nerve	a. Peristalsis
ii. Motor nerve	b. Spinal nerve
iii. Mixed nerve	c. Optic nerve
iv. Medulla oblongata	d. Nerve from brain to eyeball muscles to rotate eye

3. Choose the best match:

Column I	Column II
i. Axon	a. Sheath enclosing myelin sheath
ii. Nodes of Ranvier	b. Myelin sheath
iii. Neurolemma	c. Gaps in myelin sheath

True or False

Read the following statements and write your answer as true or false.

1. Stimulus is a sudden change in the external environment that leads to change in an organism or its body part.
2. The area between two neurons at the synapse is termed synaptic cleft, and the joining of two neurons is called the synapse.
3. The digestive system helps us to think, reason and remember things.
4. The outermost part of this structure contains cell bodies of neurons which are greyish in colour and known as grey matter.
5. The optic nerve is a type of mixed nerve.
6. Pia mater is the outermost tough fibrous membrane.
7. The cerebellum is responsible for the proper coordination and timing of muscle contraction and relaxation.
8. The pituitary gland is present below the neck on either side of the windpipe.
9. Thigmotropism is defined as the movement of plant organs in response to stimuli induced by physical contact with solid objects.
10. The hypothalamus is the reproductive gland present in the male and female body.

Fill in the Blanks

Complete the following statements with an appropriate word/term to be filled in the blank spaces.

1. _______ are cytoplasmic extensions from the cell body that are branched.
2. _______ are the neurons that carry impulses from the main nervous system to an effector such as a muscle or gland.
3. In most neurons, the axon has ______surrounding the body which is enclosed by thin sheath called _________and there are gaps between these sheaths along the length known as___________.
4. ______are the nerves that contain both motor and sensory fibres like the spinal nerve.
5. The pathway taken by nerve impulses in a reflex action is called ______.
6. The nervous system is divided into two parts: ________ and ________.
7. The human brain is enclosed in a______.
8. _______is responsible for the proper coordination and timing of muscle contraction and relaxation.
9. _______ transmits sensory signals from the skin and muscles to the brain.
10. The _________is called the master gland.
11. ________ is an endocrine as well as an exocrine gland.
12. Adrenal glands have two regions known as the ________and ________.

Assertion–Reason Questions

The following questions consist of two statements: Assertion (A) and Reason (R). Answer these questions selecting the appropriate option given below:

(a) Both A and R are true and R is the correct explanation of A
(b) Both A and R are true but R is not the correct explanation of A
(c) A is true but R is false
(d) A is false but R is true

1. Assertion (A): Pancreas produces two types of hormones for balancing blood glucose levels: glucagon and insulin.
   Reason (R): Diabetes mellitus is a condition in which the blood glucose level is high because of insufficient secretion of insulin.
2. Assertion (A): The pituitary gland is called the master gland.
   Reason (R): The pituitary gland controls and regulates other glands in the body.
3. Assertion (A): The lacrimal glands are located above the upper eyelids.
   Reason (R): The lacrimal glands only secrete and do not create the fluid every time we blink.
4. Assertion (A): Roots are negatively hydrotropic.
   Reason (R): Hydrotropism is the movement of plant organs in reaction to water.
5. Assertion (A): Myelin sheath protects the neurons from damage.
   Reason (R): It also stops leakage and weakening of signals.

One-word Answers

1. What is the ability of a person to think with logic and reason?
2. What does excess secretion of growth hormone by the pituitary cause?
3. Which region of the adrenal cortex secretes the hormones adrenaline and noradrenaline?
4. Which hormone stimulates glycogen break-down in the liver to form glucose?
5. Which hormone is secreted by the pineal gland?
6. What are buds of cell membranes that break off into the duct?
7. Which gland processes mucus to protect the duodenum from acid?
8. Which gland produces ear wax?
9. Name the region that secretes the hormones adrenaline and noradrenaline.
10. Name the condition in which blood glucose level is high because of insufficient secretion of insulin.

Define the Term

1. Gonads
2. Adrenal glands
3. Pituitary
4. Hypothalamus
5. Pancreas

Groups of Terms

Given below are groups of terms. In each group the first pair indicates the relationship between the two terms. Rewrite and complete the second pair on a similar basis.

1. Phototropism: light :: Hydrotropism: _____
2. Pineal: melatonin :: Thyroid:_____
3. Neurons: three types :: Cranial nerves: _____
4. Ductless: endocrine :: Duct: _____
5. Outermost: dura mater :: Innermost: ____

Identify the Odd One Out

1. Pituitary, pineal, sweat, thyroid
2. Apocrine, Stereocrine, Holocrine, Merocrine
3. Sebum, thyroxine, insulin, adrenaline
4. Soma, axon, dendrites, fibres

Image-Based Questions

1. Identify the part:

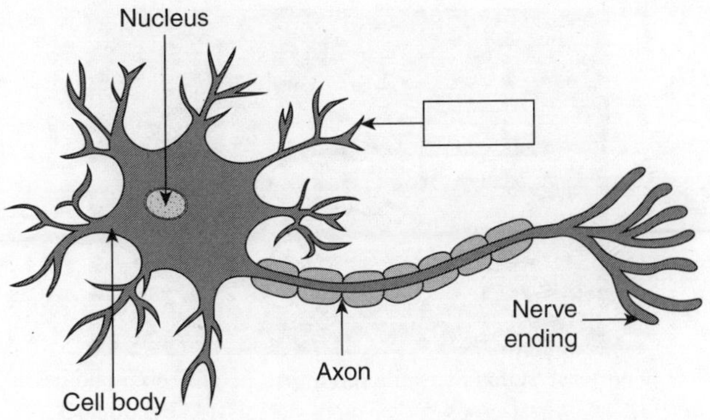

(a) Synapse (b) Myelin sheath
(c) Neurolemma (d) Dendrite

2. Identify the part:

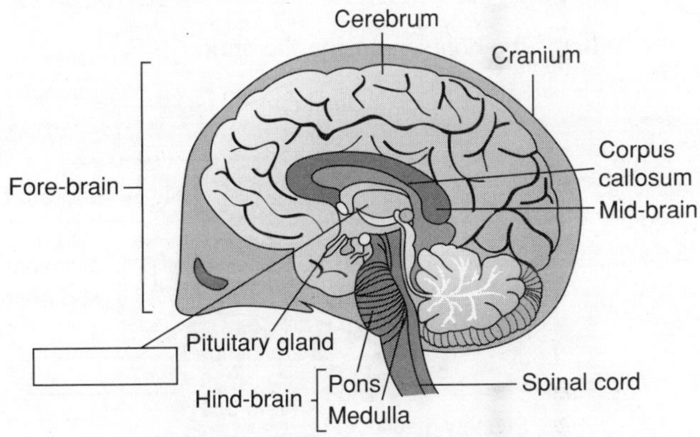

(a) Pituitary (b) Pineal
(c) Hypothalamus (d) Thyroid

3. Identify the part:

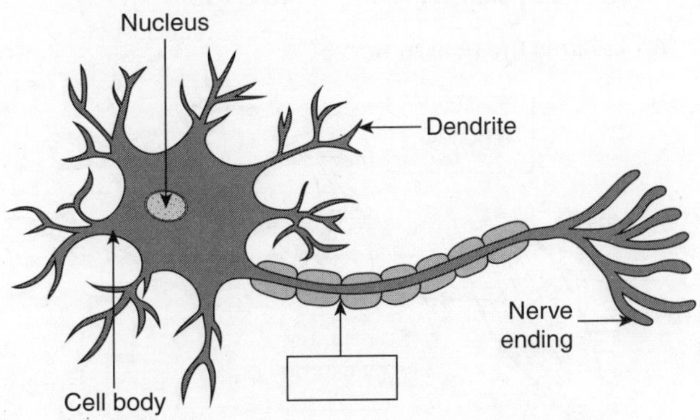

(a) Axon (b) Myelin sheath
(c) Neurolemma (d) Synapse

4. Which component causes growth in the tips of shoots?

(Courtesy: Stefan Schweihofer; https://pixabay.com/photos/wheat-crops-barley-crops-cereals-143373/)

(a) Gibberellin
(b) ABA
(c) Auxin
(d) Ethylene

5. What is the bone enclosing the brain:

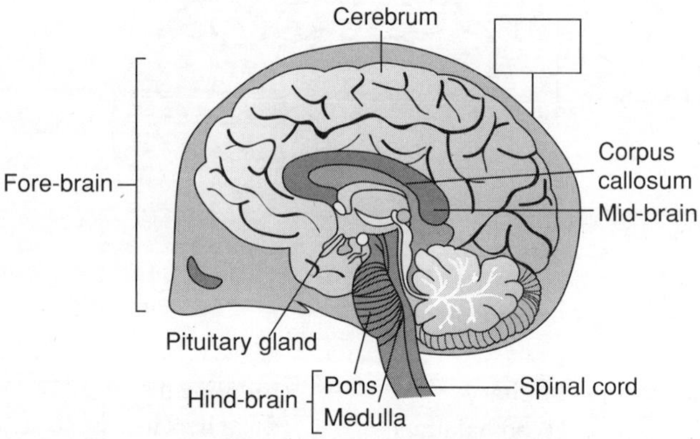

(a) Cranium (b) Femur
(c) Radicle (d) Rib

6. What is the type of nerve?

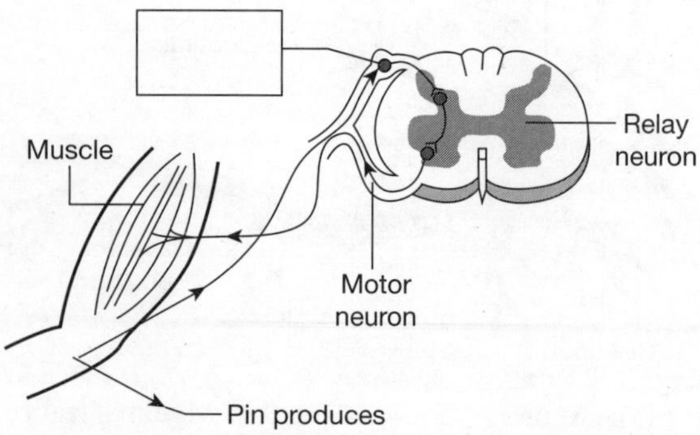

(a) Motor (b) Sensory
(c) Mixed (d) None

7. What is this organ?

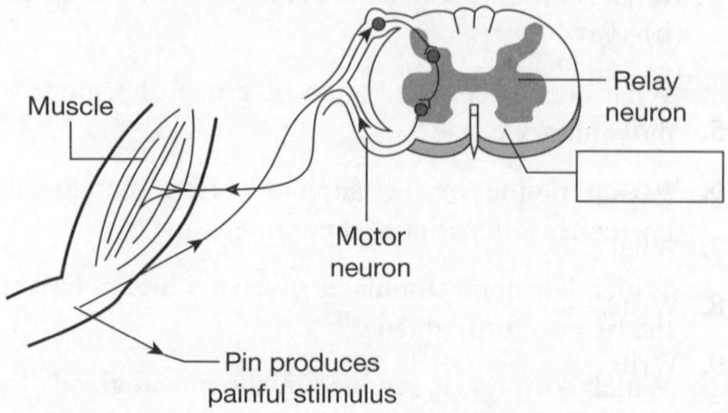

(a) Brain (b) Thalamus
(c) Medulla (d) Spinal cord

8. Choose the growth stunting hormone:

(a) Ethylene (b) ABA
(c) Auxin (d) Cytokine

Very Short Answer Questions

1. Define the nervous system.
2. What is stimulus?
3. Give any three examples of endocrine glands.
4. What is a nerve?
5. What function does the cerebellum play?
6. What are the ways of signal transduction?
7. What is a hormone?
8. Name the coverings of the brain.
9. What is thigmotropism?
10. Write a line about nerve fibres.

Short Answer Questions

1. Explain the types of neurons.
2. Explain the role of the spinal cord.
3. What is a reflex action?

4. Write about
   (a) Medulla oblongata
   (b) Pancreas
   (c) Glucagon
5. Write the relation between cerebrum and cerebellum.
6. Write about the nervous system.
7. What are the three ways of exocrine gland secretion?
8. Write a short note on the thyroid gland.
9. Write a short note on diabetes mellitus.
10. Write any two ways to tackle stress.

Long Answer Questions

1. Explain the structure of the nervous system.
2. Explain neurons and nerve fibres in detail.
3. Write an essay on the brain and spinal cord.
4. Explain the endocrine system.
5. Define the types of plant nervous systems.
6. Write a brief account of the exocrine system.
7. Write an essay on the different glands of the endocrine system.
8. Explain the disorders of endocrine glands.
9. Write a note on the relation between the endocrine system and the nervous system.
10. Write an essay on reflex action.
11. Write the methods to tackle stress.

Case Study-Based Questions

1. Pratibha was talking to Karuna about her childhood memories and they both began to wonder which part of the brain stores the memories. Can you help them understand the structure of the brain?
   (a) What are the three parts of the brain?
   (b) Which part of the brain is responsible for memory?
2. Sandeep has exams coming up and his parents have noticed a few changes in him like insomnia, palpitations and hypertension. Can you help his parents understand his situation of stress?
   (a) What is stress?
   (b) What are the ways he can curb his stress?
3. Laila has been showing symptoms of goitre as observed due to a swollen neck. Can you help her understand the disease?
   (a) What is goitre?
   (b) Why does goitre occur?
4. Kishore has been seeing his father taking insulin injections for diabetes. He has been reading about it. Can you help him to know more?
   (a) What is the function of insulin?
   (b) Which gland and cells secrete insulin?
5. In summers, Ghazal noticed that the production of sweat increases and has been wondering about it. Can you help him understand?
   (a) Which glands secrete sweat?
   (b) What is the function of sweat?

COMPETITION WINDOW (OBJECTIVE TYPE)

1. Select the correct statement:
   (a) Glucagon is associated with hypoglycaemia
   (b) Insulin acts on pancreatic cells and adipocytes
   (c) Insulin is associated with hyperglycaemia
   (d) Glucocorticoids stimulate gluconeogenesis
2. Which of the following is the correct match:

Column I	Column II
(A) Pituitary gland	(i) Grave's disease
(B) Thyroid gland	(ii) Diabetes mellitus
(C) Adrenal gland	(iii) Diabetes
(D) Pancreas	(iv) Addison

   (a) (A) (iii), (B) (ii), (C) (i), (D) (iv)
   (b) (A) (iii), (B) (i), (C) (iv), (D) (ii)
   (c) (A) (ii), (B) (i), (C) (iv), (D) (iii)
   (d) (A) (iv), (B) (iii), (C) (i), (D) (ii)
3. Which of the following hormones is responsible for milk ejection and foetal ejection?
   (a) Oestrogen (b) Prolactin
   (c) Oxytocin (d) Relaxin
4. Artificial light and extended work time with reduced sleep will disrupt the activity of:
   (a) Thymus gland
   (b) Pineal gland
   (c) Adrenal gland
   (d) Pituitary gland

5. Which of the following ecological pyramids is generally inverted?
(a) Pyramid of biomass in sea
(b) Pyramid of numbers in grassland
(c) Pyramid of energy
(d) Pyramid of biomass in forest

6. Which of the following is an amino acid-derived hormone?
(a) Epinephrine (b) Ecdysone
(c) Estradiol (d) Estriol

7. Which of the following is a temporary endocrine gland in the body?
(a) Pineal (b) Corpus luteum
(c) Corpus cardiacum (d) Corpus allum

8. Receptor sites for neurotransmitters are present on:
(a) Membranes of synaptic vesicles
(b) Pre-synaptic membrane
(c) Tips of axon
(d) Post-synaptic membrane

9. Myelin sheath is produced by:
(a) Schwann cells and oligodendrocytes
(b) Astrocytes and Schwann cells
(c) Oligodendrocytes and osteoclasts
(d) Osteoclasts and astrocytes

10. The human hind brain comprises three parts, one of which is the:
(a) Cerebellum (b) Hypothalamus
(c) Spinal (d) Corpus callosum

Answer Keys

Multiple Choice Questions

1. (a) 2. (c) 3. (b) 4. (d) 5. (a)
6. (d) 7. (a) 8. (b) 9. (a) 10. (d)

Match the Following

1. i (b), ii (a), iii (d), iv (c)
2. i (c), ii (d), iii (b), iv (a)
3. i (b), ii (c), iii (a)

True or False

1. T 2. T 3. F 4. T 5. F
6. F 7. T 8. F 9. T 10. F

Fill in the Blanks

1. Dendrites
2. Motor neurons
3. Myelin sheaths, neurolemma, nodes of Ranvier
4. Mixed nerves
5. Reflex arc
6. Central Nervous System (CNS), Peripheral Nervous System (PNS)
7. Cranium
8. Cerebellum
9. Spinal cord
10. Pituitary gland
11. Pancreas
12. Adrenal cortex, adrenal medulla.

Assertion-Reason Questions

1. (b) 2. (a) 3. (c) 4. (d) 5. (b)

One-word Answers

1. Rational thinking
2. Acromegaly
3. Medulla
4. Glucagon
5. Melatonin
6. Apocrine gland
7. Brunner's gland
8. Ceruminous gland
9. Medulla
10. Diabetes mellitus

Define the Term

1. Gonads are reproductive glands present in the male and female bodies. The male gonad comprises a pair of testes which release testosterone. This is responsible for the secondary sexual characteristics in males. The female gonad consists of a pair of ovaries. It secretes oestrogen and progesterone.
2. The adrenal glands have two regions, known as the adrenal cortex and adrenal medulla. The cortex region secretes cortisol and androgens. These hormones influence the metabolism of fats, proteins and carbohydrate. It is also known to regulate the water and salt balance in the body. The medulla region secretes adrenaline and noradrenaline. They are secreted under conditions of stress, anger or fear. Adrenaline is known as the "hormone of fight or flight". It provides extra energy and strength to the body in unfavourable conditions. It also increases the rate of respiration and heartbeat by raising the blood pressure and increasing the supply of blood to the muscles, so more glucose is released into the blood by the liver.
3. The pituitary gland is called the master gland. It is located below the brain. It controls and regulates other glands in the body. The hormones released by this gland are the growth hormone, thyroid-stimulating hormone, etc.
4. This gland is a part of the brain. The hypothalamus secretes various hormones like gonadotropin-releasing hormones and growth hormone-releasing hormones. These hormones work on the pituitary glands to stimulate other glands.
5. The pancreas is an endocrine as well as an exocrine gland. It is situated behind the stomach. It plays a vital role in digestion and secretes hormones like glucagon and insulin from Islets of Langerhans.

Groups of Terms

1. Water
2. Thyroxine
3. Twelve
4. Exocrine
5. Pia mater

Identify the Odd One Out

1. Sweat
2. Stereocrine
3. Sebum
4. Soma

Image-Based Questions

1. (d) 2. (c) 3. (a) 4. (c) 5. (a)
6. (b) 7. (d) 8. (b)

Very Short Answer Questions

1. The brain, spinal cord, sense organs and nerves make up the nervous system.
2. A sudden change in the external environment that leads to a change in organisms or their body parts is known as a stimulus.
3. Hypothalamus, pituitary gland and pineal gland.
4. A bundle of nerve fibres or axons of different and separate neurons that are enclosed in a tubular sheath is known as nerves.
5. The cerebellum is responsible for the proper coordination and timing of muscle contraction and relaxation.
6. Reflex action is an automatic response of the body to a stimulus. The pathway taken by nerve impulses in a reflex action is called a reflex arc.
7. Hormones are the chemical substances secreted by endocrine glands which are poured into the bloodstream that act on target cells or organs to coordinate the body's functions.
8. Dura mater: outermost tough fibrous membrane

   Arachnoid: thin delicate middle layer

   Pia mater: innermost highly vascular layer
9. The movement of plant organs in response to stimuli induced by physical contact with solid objects.
10. One of the cell body's branches is unusually long in comparison to the others. The axon, or nerve fibre, is the name for this branch.

Short Answer Questions

1. Sensory neuron is the neuron that carries impulses from the sensor organ receptor to the main nervous system which is the brain or spinal cord.

   Motor neurons are the neurons that carry impulses from the main nervous system to an effector such as a muscle or gland.

   Association neurons are the neurons located in the brain and spinal cord and also interconnected with sensory and motor neurons.
2. The following functions are performed by the spinal cord:
   (i) It transmits sensory signals from the skin and muscles to the brain.
   (ii) It transmits motor responses from the brain to the trunk and limb muscles.
3. A reflex mechanism involves a receptor organ, an effector organ and a type of communication pathway. When a sensory receptor is activated, a signal passes from it along a sensory neuron to the spinal cord. The message travels out of the spinal cord along a motor neuron to the effector organ, which shows the response. Such a pathway is called a reflex arc.
4. (a) It is the lowermost portion of the brain which is located at the base of the skull and controls the activities of internal organs like peristalsis, heartbeat, etc.
   (b) Pancreas is an endocrine as well as an exocrine gland. It is situated behind the stomach. It plays a vital role in digestion and secretes hormones like glucagon and insulin from the Islets of Langerhans.
   (c) It stimulates glycogen break-down in the liver to form glucose and raises sugar levels in blood.
5. The cerebellum, which lies beneath the cerebrum, is a considerably smaller part of the brain. It does not have any convolutions, but it does have a lot of furrows. It also has a grey matter outer cortex and a white matter inner cortex. The cerebrum and cerebellum function together closely.
6. The nervous system is a complex regulatory system in living organisms. It not only coordinates actions but also transmits sensory information and signals to and from the different parts of the body. The nervous system's ability to function is dependent on its ability to perceive and respond to stimuli in an environment.

   Importance of the nervous system:
   - It controls and harmonises the various voluntary activities of our body like running
   - It keeps us informed about the surroundings through the sense organs
   - It regulates involuntary activities as well as breathing
   - It helps us to think, reason and remember things
7. There are three ways in which exocrine glands secrete their substances:
   - Apocrine gland: Buds of cell membranes that break off into the duct and cause the glands to lose

part of the membrane in the process, like the mammary gland

- Merocrine gland: These glands release their substances by exocytosis where cells are not damaged, like sweat glands
- Holocrine gland: In order to release its secretions, the cell membranes burst, like sebaceous glands.

8. This is a bilobed gland that is present below the neck on either side of the windpipe. It releases thyroxine (T4) which is an iodine-containing protein. It regulates body metabolism. Iodine is vital for thyroxine release in the body. Thyroxine stimulates growth in young ones. The over-secretion of this hormone can cause enlargement of glands which increases the heartbeat, create restlessness as well as increase in the metabolic rate.
9. The pancreas produces two types of hormones for balancing blood glucose levels: glucagon and insulin. Diabetes mellitus, also known as diabetes, is a condition in which blood glucose level is high because of insufficient secretion of insulin. Excessive thirst, weight loss and frequent urination are symptoms of the disease. A diabetic person has a high concentration of sugar in the blood and excretes a large volume of urine which is loaded with sugar and so feels thirsty due to too much urination. The patient also loses weight and becomes weak. There are cases where the glucose level of a patient comes down due to over-secretion of insulin, which is very harmful.
10. Yoga: It refers to posture-based ways to maintain physical fitness, and relax and relieve stress; it consists mainly of asanas. A number of studies have shown the role of yoga in reducing anxiety and stress by enhancing mood and overall well-being which also helps to manage depression.

    Meditation: It is a practice where the individual uses mindfulness or focus on a particular thought or object or activity in order to make oneself aware and attentive to achieve mental peace and clarity. There are five essential points in meditation:

    - Awareness of the current moment
    - Ignoring and suppressing the negative thoughts of the mind
    - Mindfulness of thought
    - Awareness, attention, calm, stability
    - Unwavering commitment

Long Answer Questions

1. Hint: Explain the nervous system components and their structure and functions with a well-labelled diagram.
2. Hint: Explain neurons and nerve fibres with a diagram.
3. Hint: Explain the components, functions and structure of the brain and spinal cord with a well-labelled diagram.
4. Hint: Explain the endocrine system, hormones, glands of the endocrine system and hormones secreted by them along with their functions.
5. Hint: Explain the types and components of the plant endocrine system with hormones and their functions.
6. Hint: Explain the exocrine system and its difference from the endocrine; explain all the glands involved in the exocrine gland and their functions.
7. Hint: Explain the endocrine glands, the hormones they secrete and the glands that secrete them with their functions and location.
8. Hint: Explain all the disorders of endocrine glands mentioned in this chapter.
9. Hint: Explain the nervous and endocrine systems with their components and their relation.
10. Hint: Explain reflex action and its mechanism.
11. Hint: Explain all the methods mentioned in this chapter.

Case Study-Based Questions

1. (a) The human brain has three major parts or regions: the forebrain, the mid-brain and the hindbrain.

   (b) The forebrain or cerebrum is responsible for memory.
2. (a) It is a response to something other than normal stimuli that requires attention and action in the near future. Everyone experiences stress to some degree and there are different ways in which an individual responds to it which makes a big difference. Both the brain and the body are affected by stress and a little amount of stress is important for people to show their best performance but too much stress acts in an antagonistic way.

   (b) Yoga, meditation, sports, time management, hobby development and rational thinking are the ways to manage stress.

3. (a) When the level of the thyroid hormone is high in blood, it shows symptoms like excessive sweating, weight loss. It leads to swelling of the thyroid gland. This condition is known as goitre.
   (b) It is caused by iodine deficiency.
4. (a) It induces cells to burn extra glucose to produce energy and heat for our body, converts extra glucose to glycogen which gets stored in the muscles and liver and causes cells to convert extra sugar to fat.
   (b) Insulin is secreted by the Islets of Langerhans of the pancreas.
5. (a) Sweat is created and secreted by these glands.
   (b) Exocrine sweat glands are a type of sweat glands that cover the entire body surface and produce sweat which is a non-oily secretion that controls the body temperature.

Competition Window

1. (d)	**2.** (b)	**3.** (c)	**4.** (d)	**5.** (a)
6. (a)	**7.** (b)	**8.** (d)	**9.** (a)	**10.** (a)

CHAPTER

6 Microorganisms: Friends or Foes?

PREREQUISITES

You must have knowledge of:

- Microbes
- Prokaryotes
- Eukaryotes
- Ecosystem

LEARNING OBJECTIVES

At the end of this chapter, you will be able to:

- Describe the different types of microbes in the ecosystem
- Explain the classification of microbes
- Differentiate between harmful and useful microorganisms
- List the applications of microbes
- Describe the human immune system
- List the different diseases in humans caused by microbes

INTRODUCTION

Different varieties of plants and animals are present in the world. Many of these living organisms cannot be observed with the unaided eye. These organisms are known as microbes or microorganisms ('*micros*' in Greek means 'small'). Microbes are diverse by nature and include different organisms like fungi, bacteria, protozoa and some algae (Fig. 6.1). Viruses are a type of microscopic organism that is generally considered as non-living. The special feature of the virus is that it only reproduces inside a host cell such as plants, animals or bacteria.

Microorganisms are mostly unicellular (made up of only one cell) like bacteria, but can be multicellular (made of many cells) like moulds. In the biosphere, microbes can be found anywhere – in the air, oceans, hot springs, deserts, soil, even inside the animal body. The study of microbes is known as **microbiology** and the person who studies microbes is called a microbiologist.

Microbes can be classified as harmful or useful, depending on their effect on the health of humans. The presence of harmful microbes can be felt by actions such as certain fungi and bacteria spoiling food or causing diseases like typhoid, diarrhoea, common cold, Covid-19, tuberculosis, cholera, AIDS, polio, measles and chickenpox. Some microbes that are useful to us are those that enable the manufacture of curd, cheese, alcohol, vinegar or medicines like antibiotics. They also help in the proper regulation of biogeochemical cycles like the nitrogen and carbon cycles in the environment.

SCIENCE IN DAILY LIFE

1. Collect water samples from different sources:
   - Drinking water
   - Pond water
   - Tap water
2. Take a few drops of each sample on different slides and spread it.
3. Observe the slides under the microscope.

You will be able to see different types of microbes of different quantities in all the three samples. The fewest microbes will be observed in drinking water, followed by tap water and the maximum number of microbes will be observed in pond water.

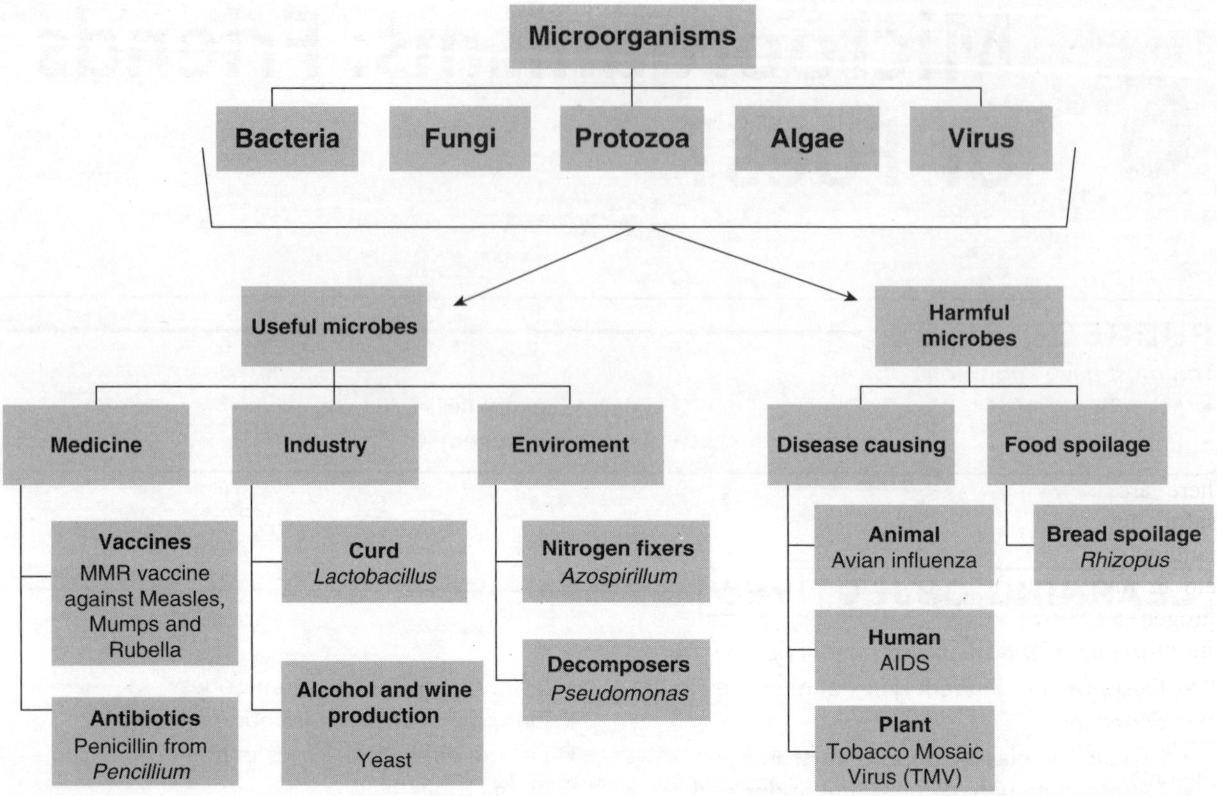

Fig. 6.1 Types of microorganisms and their classification based on whether they are useful or harmful to humans

Fun Fact!

In 1676, Antoni van Leeuwenhoek discovered bacteria and so is known as the "Father of Microbiology".

SCIENCE IN DAILY LIFE

1. Take a beaker and fill it with moist soil collected from a field.
2. Add water to the beaker and wait till the soil particles settle. Take a drop from it and observe it under a microscope.

You will observe organisms, which proves the presence of microbes in soil.

These activities show evidence of the presence of tiny organisms in the soil and water, but all of them cannot be described as microbes. Microorganisms are so small that they are only visible under a microscope. Some are visible under a simple magnifying lens, like moulds (a type of fungus) on bread.

CLASSIFICATION OF MICROORGANISMS

Microorganisms can be classified as bacteria, fungi, protozoans, some algae and viruses. We will now examine each type in detail.

Bacteria

Bacteria are single-celled microorganisms enclosed in a cell wall but without a nucleus and other structures. Bacteria are small in size but bigger than viruses. They are ubiquitous, which means that they are found everywhere – in the soil, water, air, every surface around us and outside and inside our bodies. Bacteria are different from viruses in that they can respire, move, feed and even reproduce on their own.

Bacteria were among the first life forms to appear on Earth. They can be found in a wide range of habitats, from favourable to extremely unfavourable: water, soil, methane-rich environments, acidic hot springs, high salty areas, radioactive waste and even in the deep crust of Earth.

Bacteria are generally classified into categories on the basis of their shape (Fig. 6.2): rod, spiral, spherical and comma. Among these, the rod shape is the most common.

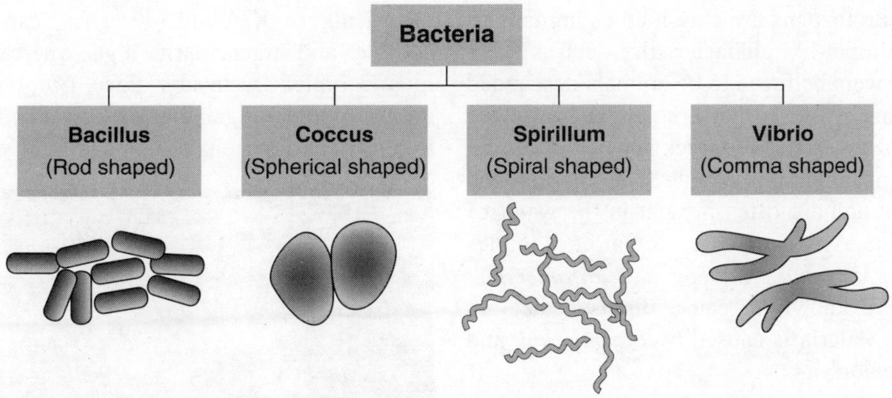

Fig. 6.2 Categorisation of bacteria based on their shape

There are two kinds of bacteria present around us: useful and harmful. Useful bacteria facilitate the manufacture of different commercial products like curd, help in maintaining biogeochemical cycles like fixing nitrogen and help in the manufacture of antibiotics. The other kind of bacteria are harmful to humans and other living organisms as they cause diseases that could even lead to death. Some of these diseases are typhoid, diarrhoea, cholera, food poisoning, diphtheria and tuberculosis.

The most common examples of bacteria are *E.coli*, *Rhizobium* and *Lactobacillus*.

Fun Fact!

Mycoplasma genitalium is the smallest bacteria, with a diameter that ranges from 200 to 300 nanometres.

Thiomargarita namibiensis is the longest/largest bacteria, with a diameter that ranges from 100 to 300 micrometres.

Fun Fact!

In 2020, the Nobel Prize for Chemistry was given for the discovery of the technique of CRISPR gene editing, which is a mechanism of the bacterial immune system that prevents viral infections.

Fungi

Fungi are microbes that are devoid of chlorophyll as well as the ability to make their own food. Fungi can be unicellular, like yeast, or multicellular, like mushroom, puffballs or toadstools. Fine, thread-like filamentous structures that arise from the fungi body are known as hyphae. All fungi except yeast have hyphae. Warm and moist conditions are the most favourable for the growth of fungi.

Fungi show different modes of nutrition: they could be parasitic and feed on living organisms, like *Candida albicans* which is present in the gut of humans and can cause disease, or they could be saprophytic and feed on dead organisms, like mushrooms and mould. Mould is very common and can be observed in everyday life, like *Rhizobium* on moist bread (bread mould), *Penicillium* and *Aspergillus*. Mould and yeast (*Saccharomyces cerevisiae*) are the only fungi that can be considered as microbes; fungi like mushroom, toadstool or puffballs are bigger in size. Some of the most common diseases caused by fungi are ringworm, candidiasis and athlete's foot.

SCIENCE IN DAILY LIFE

During the rainy season, bread that is moist gets spoiled and its surface is covered with patches of a greyish-white substance.

Observe these patches under a magnifying glass. You will be able to see tiny, round black structures. Study the structures carefully.

Stale bread under moist conditions is usually attacked by fungi, which form white patches. These patches are generally referred to as bread mould.

Protozoa

There has been a lot of debate about protozoa as they are single-celled organisms like prokaryotes but have the characteristics of eukaryotes. So, protozoa can be defined as single-celled eukaryotes that are either parasitic or free living (Fig. 6.3). Protozoans are classified as animals as they often shows animal-like characteristics, such as a lack of cell wall. They were believed to be animals descended from the first forms of life or the 'first animal' and were thus named protozoa (from the Greek '*protos*' meaning 'first' and '*zoon*' meaning 'animal'). Free-living protozoa are more common and are often present in the water of ponds, lakes, drains, rivers and seas. Common examples are *Euglena*, *Trypanosoma*, *Amoeba* and *Paramecium*. Some protozoa are known to cause diseases that are lethal to humans. Malaria is caused by *Plasmodium* and dysentery by *Entamoeba*.

Fun Fact!

Entamoeba histolytica, a protozoan known to cause amoebic dysentery, can remain in the human body for as long as 40 years!

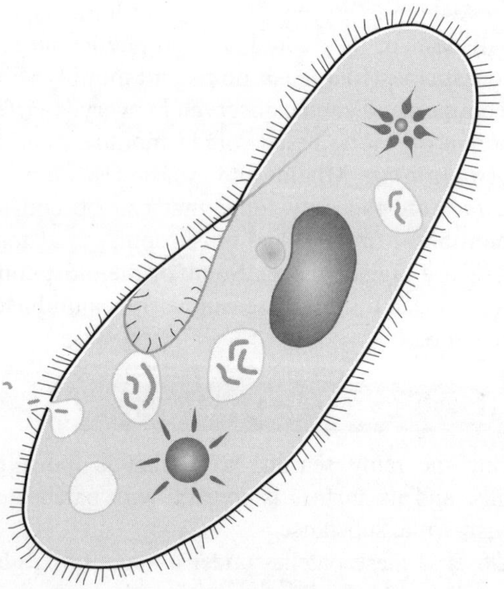

Fig. 6.3 *Paramecium*

Algae

This is a large group of simple, plant-like organisms. They are autotrophs (can make their own food) as they contain chlorophyll. However, they differ from plants in that they lack stems, roots or leaves. Most algae are multicellular (Fig. 6.4). Common examples are diatoms, blue-green algae (BGA), *Chlamydomonas*, seaweeds and *Spirogyra*. BGA has the ability to fix nitrogen gas from the atmosphere. Examples of unicellular algae are diatoms and *Chlamydomonas* and examples of multicellular organisms are BGA and *Spirogyra*. Seaweeds are the most complex and largest marine algae whereas *Spirogyra* is the most complex freshwater algae. *Chlorella* and *Spirulina* are two popular algae that are consumed in many nations as a source of protein and omega-3 fatty acids.

Fig. 6.4 Algae in water
(Algae (*Cladophora glomerata*) on a rope in...; W.carter; 2019; https://commons.wikimedia.org/wiki/File:Algae_on_a_rope_in_Norra_Grundsund_5.jpg. Licensed under: CC0)

Viruses

There has been a lot of discussion about whether viruses should be considered as biotic or abiotic. This is because of the inability of viruses to feed, grow, respire, excrete or move. Today, viruses are considered as living organisms that can develop and reproduce only inside living organisms (Fig. 6.5). The living organism is termed the host and could be a plant, animal or bacteria. Thus, viruses show a dual character as they behave like a non-living organism outside the host and like a living organism inside the host. Viruses are known to be agents of diseases in humans, plants and other animals. Polio, chickenpox, AIDS,

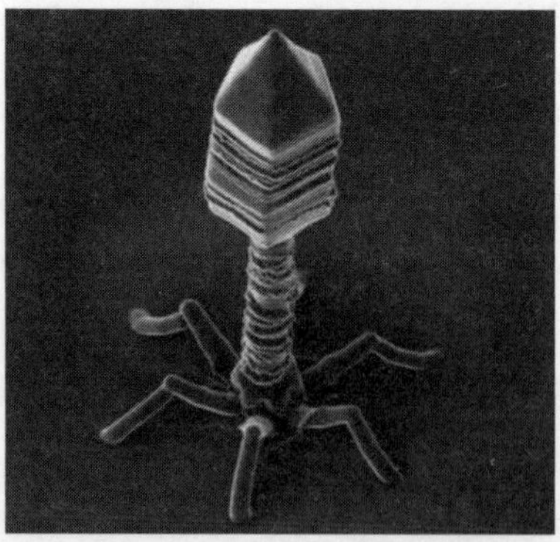

Fig. 6.5 A bacteriophage virus
(SEM images of 3-D nanostructures fabricated by FIB-CVD. "Bacteriophage"; Reo Kometani and Sunao Ishihara; 2009; https://commons.wikimedia.org/wiki/File:IBID_bacteriophage.jpg. Licensed under: CC-BY-4.0)

herpes, influenza and smallpox are caused by viruses. Antibiotics have no effect on viruses.

Fun Fact!

Did you know that viruses have been responsible for many pandemics in history:

- 1889–1890: Flu pandemic
- 1918–1920: Spanish flu
- 1957–1958: Asian flu
- 1968: H3N2 virus
- 1981–Present: AIDS pandemic
- 2009–2010: H1N1 virus
- 2019–Present: COVID-19

WHERE DO MICROORGANISMS LIVE?

Microorganisms live in all types of environments, whether they are single-celled like bacteria, protozoa and some algae or multicellular like fungi and most algae. Microbes are found in the soil, air and in water bodies. The living conditions of microbes is vast and could range from hot springs and icy regions to marshy lands and deserts. Some microorganisms have the ability to grow on other living organisms, like a parasite. The dead and decomposed bodies of organisms also serve as hosts for microbes. Human and other animal bodies also serve as a place for microbes to grow.

Fun Fact!

Deinococcus radiodurans has earned a place in the Guinness Book of World Records as the world's toughest organism. Found inside nuclear reactor walls, it has been known to withstand 1500 times the dose of radiation that would kill a human!

MICROORGANISMS AND US

Microorganisms play a very important role in our everyday life. Over the years, scientists have discovered that the majority of microorganisms are useful to humans and the environment while only a handful are harmful to us.

Useful Microorganisms

Microbes are used by humans for different purposes, such as:

- Making curd, cake and bread
- Producing acetic acid (vinegar), alcohol and wine
- Preparing antibiotics
- Fixing nitrogen from the atmosphere to increase soil fertility
- Decomposing dead organic matter into simple forms

In Everyday Life

Microbes are used in everyday life for many purposes. The most common is for the preparation of bread, curd and cake.

- **Curd:** Curd results from fermentation of milk by *Lactobacillus* (Fig. 6.6). Curd contains various microbes but *Lactobacillus* is responsible for promoting the formation of curd by multiplying in milk. Curd is rich in Vitamin B12. To prepare curd, add a small quantity of pre-made curd to warm milk. Curd is also one of the main ingredients in idli and bhatura and makes them spongy and soft.

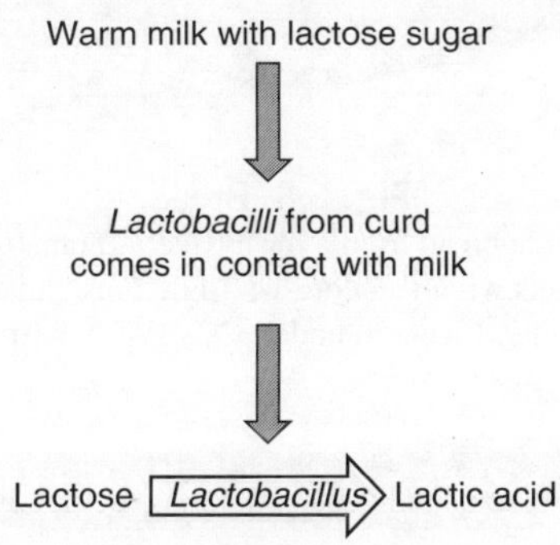

Fig. 6.6 Steps in the preparation of curd from milk

- **Cheese and pickles:** Cheese (Fig. 6.7) and pickles are made using bacteria. *Lactobacillus* belongs to the category of lactic acid bacteria (LAB), which is involved in cheese making. Different types of cheese are made using different types of bacteria. For example, the characteristic holes of Swiss cheese are obtained by the carbon dioxide produced by *P.shermanii*.

Fig. 6.7 Types of cheese

(Eight cheese collection of cheese platter in...; Geoffreyrabbit; 2020; https://commons.wikimedia.org/wiki/File:Cheese_selections_in_the_buffet.jpg. Licensed under: CC-BY-SA-4.0)

- **Bread:** Yeast is used in the baking industry. It is known to undergo anaerobic respiration to produce carbon dioxide. When bread is baked, the yeast in it reproduces rapidly and releases carbon dioxide that fills up the bread and enhances its volume. This results in soft, light and spongy bread with multiple holes of aeration (Fig. 6.8). The same mechanism is involved in the making of pastry and cake.

Fig. 6.8 Bread
(Freshly made bread from a local bakery; FranHogan; 2018; https://commons.wikimedia.org/wiki/File:Fresh_made_bread_05.jpg. Licensed under: CC-BY-SA-4.0)

SCIENCE IN DAILY LIFE

Let's bake bread!

1. Mix half a kilogram of white flour (maida) with warm water while adding sugar.
2. Add a small amount of yeast powder to the batter and knead it to make a soft dough.
3. Set the dough aside for two hours. The dough is observed to increase in volume. Now bake it to get soft and spongy bread.

In Commercial Production

Microorganisms are utilised for the large-scale production of alcohol and vinegar (Table 6.1). Yeast serves as one of the main bacteria involved in the production of alcohol and vinegar (acetic acid). The alcohol that is produced is further utilised in the production of beer, whisky, wine and industrial spirit. Yeast undergoes anaerobic respiration to form carbon dioxide and alcohol by fermenting sugar. Substances like fruit and cane juice along with barley, maize, paddy, wheat and rice act as substrates for the growth of yeast in the production of alcohol. The crops involved contain starch that is eventually converted into sugar. Fermentation was discovered by Louis Pasteur in 1857 and is defined as the conversion of sugar into alcohol by the action of yeast.

Table 6.1 Microorganisms used in commercial production of food items

Type of microorganism	Microorganism	Commercial use
Fungi	Yeast	Alcohol production
Bacteria	*Lactobacillus*	Curd production
Bacteria	*Acetobacter malorum*	Vinegar production
Bacteria	*Gluconacetobacter azotocaptans*	Coffee production
Bacteria	*Propionibacterium shermanii*	Cheese production

Acetic acid is a weak acid and its diluted form is known as vinegar. In order to obtain vinegar, alcohol is produced first by the yeast. Then, bacteria known as *Acetobacter* is added to the mixture with air bubbled through it. By the action of this bacteria, alcohol is converted to vinegar.

SCIENCE IN DAILY LIFE

1. Take 2 spoonfuls of curd.
2. Warm 2 cups of milk in a pan and pour the milk into a bowl.
3. Add the 2 spoonfuls of curd and stir it until it is mixed well. Close the bowl with a lid and keep it undisturbed for the entire night. The next day, you will notice that the two spoonfuls of curd have converted the milk to curd.

For the Production of Medicines

- **Antibiotics:** An antibiotic is a medicine that stops the growth of or kills the microorganisms that cause disease. The source of antibiotics is a microorganism. Some common antibiotics that have been manufactured and are still in high demand are streptomycin, penicillin, tetracycline and erythromycin. Antibiotics are manufactured either using bacteria and fungi or synthetically. The disease-causing microbes are known as pathogens and antibiotics kill these pathogens without harming any other human cells.

Fun Fact!

Did you know that the first antibiotic, penicillin, was discovered 'accidently' by Alexander Fleming in 1929?

Alexander Fleming was working on a disease-causing microbe and cultivated a culture of it. A small, green fungus growth was noticed by Fleming in one of his cultures and he observed that it stopped the growth of culture in that area. He studied it further and observed that the growth of many pathogens was restricted by this fungus. This fungus, known as *Penicillium*, was then found to stop many bacterial and fungal infections and the antibiotic called penicillin was developed. In the Second World War, this antibiotic saved the lives of thousands of soldiers. Fleming was awarded the Nobel Prize for Medicine in 1945, along with Florey and Chain, who invented techniques to produce penicillin in large quantities.

Antibiotics are prescribed to us by doctors every time we fall ill and are very effective against a large range of bacteria and fungi. Antibiotics may be given in many forms such as capsules, tablets or injections (Fig. 6.9). They may kill or stop bacterial and fungal growth but are ineffective against viruses. The common viral infections like flu or common cold cannot be treated by antibiotics. Antibiotics are also given to plants and other animals mixed with feedstock to prevent them from getting infected by different bacteria and fungi.

Various points should be considered while prescribing antibiotics, as every antibiotic has a different mechanism to stop infections: Valinomycin disrupts the cellular membrane by creating pores in it, penicillin stops cell wall synthesis, while streptomycin and erythromycin inhibit protein formation.

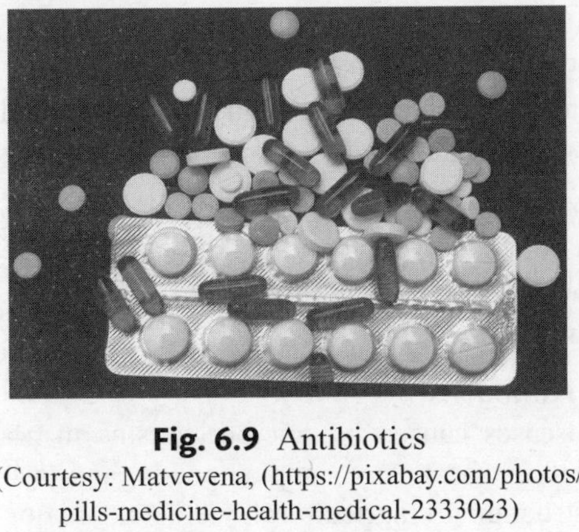

Fig. 6.9 Antibiotics
(Courtesy: Matvevena, (https://pixabay.com/photos/pills-medicine-health-medical-2333023)

- The full course of the antibiotics should be completed as prescribed.
- Do not consume antibiotics unless prescribed by as doctor.
- Antibiotics should not be taken unnecessarily as it could result in killing of useful bacteria inside the human body.
- Antibiotics should be taken at the proper dose.

- **Vaccines:** Apart from the production of antibiotics, microbes can also be used in preparing vaccines. A vaccine is a medicine that is prepared to provide strong immunity against a specific disease. Vaccines are manufactured by different microbes on a large scale and given to humans to protect them from future onset of the disease. The way a vaccine works is very different from how an antibiotic works (Fig. 6.10).

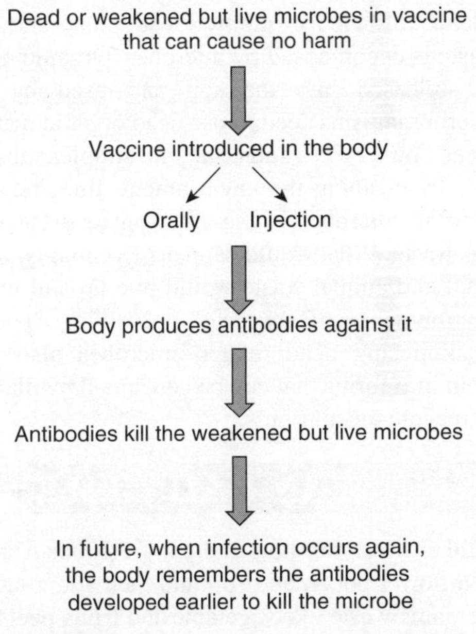

Fig. 6.10 Steps in the action of vaccines

The process of giving a vaccine in order to provide protection against a specific disease orally or by injection is known as vaccination. To maintain good health, vaccination should be carried out in a particular time period. Many vaccines are available for children as, at a certain age, some diseases attack children and proper prevention is required. The vaccines that are available for children to be taken at a certain age are for polio, measles mumps rubella (MMR), diphtheria, tuberculosis, cholera and typhoid. The child is immunised against a particular disease after vaccination and this process of attaining immunity is known as immunisation. In 1979, Edward Jenner invented the vaccine for smallpox; this led to the eradication of this disease in most parts of the world. In India, under the National Health Programme, several vaccines are provided free of cost. In the Pulse Polio Programme, children are given polio vaccines orally.

Environmental Uses

- **Increase soil fertility:** Some bacteria in the environment help to fix nitrogen from the atmosphere and increase the fertility of soil. These microbes are known as nitrogen fixers. Example are blue-green algae and some bacteria. BGA stores compounds that are rich in nitrogen while fixing nitrogen and this is why they are used as fertiliser in agriculture fields. Addition of BGA increases soil fertility. Peas, beans and other pulses that are known as leguminous plants have *Rhizobium* living in the root nodules that helps to fix nitrogen in the soil.
- **Clean the environment:** Dead plant and animal waste is decomposed by microbes. Organic nutrients are released into the soil and used by plants. Microorganisms decompose dead organic matter and break down faeces and urine into simpler substances, thereby cleaning the environment. Imagine if there were no microbes in the atmosphere to decompose this waste. What would happen? Eventually, the dead plant and animal waste would pile up and make the environment unfavourable to live in. Apart from decomposing dead matter, microbes also convert them into forms that can be consumed by plants, and so recycle the nutrients.

SCIENCE IN DAILY LIFE

1. Take two flower pots and mark them as A and B. In flower pot A, add soil and then add a layer of organic waste like vegetable and fruits peels with other kitchen waste.
2. In pot B, add soil and then add a layer of polythene bags and other plastics.
3. After 3 or 4 weeks, observe pot A and you will see that microbes have converted the waste into manure and decomposed forms. These nutrients could be used to grow plants.
4. In pot B, the plastic waste could be observed as unchanged and no decomposition would have occurred. This is because microorganisms cannot decompose inorganic matter. This shows the activity of microbes in decomposing organic matter.

Fun Fact!

Alcanivorax borkumensis is a bacillus that is known to break down oil and gas. It is known as an 'oil-eater' and could be the future of water pollution treatment.

Harmful Microorganisms

Microorganisms can cause many diseases in humans, animals and plants. These disease causing microorganisms are known as pathogens. The common diseases known to be caused by microbes are the common cold, tuberculosis, tetanus, food poisoning, typhoid and measles. Some microbes also spoil food, leather and clothing.

Pathogens

Pathogens enter our body through water, air and the food we consume. The diseases caused by pathogens are said to be communicable when they are transmitted by direct contact of an infected individual with a healthy person. Examples of communicable diseases are the common cold, cholera and chickenpox. The individual suffering from the disease is termed the infected person. Communicable diseases can spread in many ways, such as:

- By consuming infected food or water
- Through the bites of insects like mosquitoes
- By breathing pathogen-infected air
- By sharing infected needles
- Through physical contact with an infected person
- Using articles handled by an infected person

In the case of the common cold, when an infected person sneezes, the droplets released carry thousands of common cold viruses that spread in the air. These droplets can enter the body of a healthy individual while breathing and cause disease. Communicable diseases can be prevented in simple ways:

- Infected persons while sneezing should always cover their mouth and nose
- Only disposable syringes and injections should be used
- Keep food and water protected from flies
- Avoid physical contact with infected persons
- Insect and mosquito bites should be prevented
- Take vaccinations at the desired doses at the desired time

Plant Pathogens

The diseases caused by microorganisms in plants and crops like wheat, rice, apple, paddy and maize lead to destruction of agricultural yield and major loss to farmers. Some common diseases are citrus canker, rust of wheat and yellow mosaic disease of wheat.

- **Rust of wheat:** This disease is caused by fungi. It is transmitted by seeds and air in wheat plants.
- **Yellow vein mosaic:** This disease is caused by the yellow mosaic virus (Fig. 6.11). It is transmitted by insects in okra or bhindi.

Fig. 6.11 Yellow mosaic virus in zucchini
(Striped zucchini with Zucchini Yellow Mosaic Virus...; Downtowngal; 2015; https://commons.wikimedia.org/wiki/File:Zucchini_with_Zucchini_Yellow_Mosaic_Virus.jpg. Licensed under: CC0)

- **Citrus canker:** This disease is caused by bacteria in citrus fruits (Fig. 6.12) like lemon or orange. It is transmitted by air.

Fig. 6.12 Citrus canker
(Citrus canker (*Xanthomonas citri* (ex Hasse 1915)...; Timothy Schubert; 2004; https://commons.wikimedia.org/wiki/File:Citrus_canker_Xanthomonas_citri_-_symptoms_on_leaves_UGA1262029.jpg. Licensed under: CC-BY-3.0)

Animal Pathogens

The diseases caused by microorganisms in animals like cow, pig, buffalo, goat, poultry bird and sheep lead to mortality of animals along with feedstock loss and should be prevented. Some common examples of animal diseases caused by pathogens are:

- **Anthrax:** In 1876, Robert Koch discovered the causative agent of anthrax, *Bacillus anthracis*. This is a bacteria that is transmitted through air and infects cattle.
- **Foot and mouth disease:** This disease is in cattle caused by a virus (Fig. 6.13). It is transmitted by air and causes blisters to occur on the feet and mouth of cattle.

Fig. 6.13 Foot and mouth disease
(When this shorthorn heifer in the United Kingdom became afflicted ...; Lawrence Livermore National Laboratory; 2006; https://commons.wikimedia.org/wiki/File:Shorthorn_heifer_foot-and-mouth_disease1.png. Licensed under: CC-PD-Mark)

- **Aspergillosis:** Poultry birds are affected by this disease, which is caused by a fungus.

Carriers of Pathogens

The insects that carry pathogens and infect other organisms without suffering from the disease are known as carriers or vectors. Common carriers are the housefly and mosquito. They bite us or contaminate our food and thus transfer the disease-causing microbes to our body. These two common carriers have different breeding grounds as the housefly lives in insanitary and filthy places while mosquitoes breed in stagnant water.

- **Housefly:** The housefly (Fig. 6.14) feeds on animal excreta, dead and decomposed organic matter, refuse, garbage and our food. It lay eggs on garbage dumps. The housefly has fine hairs on the legs; these act as the main microbial travel system. When a fly sits on garbage, particles of the garbage along with microbes stick to the fine hair of the legs. The fly then sits on uncovered food and thus transfers the microbes to the food. Food is thus contaminated and when it is

Fig. 6.14 A housefly
(Common house fly (*Musca domestica*); Alvesgaspar; 2007; https://commons.wikimedia.org/wiki/File:Musca_domestica_September_2007-1.jpg. Licensed under: CC-BY-SA-3.0)

consumed, the disease-causing microbes enter our body and cause disease. The common housefly carries microbes that cause diarrhoea, typhoid, tuberculosis and cholera.

- **Prevention of disease:**
  - Do not litter garbage
  - Always keep the garbage in covered-lid bins
  - Always cover the food so that flies cannot sit on the food
  - Kill the flies using insecticides
  - Do not eat uncovered food from unhygienic places
  - Some diseases caused by flies can be prevented by proper vaccination

- **Mosquito:** Unlike the housefly, the mosquito does not carry microbes outside its body but does so inside its body. The pathogens are transmitted by the action of biting. Mosquitoes breed in still waters like lakes, ponds, drains and sewage. One of the most common and lethal diseases spread by the mosquito is malaria. The female *Anopheles* mosquito carries the parasite of malaria, *Plasmodium*, which survives by using blood as its food. The life cycle of *Plasmodium* is completed in mosquitoes and humans in the following stages:
  1. The female *Anopheles* mosquito bites an infected person and sucks in their blood.
  2. The parasite enters the mosquito and when the same mosquito bites a healthy individual, the parasite is transferred to the bloodstream of the healthy individual.
  3. The healthy person is now infected with malaria. (Please see chapter 7 for more information on malaria)

Mosquitoes also spread diseases such as dengue and encephalitis. Dengue is caused by the female *Aedes* mosquito.

Fun Fact!

In 2015–2016, WHO declared the Zika virus epidemic that started in Brazil and spread to other countries like America. Zika virus is a disease spread by the *Aedes* mosquito with no vaccine currently available.

- **Prevention of mosquito-borne diseases:**
  - Do not let water to get collected near the house
  - Drain out the stagnant water pools near the house
  - The doors and window panes should have iron mesh or net
  - Kill the mosquitoes regularly inside the home by insecticides
  - Mosquito nets or mosquito repelling creams should be used while sleeping at night

Causative Organisms

The microorganisms that are responsible for a disease are known as causative agents or causative microorganisms. Table 6.2 lists the causative microorganisms which are common in everyday life.

Table 6.2 Causative microbe, mode of transmission and preventive measures of common human diseases

Human disease	Causative microorganism	Mode of transmission	Preventive measures
Tuberculosis	Bacteria	Air	Keep the patient in complete isolation. Keep the personal belongings of the patient away from those of the others. Vaccination to be given at suitable age.
Measles	Virus	Air	
Chickenpox	Virus	Air/Contact	
Polio	Virus	Air/Water	
Cholera	Bacteria	Water/Food	Maintain personal hygiene and good sanitary habits. Consume properly cooked food and boiled drinking water. Vaccination.
Typhoid	Bacteria	Water	
Hepatitis A	Virus	Water	Drink boiled drinking water. Vaccination.
Malaria	Protozoa	Mosquito	Use mosquito net and repellents. Spray insecticides and control breeding of mosquitoes by not allowing water to collect in the surroundings.

HUMAN IMMUNE SYSTEM

Health can be defined as an individual's wellbeing that includes mental, physical and social aspects. Immunity refers to a person's ability to live disease-free with no infections (Fig. 6.15).

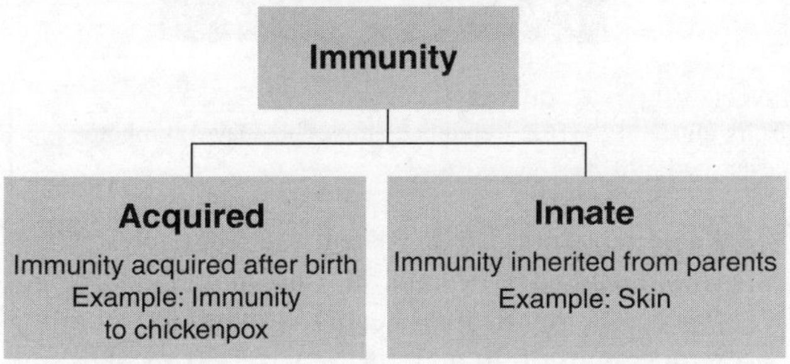

Fig. 6.15 Different types of immunity in humans

The human immune system is a well-developed system composed of a complex network of cells, tissues, organs and organ systems. This complex network functions together to defend the human body against pathogens. The process and involvement of the immune system against foreign particles is known as the immune response of the body. There are two components to the immune response: lymphatic organs and white blood cells (Fig. 6.16).

The first line of defence works against foreign particles that try to enter the body. For example, the skin of the human body acts as the first line of defence.

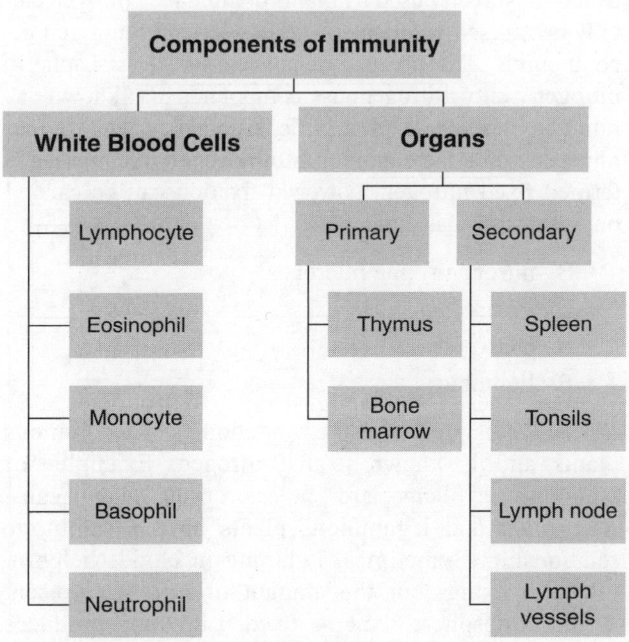

Fig. 6.16 Components of immunity

Once the pathogen enters the body, it comes across cells specialised in killing as well as neutralising pathogens.

The immune system can differentiate between self and foreign cells. It accomplishes it by using molecules like proteins which are located on the cell surface as a detector. It helps the cells to recognise specific cells and ignore its own cells. The substance that can initiate the immune response is known as the immunogen. Immunogens could be fungi, bacteria, toxins or viruses.

FOOD POISONING

Food that is not taken care of and is either left uncovered, not stored properly or not well-preserved can come in contact with microorganisms that cause spoilage of food. The microorganisms growing on food also release toxins. These toxins leads to spoiling of food by giving it a strange taste and changing the colour of the food. If this food is consumed by a person, it could lead to food poisoning.

Food poisoning is the disease that is caused by the growth of microorganisms on food items and which causes illness due to the toxins released. The main symptoms of food poisoning are fever, headache, abdominal pain, diarrhoea and vomiting. The microorganisms come in contact with food through different sources like air, cockroaches, flies, dirty containers, unclean hands, insects, sick farm animals and rats. The two most common microbes known to cause food poisoning are *Clostridium botulinum* and *Salmonella*. The fungus that is responsible for the food poisoning is *Aspergillus.* Spoiling of food can be prevented by preserving it using various techniques.

Food Preservation

Foods like milk, meat, vegetables and fruits undergo spoilage very easily due to the presence of water in them that facilitates the growth of microbes. To prevent food from being spoiled by the action of microbes, the food should be preserved using various techniques. These techniques are as follows:

- **Dehydration:** Dehydration or drying is a method of preserving food materials by removing water from the food. Water removal acts as a preservation mechanism as water facilitates the growth of microbes. Vegetables like peas and spinach leaves are sun-dried to preserve them for future use. A major example of dehydration is grapes being dehydrated to raisins (Fig. 6.17).
- **Heating:** Heating is a way to preserve food as it kills microbes that are responsible for spoiling food. Boiling milk is an example of preserving it for longer use.

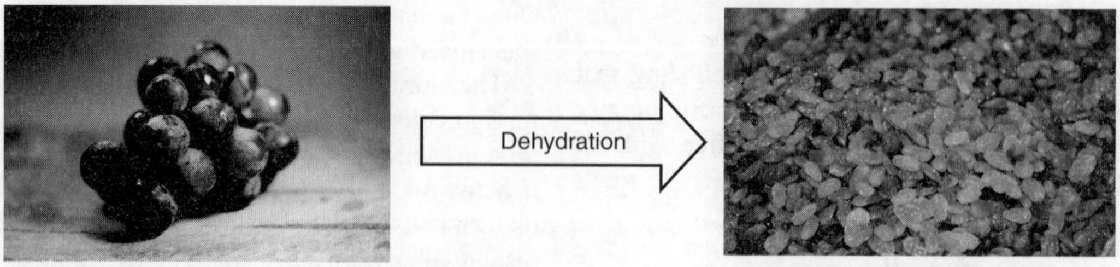

Fig. 6.17 Dehydration converts grapes to raisins
(Courtesy: Pexels, https://pixabay.com/photos/grapes-red-grapes-fruits-cluster-1283162; Republica, https://pixabay.com/photos/raisins-grapes-dried-desiccated-89013)

- **Refrigeration:** Low temperature provides unfavourable conditions for microbes to grow on food and so preserves it. Refrigerators at home preserve perishable food items like milk, meat, fruits, vegetables, cooked food, etc.
- **Deep freezing:** Preservation of food at a temperature lower than the freezing point of water in excessively cold conditions is known as deep freezing. This can be done in the freezer compartment of a refrigerator or by using special deep-freezing machines. Meats, vegetables and fruits are preserved using deep freeze and can remain fresh and spoilage-free for longer periods of time.
- **Common salt:** Since ancient times, common salt has been used to preserve meat and fish as it prevents the spoilage of food. Common salt has also been employed to preserve vegetables in our homes in the form of chutney or pickles. The only point of preservation to be considered using common salt is that only raw vegetables or fruits should be preserved.
- **Sugar:** The water content from fruits is removed by sugar as it absorbs it and inhibits the growth of microbes. Certain fruits that have been preserved by sugar in the form of jams and jellies are apples, strawberry, guava, mango, etc.
- **Oil and vinegar:** Oil and vinegar have been used to preserve certain vegetables and fruits like mango, gooseberry, lemon and garlic in the form of pickles. It is successful in preventing food from spoiling as it creates conditions unfavourable for the growth of microbes.
- **Chemical preservatives:** Sodium benzoate, citric acid and sodium metabisulphite are three special chemicals that are employed to preserve food. These chemicals only kill bacteria and do not harm the human body. Sodium metabisulphite and sodium benzoate are used in jams and jellies to preserve fruits while citric acid is used in confectionary.
- **Pasteurisation:** Louis Pasteur was a French scientist who invented a wonderful way to preserve food. Pasteurisation involves heating milk up to its boiling temperature of around 70°C for 15–30 seconds and then suddenly exposing it to cold temperature. This sudden change kills microbes that are left over after heating. This method of preserving milk is used by big dairies and industries. Pasteurised milk can be consumed directly from packets as it is microbe-free and has a longer shelf life.
- **Air-tight packets:** Fruits, vegetables and other food items are sold in markets in air-tight packets as they are protected from exposure to microbes.

NITROGEN FIXATION

About 78% of the atmosphere is composed of nitrogen. However, this abundance of nitrogen in the atmosphere cannot be directly used by plants or animals. Nitrogen can only be utilised by plants if it is converted to nitrogenous compounds. The process of conversion of atmospheric nitrogen into nitrogenous compounds is known as nitrogen fixation. Atmospheric nitrogen is termed free nitrogen while the compounds of nitrogen like nitrates is termed fixed nitrogen. Nitrogen fixation can be carried out using four basic methods:

- By nitrogen fixing microbes
- By *Rhizobium*
- By BGA
- By lightning

Rhizobium is present in the root nodules of leguminous plants and is known to fix nitrogen. Examples of leguminous plants are pulses, peas and beans. *Rhizobium* and leguminous plants have a symbiotic relationship. Sometimes, lightning in clouds helps in nitrogen fixing, but the amount of nitrogen present in the atmosphere remains fixed. Lightning produces a tremendous amount of heat, which makes the free atmospheric nitrogen bind to the oxygen present in the environment and results in nitrogenous compounds. The nitrogenous compounds formed are mixed with rainwater and fall on the earth and mix with the soil. The maintenance of the biogeochemical cycle involves many microorganisms and processes and is known as the nitrogen cycle.

Nitrogen Cycle

For the growth of plants and animals, nitrogen plays a vital role. Nitrogen is involved in the growth and development of the body as it is a key element in proteins, nucleic acids, vitamins and chlorophyll. This nitrogen element is continuously exchanged between living organisms like microbes, plants and animals and non-living things like soil, water and air. The nitrogen cycle can be defined as the continuous exchange of the nitrogen element from living organisms to the non-living environment.

The main steps of nitrogen cycle are as follows (Fig. 6.18):

1. The free atmospheric nitrogen is converted into nitrogenous compounds in the soil by the action of *Rhizobium*, nitrogen fixing microbes, BGA or lightning.
2. Plants that are growing in the nitrogen-rich soil take up the nitrogen and utilise it for their growth and development by using it to form organic compounds.
3. These plants are eaten by animals that convert the plant protein into animal protein. The nitrogen of the protein is utilised by animals in various ways and the source of this protein is not only plants but also different animals that are eaten by other animals.
4. After the death of animals or plants, the nitrogenous compounds from the dead and decomposed bodies are converted into simpler forms by bacteria. The nitrogen that plants took up are now returned to the soil in simple form and used by other plants, and so the cycles continues.
5. There are some denitrifying bacteria that convert the decomposed nitrogenous compounds into nitrogen gas. This allows the nitrogen to return to the atmospheric in free nitrogen form.

By this process, the level of nitrogen is maintained in the environment.

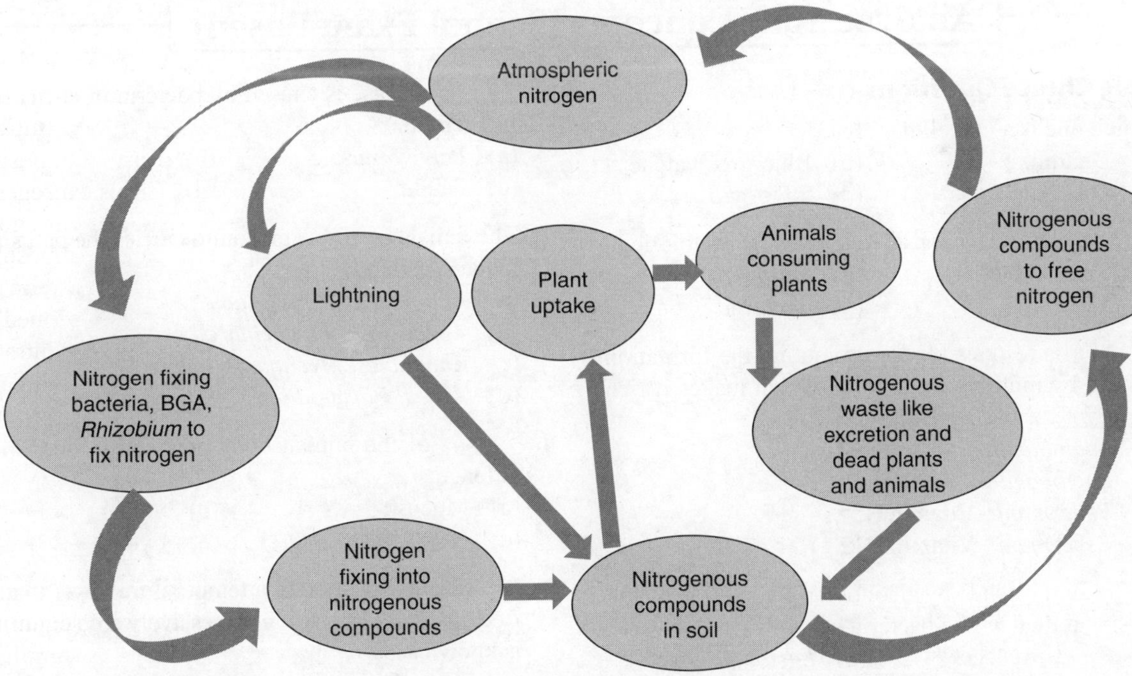

Fig. 6.18 The nitrogen cycle

SUMMARY

- Microorganisms are not visible to the unaided eye but are present all around us.
- They are divided into different groups: bacteria, protozoa, fungi, algae and viruses.
- Microorganisms can be beneficial to humans and the environment in different ways: in curd and cheese making, in the food industry where alcoholic beverages and vinegar are made, in the manufacture of medicines to make vaccines and antibiotics, in cleaning up the environment and maintaining important biogeochemical cycles like the nitrogen cycle.
- Microorganisms can be harmful to us as well in different ways, like spoiling food by growing on it and causing food poisoning by releasing toxins in the food, which makes food preservation important, or by spreading infectious diseases using different carriers like air, water, mosquito and housefly.
- Microorganisms are present all around us as friend or foe and small observations and detailed studies help in identifying the occurrence of microbes in everyday life.
- The process and involvement of the immune system against foreign particles is known as the immune response of the body.
- There are two components to the immune response: lymphatic organs and white blood cells.
- The immune system can differentiate between self and foreign cells.

PRACTICE QUESTIONS

Analyse Your Concepts (School Exam Based)

Multiple Choice Questions

1. Which one is a unicellular alga?
(a) Diatoms (b) Blue-green algae
(c) Dinoflagellates (d) *Spirogyra*

2. _______ should be considered as biotic or abiotic.
(a) Slime moulds (b) *Spirogyra*
(c) Viruses (d) Bacteria

3. _______ is responsible for promoting the formation of curd by multiplying in milk, which makes it rich in ______.
(a) *Lactobacillus*, Vitamin B5
(b) *Lactobacillus,* Vitamin B12
(c) *P.shermanii,* Vitamin B5
(d) *P.shermanii,* Vitamin B12

4. Acetic acid is made by alcohol fermented first by the yeast and then by the bacteria known as:
(a) *Saccharomyces* (b) *Lactobacillus*
(c) *P.shermanii* (d) *Acetobacter*

5. Leguminous plants have ______ living in the root nodules that helps to fix _______in the soil.
(a) *Rhizobium*, nitrogen (b) *Rhizobium,* oxygen
(c) BGA, nitrogen (d) BGA, oxygen

6. ______ disease is caused by bacteria in citrus fruits like lemon.
(a) Yellow mosaic (b) Rust
(c) Blight (d) Citrus canker

7. The female _______ mosquito carries the parasite of malaria called ______.
(a) *Anopheles, Plasmodium*
(b) *Aedes, Mycobacterium*
(c) *Anopheles, Mycobacterium*
(d) *Aedes, Plasmodium*

8. ______ of the human body acts as the first line of defence.
(a) Mucus (b) Skin
(c) Hair (d) Eye

9. Preservation of food at a temperature lower than the freezing point of water in excessively cold conditions is known as:
(a) Deep freezing (b) Refrigeration
(c) Pickling (d) Pasteurisation

10. The substance that can initiate the immune response is known as the:
(a) Immunity (b) Antibody
(c) Microbe (d) Immunogen

Match the Following

Each question contains statements given in two columns which have to be matched. Statements in column I have to be matched with statements in column II.

1. Choose the correct option:

Column I	Column II
i. Malaria	a. Virus
ii. Anthrax	b. *Plasmodium*
iii. AIDS	c. Toxins
iv. Food poisoning	d. Bacteria

2. Choose the correct option:

Column I	Column II
i. Coccus	a. Comma shaped
ii. Bacillus	b. Spiral shaped
iii. Vibrio	c. Spherical shaped
iv. Spiral	d. Rod shaped

3. Choose the best match:

Column I	Column II
i. Cholera	a. Water
ii. Malaria	b. Air
iii. Tuberculosis	c. Water/Food
iv. Typhoid	d. Mosquito

4. Choose the best match:

Column I	Column II
i. WBC	a. Spleen
ii. Primary lymphoid organ	b. Lymphocyte
iii. Secondary lymphoid organ	c. Bone marrow

5. Choose the correct option:

Column I	Column II
i. Dehydration	a. Fish
ii. Common salt	b. Milk
iii. Pasteurisation	c. Grapes
iv. Oil and vinegar	d. Vegetables

True or False

Read the following statements and write your answer as true or false.

1. Citric oxide is used as a chemical preservative.
2. Louis Pasteur invented a milk preservation technique called pasteurisation.
3. *Rhizobium* is a free-living nitrogen fixer.
4. Sodium metabisulphite is a special chemical preservative for food.
5. Microorganisms that cause disease are called host microorganisms.
6. Vaccines are made using live microorganisms that can infect the human body.
7. Polio and rabies are bacterial diseases.
8. *Aedes* is responsible for the spread of malaria.
9. Malaria is caused by *Plasmodium*.
10. Food poisoning occurs when microbes release spores into food.
11. Housefly-carried diseases can be prevented by keeping the food covered.
12. Fungi have a filamentous structure called hyphae.
13. Edward Jenner invented the vaccine for chickenpox.
14. Lightning can fix atmospheric nitrogen.
15. Skin is a type of adaptive immunity.

Fill in the Blanks

Complete the following statements with an appropriate word/term to be filled in the blank spaces.

1. ______ is produced with the help of yeast.
2. Blue-green algae fix __________directly from air.
3. ________ is used to observe microbes.
4. Typhoid is caused by_________.
5. ________ has been used to preserve meat and fishes for ages.
6. Louis Pasteur discovered for _________ milk preservation.
7. Two human diseases caused by bacteria are ______ and _______.
8. ________ invented the vaccine for smallpox.

9. _________ and _______ are two commonly known antibiotics.

10. Food materials that can be preserved by using sugar are _____ and ______.

Assertion–Reason Questions

The following questions consist of two statements: Assertion (A) and Reason (R). Answer these questions selecting the appropriate option given below:

(a) Both A and R are true and R is the correct explanation of A
(b) Both A and R are true but R is not the correct explanation of A
(c) A is true but R is false
(d) A is false but R is true

1. Assertion (A): All fungi except yeast have hyphae.
Reason (R): Warm and moist conditions are most favourable for the growth of fungi.

2. Assertion (A): When bread is baked, the yeast in it reproduces rapidly and releases carbon dioxide.
Reason (R): Carbon dioxide fills up the bread and enhances its volume.

3. Assertion (A): Every antibiotic has a different mechanism to stop infections.
Reason (R): Penicillin inhibits protein formation.

4. Assertion (A): Louis Pasteur invented the vaccine for smallpox.
Reason (R): This led to the eradication of this disease in most parts of the world.

5. Assertion (A): Lightning in clouds helps in nitrogen fixing.
Reason (R): There are some denitrifying bacteria that convert the decomposed nitrogenous compounds into nitrogen gas.

One-Word Answers

1. Which type of microbe causes typhoid?
2. What can be defined as single-celled eukaryote that is either parasitic or free living?
3. What is the most complex freshwater alga?
4. The characteristic holes of Swiss cheese are obtained by which bacteria?
5. Who discovered fermentation in 1857?
6. Which antibiotic stops cell wall synthesis?
7. What is the causative agent of anthrax?
8. Dengue is caused by which mosquito?
9. What is the result of eating spoiled food?
10. Which method of food preservation was given by Louis Pasteur?

Define the Term

1. Protozoa
2. Increase in soil fertility by microbes
3. Antibiotic
4. Anthrax
5. Nitrogen cycle

Groups of Terms

Given below are groups of terms. In each group the first pair indicates the relationship between the two terms. Rewrite and complete the second pair on a similar basis.

1. *Bacillus*: bacteria :: *Aspergillus*: _____
2. Wheat: rust :: Okra:_____
3. Diarrhoea: housefly :: Dengue:_____
4. Jams: sodium benzoate :: Confectionary: _____
5. Measles: virus :: Malaria: ____

Identify the Odd One Out

1. Cheese, curd, bread, ice cream
2. BGA, *Spirogyra*, *Chlamydomonas*, *Candida*
3. *Rhizobium*, BGA, *Nitrosomonas*, *Penicillium*
4. Pox, Malaria, AIDS, Typhoid
5. Refrigeration, jam, common salt, chilli

Image-Based Questions

1. What is the term used to describe this shape of bacteria?

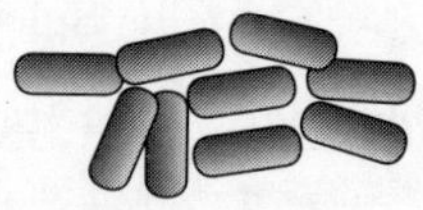

(a) Coccus (b) Spiral
(c) Vibrio (d) Bacillus

2. To which category does this microbe belong?

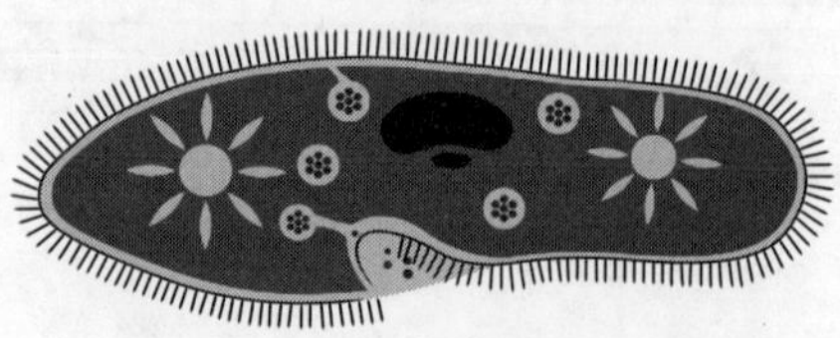

(a) Bacteria (b) Virus
(c) Protozoa (d) Algae

3. What is the term used to describe this shape of bacteria?

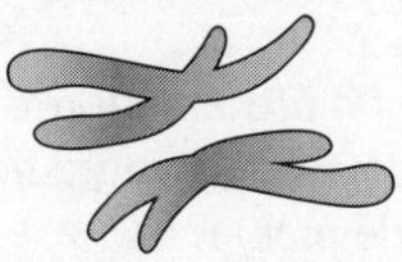

(a) Bacillus (b) Vibrio
(c) Coccus (d) Spiral

4. What is the name of this virus?

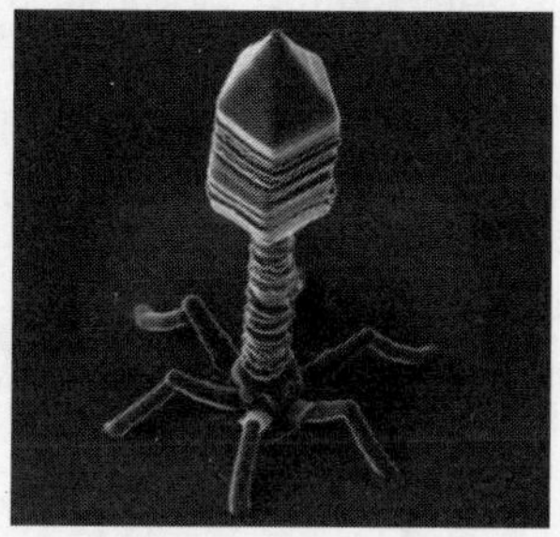

(a) HIV (b) TMV
(c) Bacteriophage (d) Polio virus

5. Choose the disease caused by this carrier:

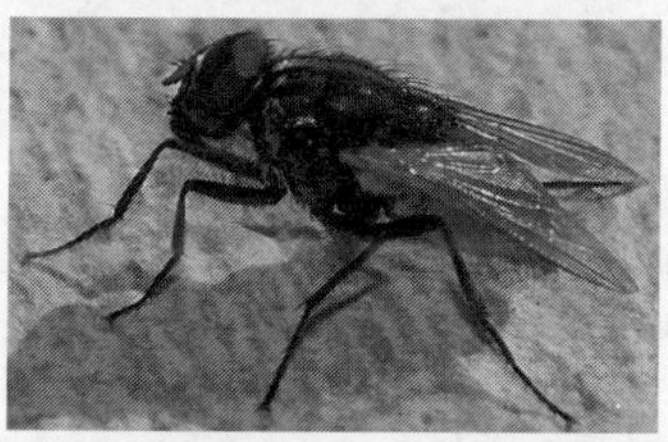

(a) Tuberculosis (b) Malaria
(c) Polio (d) HIV

6. What is the method of preservation involved in the conversion of grape to raisin?

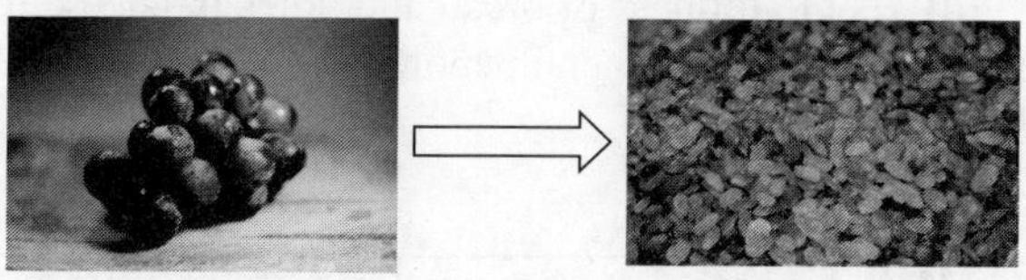

(a) Deep freezing
(b) Chemical preservation
(c) Sugar
(d) Dehydration

7. Choose the disease caused by this carrier:

(a) Malaria (b) Tuberculosis
(c) Cholera (d) Typhoid

Very Short Answer Questions

1. What is the instrument used to view microorganisms?
2. Name all the major groups of microbes.
3. Which organism is responsible for curd formation from milk?
4. Which microbe is responsible for the production of alcohol?
5. Which microorganism reproduces only inside the living cells of other organisms?
6. Who discovered penicillin?
7. What are different ways of nitrogen fixation?
8. Name two carriers of disease-causing microorganisms.
9. Name two leguminous plants that can fix nitrogen.
10. Name three special chemicals preservatives.

Short Answer Questions

1. Name any two human diseases caused by
   (a) Bacteria (b) Fungi
   (c) Protozoa (d) Virus
2. Give a brief account of the ways of nitrogen fixation and the nitrogen cycle.
3. Name any two antibiotics with their mechanism of action.
4. How do microorganisms increase soil fertility?
5. Explain the process of curd formation with a flow diagram.
6. Give a brief account of the diseases caused by mosquitoes with its species.
7. In how many ways can a communicable disease spread?
8. Discuss one disease each caused by microbes in
   (a) Animals (b) Plants
   (c) Humans
9. Write three ways of preventing carrier spread disease by
   (a) Mosquito (b) Housefly
10. Describe two ways of meat preservation.

Long Answer Questions

1. Microorganisms can be harmful to humans. Justify the statement.
2. Describe the use of microorganisms to humans.
3. What is the nitrogen cycle? Explain the steps with a labelled diagram.
4. Explain the mechanism of vaccination.
5. What are the ways to preserve food?
6. Explain the steps involved in preventing communicable diseases.
7. Explain human diseases that involve all five microbe categories.
8. What is immunity? What are the two types of immunity? What are the components of immunity?
9. Explain the role of microorganisms in
   (a) Food industry (b) Medicine industry
   (c) Everyday life (d) Environment
10. Write a short note explaining the five groups of microorganisms with examples.

Case Study-Based Questions

1. Reema visited a lake and saw it was covered with a green slimy film. She wanted to see and read more about it, but she got confused. Can you help Reema?
   (a) What are algae?
   (b) What are the types of algae?
   (c) What are different functions of algae?
2. Sonu and Rinku live with their grandparents. Every summer, they witness their grandmother drying raw mangoes in the sun and then making pickles with them by adding different ingredients. Sonu and Rinku love the taste but can you tell them the science behind it?
   (a) What is food preservation?
   (b) What preservative ingredients are added in home-made pickle?
3. Recently, Ahmed noticed that most of his friends were suffering from malaria and dengue. After gathering information, he got to know that the causative agent behind this is mosquitoes. In order to prevent other people from getting infected, Ahmed decided to raise awareness in his society to prevent the breeding of mosquitoes. Can you help Ahmed?
   (a) What are the major breeding grounds of mosquito?
   (b) What are the ways to prevent mosquito-borne diseases?
4. Suman has a baby sister. She often sees her mother taking her sister regularly to hospital. Her mother told her that to vaccinate her sister is an important process. Can you help Suman understand this concept?
   (a) What is vaccination?
   (b) Why is vaccination important?
   (c) Which vaccines are available for children?
5. Ankita along with her father walks to a bakery every other day to purchase bread. This time she saw a block of Swiss cheese that attracted her attention due to its characteristic holes. She started wondering different things related to it. Can you help her to find the answers?
   (a) What is the organism behind the making of cheese?
   (b) Why are holes present in cheese?
   (c) Is the fullness of bread and holes in cheese due to the same phenomenon?

NCERT Corner

1. Select the correct option:
   (a) Yeast is used in the production of:
      (i) Alcohol
      (ii) Sugar
      (iii) Hydrochloric acid
      (iv) Oxygen
   (b) The following is an antibiotic:
      (i) Alcohol
      (ii) Sodium bicarbonate
      (iii) Yeast
      (iv) Streptomycin
   (c) Carrier of malaria-causing protozoan is:
      (i) Cockroach
      (ii) Housefly
      (iii) Female *Anopheles* mosquito
      (iv) Butterfly
   (d) The most common carrier of communicable diseases is:
      (i) Ant
      (ii) Spider
      (iii) Dragonfly
      (iv) Housefly

(e) Bread rises because of:
   (i) Heat
   (ii) Growth of yeast cells
   (iii) Grinding
   (iv) Kneading

(f) The process of conversion of sugar into alcohol is called:
   (i) Nitrogen fixation (ii) Fermentation
   (iii) Moulding (iv) Infection

2. Match the organisms in Column A with their action in Column B.

Column A	Column B
i. Bacteria	a. Fixing nitrogen
ii. *Rhizobium*	b. Setting of curd
iii. Yeast	c. Causing malaria
iv. Protozoan	d. Causing AIDS
v. *Lactobacillus*	e. Baking of bread
vi. Virus	f. Causing cholera
	g. Producing antibodies

3. Can microorganisms be seen with the naked eye? If not, how can they be seen?

4. What are the major groups of microorganisms?

5. Name the microorganisms which can fix atmospheric nitrogen in the soil.

6. Write 10 lines on the usefulness of microorganisms in our lives.

7. Write a short paragraph on the harmful effects of microorganisms.

8. What are antibiotics? What precautions must be taken while taking antibiotics?

COMPETITION WINDOW (OBJECTIVE TYPE)

1. Match the following

Column A	Column B
a. *Lactobacillus*	i. Cheese
b. *Saccharomyces*	ii. Curd
c. *Aspergillus*	iii. Citric acid
d. *Acetobacter*	iv. Bread
	v. Acetic acid

(a) a (ii), b (iv), c (v), d (iii)
(b) a (ii), b (iv), c (iii), d (v)
(c) a (iii), b (iv), c (v), d (i)
(d) a (ii), b (i), c (iii), d (v)

2. Conversion of milk to curd improves its nutritional value by increasing the amount of:
(a) Vitamin B12 (b) Vitamin A
(c) Vitamin D (d) Vitamin E

3. Which of the following is correctly matched by the product produced by them?
(a) *Acetobacter aceti*: antibiotic
(b) Methanobacterium: lactic acid
(c) *Penicillium notatum:* acetic acid
(d) *Saccharomyces cerevisiae:* ethanol

4. Which one of the following produces alcohol?
(a) *Saccharomyces cerevisiae*
(b) *Torulopsis utilis*
(c) *Clostridium botulinum*
(d) *Leuconostoc*

5. Baker's yeast is:
(a) *S. cerevisae*
(b) *S. ludwigii*
(c) *S. octosprus*
(d) *Schizosaccharomyces*

6. Nitrogen fixation is:
(a) Nitrogen to ammonia
(b) Nitrogen to nitrate
(c) Nitrogen to amino acids
(d) Both (a) and (b)

7. Which is false about antibiotics?
(a) Cure all diseases
(b) Produced by microbes
(c) Produce allergy in some
(d) Coined by Waksman

8. Which of the following can use molecular nitrogen as a nutrient?
   (a) *Methanomonas* (b) *Mucor*
   (c) *Rhizobium* (d) *Spirogyra*

9. Substrate for alcohol formation:
   (a) Sucrose (b) Galactose
   (c) Maize (d) Bajra

10. Antibiotic are drugs commonly used to cure:
    (a) Fungi (b) Virus
    (c) Protozoa (d) Bacteria

11. Yogurt is produced with the help of:
    (a) *Lactobacillus bulgaricus*
    (b) *Lactobacillus thermophiles*
    (c) *Streptococcus thermophiles*
    (d) Both (a) and (b)

12. Antibiotics are mostly obtained from:
    (a) Fungi
    (b) Actinomycetes/bacteria
    (c) Both (a) and (b)
    (d) Cyanobacteria

13. The first antibiotic was discovered by:
    (a) A. Fleming (b) R. Koch
    (c) W. Fleming (d) Louis Pasteur

14. During fermentation from yeast, alcohol is formed from:
    (a) Protein (b) Sugar
    (c) Mucopolysaccharide (d) Fats

15. If a wheat field is inoculated with *Rhizobium*:
    (a) Soil will become rich in calcium
    (b) Soil become nitrogen rich
    (c) No effect
    (d) Soil become rich in nitrogen

16. Fleming, Chain and Florey were awarded the Nobel Prize in 1945 for the discovery of:
    (a) HIV (b) CT scan
    (c) *Penicillium* (d) *Staphylococcus*

17. Vinegar is produced by the help of:
    (a) *Lactobacillus* (b) *Azotobacter*
    (c) *Acetobacter* (d) *Rhizobium*

18. Ethanol is produced by starch as a raw material by the action of
    (a) Yeast (b) Protozoa
    (c) Virus (d) Rickettsia

19. Chickenpox : virus:: Malaria :
    (a) Fungus (b) Bacteria
    (c) Protozoa (d) Virus

20. Riya is doing an activity in the biology lab. She and her team-mates grew yeast on natural sugars present in grain like barley, wheat, rice, fruit juices. The result of the activity would be:
    (a) Formation of beer (b) Formation of wine
    (c) Formation of cider (d) All of the above

21. Refer to the shape of bacteria and read the statements below. Based on the information, identify the bacteria and select the correct option:

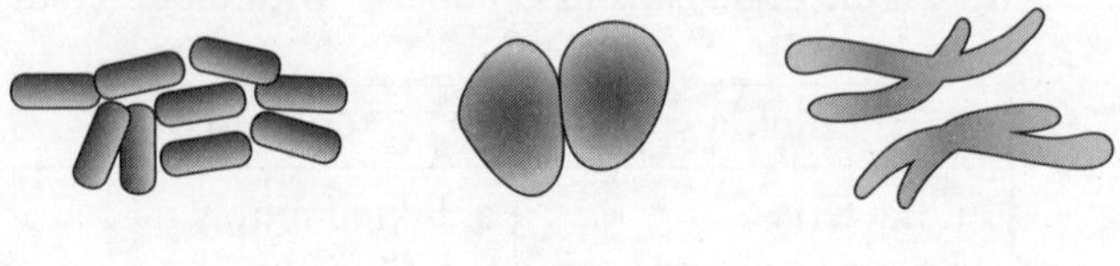

X Y Z

I. Bacteria Z causes cholera
II. Bacteria Y causes a disease that disrupts proper exchange of gases
III. Bacteria X converts lactose sugar of milk to lactic acid of curd

    (a) X is a lactobacillus.
    (b) Y is *Vibrio cholerae*
    (c) Y is *Salmonella typhii*
    (d) Both (a) and (b)

22. Latika observed her grandmother adding lots of salt to unripe mangoes to prepare pickles. Her grandmother also added a lot of oil to the jar. What could be the reason for it?
    (a) Salt prevents growth of microbes by retaining water in the food item and oil provides an environment where microbes cannot grow
    (b) Salt prevents growth of microbes by forcing them to lose water by osmosis and oil provides an environment where microbes cannot grow
    (c) Salt prevents growth of microbes by retaining water in the food item and oil prevents growth of microbes by forcing them to lose water by osmosis
    (d) Salt prevents growth of microbes by forcing them to lose water by osmosis and oil prevents growth of microbes by retaining water in the food item

23. One day you wake up with a sore throat and runny nose. Your doctor takes a swab of your throat and sends it to a lab and telephones you the next day to say that antibiotics will not help you. Which of the following is most likely the reason behind the doctor's statement?
    (a) Having waited a day, it is too late to take an antibiotic

(b) You need an antiseptic, not an antibiotic
(c) You need to be vaccinated instead of taking an antibiotic
(d) You are infected by a virus

24. Cyanobacteria:

A. Also referred as blue-green algae
B. Lacks cell wall
C. Fixes atmospheric nitrogen
D. Lacks chlorophyll

Choose the correct option

(a) A and B
(b) D and C
(c) A and C
(d) D and B

Answer Keys

Multiple Choice Questions

1. (a)	**2.** (c)	**3.** (b)	**4.** (d)	**5.** (a)
6. (d)	**7.** (a)	**8.** (b)	**9.** (a)	**10.** (d)

Match the Following

1. i (b), ii (d), iii (a), iv (c)
2. i (c), ii (d), iii (a), iv (b)
3. i (c), ii (d), iii (b), iv (a)
4. i (b), ii (c), iii (a)
5. i (c), ii (a), iii (b), iv (d)

True or False

1. F	**2.** T	**3.** F	**4.** T	**5.** F
6. F	**7.** F	**8.** F	**9.** T	**10.** F
11. T	**12.** T	**13.** F	**14.** T	**15.** F

Fill in the Blanks

1. Alcohol
2. Nitrogen
3. Microscope
4. Bacteria
5. Common salt
6. Pasteurisation
7. Cholera and typhoid
8. Edward Jenner
9. Streptomycin and erythromycin
10. Jellies and jams

Assertion–Reason Questions

1. (b)	**2.** (a)	**3.** (c)	**4.** (d)	**5.** (b)

One-word Answers

1. Bacteria	**6.** Penicillin
2. Protozoa	**7.** *Bacillus anthracis*
3. *Spirogyra*	**8.** *Aedes*
4. *P.shermanii*	**9.** Food poisoning
5. Louis Pasteur	**10.** Pasteurisation

Define the Term

1. Protozoa are single-celled organism like prokaryotes but have the characteristics of eukaryotes. They are often classified as animals as they shows animal-like behaviour such as lack of cell wall. Free-living protozoa are more common and are often present in the water of ponds, lakes, drains, rivers and seas. Common examples are *Euglena*, *Trypanosoma*, *Amoeba* and *Paramecium*.
2. Some bacteria in the environment fix nitrogen from the atmosphere to increase the fertility of soil. These microbes are known as nitrogen fixers. Example are blue-green algae and some bacteria. BGA store compounds that are rich in nitrogen while fixing nitrogen and this is why they are used as fertiliser in agriculture fields. Peas, beans and other pulses known as leguminous plants have *Rhizobium* living mutually in the root nodules that helps to fix nitrogen in the soil.
3. An antibiotic is a medicine that stops the growth of or kills microorganisms that cause disease. The source of antibiotics is another microorganism. Some common antibiotics are streptomycin, penicillin, tetracycline and erythromycin.
4. The diseases caused by microorganisms in animals like cows, pigs, buffalo, goat, poultry bird and sheep lead to mortality along with feedstock loss. In 1876, Robert Koch discovered the organism that causes anthrax, *Bacillus anthracis*, which is transmitted through air and infects cattle.
5. Nitrogen is involved in growth and development of organisms as its is a key element in proteins, nucleic acids, vitamins and chlorophyll. It is continuously exchanged between living organisms like microbes, plant and animals and non-living like soil, water and air. The nitrogen cycle is the continuous exchange of nitrogen between living organisms and the non-living environment.

Groups of Terms

1. Fungi	**4.** Citric acid
2. Yellow mosaic virus	**5.** Protozoa
3. Mosquito/*Aedes*	

Identify the Odd One Out

1. Ice cream	**4.** Malaria
2. *Candida*	**5.** Chilli
3. *Penicillium*	

Image-Based Questions

1. (d)	**2.** (c)	**3.** (d)	**4.** (c)	**5.** (a)
6. (d)	**7.** (a)			

Very Short Answer Questions

1. The microscope is used to view microorganisms.
2. The major groups of microbes are bacteria, protozoa, fungi, viruses and algae.
3. *Lactobacillus* is responsible for curd formation.
4. Yeast is one of the main bacteria involved in the production of alcohol.
5. Viruses reproduce only inside the living cells of other organisms.
6. Alexander Fleming discovered penicillin.
7. Biological and physical ways of nitrogen fixation can occur (microbes, *Rhizobium*, blue-green algae, lightning)
8. *Plasmodium*, poxvirus
9. Pulses and soybean
10. Sodium benzoate, citric acid and sodium metabisulphite are the three special chemicals used to preserve food.

Short Answer Questions

1. (a) Diphtheria and tuberculosis; (b) Ringworm, candidiasis; (c) Malaria and dysentery, (d) AIDS, Herpes
2. The nitrogen fixation can be done by four methods: nitrogen fixing microbes, *Rhizobium*, blue-green algae, lightning. The main steps of nitrogen cycle are: The free atmospheric nitrogen is converted into nitrogenous compounds in the soil by the action of *Rhizobium*, nitrogen fixing microbes, blue-green algae or lightning. Plants that are present in the nitrogen-rich soil take up the nitrogen and utilise it for growth and development in organic compounds formation. The plants are then eaten by animals that convert the plant protein into animal proteins. The nitrogen of the protein is utilised by animals in various ways and the source of this protein is not only plants but also different animals that are eaten by other animals. After the death of animals or plants, the nitrogenous compounds from the dead and decomposed bodies are converted into simpler forms by bacteria. The nitrogen that plants take up are now returned to the soil in simple form and used by other plants and so the cycles continues; There are some denitrifying bacteria that convert the decomposed nitrogenous compounds into nitrogen gas. This allows the nitrogen gas to return to the atmosphere in free nitrogen form.
3. Valinomycin disrupts the cellular membrane by creating pores in it, penicillin stops cell wall synthesis.
4. Microbes that have the ability to fix nitrogen in the soil and make it fertile are blue-green slgae and some bacteria. BGA store compounds that are rich in nitrogen while fixing nitrogen and this is the reason why they are used as fertiliser in agriculture fields. Addition of BGA increases the soil fertility. Peas, beans and other pulses that are known as leguminous plants have *Rhizobium* living mutually with the root nodules that helps to fix nitrogen in the soil.
5. Hint: Refer to the figure given in the chapter.
6. Dengue is caused by the female *Aedes* mosquito; the female *Anopheles* mosquito causes malaria.
7. Taking infected food or water. By insect bites from mosquitoes. By breathing pathogen-infected air. Sharing infected needles. Physical contact with an infected person. Using articles of an infected person.
8. (a) Rust of wheat: This disease is caused by fungi. It is transmitted by seeds and air in wheat plants.
   (b) Foot and mouth disease: This disease is caused by virus in cattle. The disease is transmitted by air and causes blisters to occur in the feet and mouth of cattle.
   (c) Malaria: One of the most common and lethal diseases spread by mosquitoes is malaria. The female *Anopheles* mosquito carries the parasite of malaria, *Plasmodium*. This parasite uses blood as its food and its life cycle is completed in mosquitoes and humans.
9. (a) Do not let water collect near the house. Drain out stagnant water pools near the house. Doors and window panes should have iron mesh or net. Kill mosquitoes regularly inside the home using insecticides. Mosquito nets or mosquito repelling creams should be used while sleeping at night.
   (b) Do not litter garbage. Always keep the garbage in covered-lid bins. Always cover the food so that flies cannot sit on them. Kill flies using insecticides. Do not eat uncovered food from unhygienic places. Some diseases caused by flies can be prevented by proper vaccination.
10. Common salt and refrigeration have been used to preserve meat.

Long Answer Questions

1. Hint: Describe the types of microbes and the pathogenicity caused by them in plants, animals and humans.
2. Hint: Explain the production of curd, cheese, vinegar, alcohol, antibiotics and vaccines.

3. Hint: Explain the nitrogen cycle and its steps with a labelled diagram, as given in the chapter.
4. Hint: Explain vaccines, their types and mechanism of vaccination.
5. Hint: Explain preservation and the different ways of preservation.
6. Hint: Mention prevention of diseases with hygiene maintenance and mosquito and housefly.
7. Hint: Explain the diseases caused by bacteria, viruses, protozoa, fungi and algae.
8. Hint: Explain immunity and the types of immunity along with components mentioned in the chapter, with a well-labelled diagram.
9. Hint: Explain (a) vinegar or cheese (b) antibiotic or vaccine, (c) curd or bread making, (d) soil fertility.
10. Hint: Explain the characteristics, usefulness and diseases caused by bacteria, viruses, protozoa, fungi and algae.

Case Study-Based Questions

1. (a) A large group of simple, plant-like organisms is known as algae. They are autotrophs (ability to make their own food) as they contain chlorophyll. They make food like plants but the feature that differentiates them is lack of stem, roots or leaves. Most algae are multicellular.
   (b) There are two types: examples of unicellular algae are diatoms and *Chlamydomonas* and examples of multicellular organisms are BGA and *Spirogyra*.
   (c) *Chlorella* and *Spirulina* are two popular forms of algae that are consumed in many nations as a source of protein and omega-3 fatty acids.
2. (a) To prevent food from spoilage by the action of microbes, it is preserved by various techniques. The process of protecting food materials from attacks by microorganisms that spoil the food is known as preservation.
   (b) Spices, common salt, oil, sugar
3. (a) Stagnant and collected water
   (b) Do not let water to collect near the house. Drain out stagnant water pools near the house. Doors and window panes should have iron mesh or net. Kill mosquitoes regularly inside the home using insecticides. Mosquito nets or mosquito repelling creams should be used while sleeping at night.
4. (a) The process of giving a vaccine in order to provide protection against a specific disease either orally or by injection is known as vaccination.
   (b) Vaccine is a medicine that is prepared to provide strong immunity against a specific disease. The vaccines are manufactured by different microbes on a large scale and given to humans in order to protect them from future onset of this disease.
   (c) The vaccines that are available for children to be taken at a certain age are polio, measles mumps rubella (MMP), diphtheria, tuberculosis, cholera, typhoid, etc.
5. (a) Lactobacillus belongs to the category of Lactic Acid Bacteria (LAB) which is involved in cheese-making.
   (b) The characteristic holes of Swiss cheese are obtained by the carbon dioxide production effects of *P.shermanii*.
   (c) Yes

NCERT Corner

1. (a) (i)
   (b) (iv)
   (c) (ii)
   (d) (iv)
   (e) (ii)
   (f) (ii)
2. i-(f), ii-(a), iii-(e), iv-(c), v-(b), vi-(d)
3. No, they are extremely small and so can only be seen under the microscope.
4. There are 5 major groups of microorganisms: bacteria, virus, fungus, protozoa and algae.
5. Blue-green algae and *Rhizobium*
6. Hint: Write about commercial use, medicinal use, daily life use, cleaning the environment, increasing soil fertility.
7. Hint: Mention the microbes that cause disease in plants, animals and humans.
8. Antibiotics are medicines produced by microorganisms to kill other microbes.

Competition Window

1. (b)	2. (a)	3. (d)	4. (a)	5. (a)
6. (d)	7. (a)	8. (c)	9. (a)	10. (d)
11. (d)	12. (c)	13. (a)	14. (b)	15. (c)
16. (c)	17. (c)	18. (a)	19. (c)	20. (d)
21. (d)	22. (b)	23. (d)	24. (c)	

CHAPTER **7**

Health, Hygiene and First Aid

PREREQUISITES

You must have knowledge of:

- Microbes
- Basic human physiology
- Organ, tissue and cell-level organisation
- Functions of organs

LEARNING OBJECTIVES

At the end of this chapter, you will be able to:

- Explain disease and its causes
- Explain the significance of health
- Differentiate between acute and chronic diseases
- Describe infectious agents
- Differentiate between infectious and non-infectious diseases
- State the principles of treatment of diseases
- Explain the prevention of infectious diseases
- Discuss first aid

INTRODUCTION

The state of complete mental, physical and social well-being is known as health. It is not merely the absence of infirmity or disease. Health can be promoted by reducing unhealthy activities and practices such as stress or smoking and instead encouraging healthy activities such as getting sufficient sleep and exercising regularly. The factors that affect health can have an individual basis depending on the choices they make such as engaging in high-risk behaviour or depend on the way society is organised, which makes healthcare services inaccessible to people.

DISEASE

Any disturbance in the normal functioning of the body is known as disease. It may affect a specific organ system or the entire body. In other words, any condition that results in deterioration of the proper and normal functioning of cells, tissues and organs is known as disease. The different types of diseases vary in their symptoms, signs and causes. The branch of biology that deals with the study of disease, its cure, nature and diagnosis is known as **pathology**.

Classification of Diseases

Diseases can be classified on the basis of different criteria:

- **Anatomical classification:** This is based on the affected tissue or organ. For example, heart disease is a type of anatomic classification.
- **Topographical classification:** This type of classification is dealt with by specialisations in medicine, such as chest disease, vascular disease, abdominal disease and gastrointestinal disease. For example, ENT specialists only treat diseases specific to the ear, nose and throat.
- **Physiological classification:** This is based on the affected function or process such as digestion, respiration and metabolism. For example, diabetes.
- **Pathological classification:** This type of classification considers the nature of the disease. For example, cancer is associated with uncontrolled cell proliferation and has many variations.

- **Epidemiological classification:** This refers to the rate of distribution, occurrence and control of the disease in a population.

Causes of Diseases

Diseases are caused by different microorganisms such as bacteria, fungi and viruses. Some diseases are caused by multicellular organisms like worms (Table 7.1).

Table 7.1 Diseases and their causative organisms

Disease	Causative agent
AIDS	Human Immunodeficiency Virus (virus)
Aspergillosis	*Aspergillus* (fungi)
Cholera	*Vibrio cholerae* (bacteria)
Meningoencephalitis	*Naegleria fowleri* (amoeba)
Hepatitis	Hepatitis A, Hepatitis B, Hepatitis C, Hepatitis D, Hepatitis E viruses

Types of Diseases

- **Congenital diseases:** Here, the condition is present from birth. The reasons for the disorder could be environmental factors such as encounters with chemicals, smoking and infection during pregnancy or genetic. The impact of these disorders on the development and health of the child is not always severe; they could be quite mild. These diseases begin at birth and last throughout the life of the person. This is why pregnant women are advised to learn about the family's health history. Some tests can help them find out if the foetus is affected by such conditions. Some common congenital disorders are congenital heart diseases such as a hole in the heart and problems with blood vessels or valves. Diagnosis is possible using routine scans during pregnancy or during screening of the newborn. Cerebral palsy is also a congenital disease, while fragile X syndrome can be diagnosed through genetic testing during pregnancy. Trisomy 21 or Down syndrome can also be diagnosed by genetic testing.

  The main **causes** of congenital diseases are:
  - Alcohol exposure
  - Toxin intake (chemicals and drugs)
  - Harmful medications and supplements
  - Smoking
  - Pathogenic infections
  - Malnutrition
  - Genetic factors
  - Complications from physical restraint such as immobility, leading to muscular atrophy
  - Age of parents

Fun Fact!

Did you know that amniocentesis is a test performed after 15 weeks of pregnancy which can provide a definite diagnosis of congenital disorders? Here, an amniotic fluid sample is taken from the pregnant woman.

- **Acquired diseases:** These diseases develop after birth. There are many types of acquired diseases:
  - **Degenerative diseases:** Malfunction of vital organs due to degeneration of cells gives rise to degenerative diseases such as osteoporosis, which shows incremental bone weakness due to degeneration of cells, leading to increase in the risk of bone fractures. Neurons also undergo degeneration, leading to neurodegenerative disorders such as Alzheimer's disease which is caused by ageing, heredity and lifestyle choices.
  - **Deficiency diseases:** Any imbalance in the regular diet of humans can lead to excess or insufficient intake of vitamins and minerals, leading to disease. Vitamins are not made by the body and need to be taken as food to ensure the normal functioning of the body. Vitamin deficiency can lead to many diseases:
    - **Vitamin A deficiency:** Vitamin A is obtained from milk, spinach, eggs, carrot and fish. It is important for normal growth, vision, reproduction and the immune system. Children below 5 years of age suffer from a serious eye disorder called xerophthalmia which places them at risk of becoming blind; night blindness is also caused by vitamin A deficiency.
    - **Vitamin B deficiency:** Different kinds of B vitamins are available, such as B1, B2 and B12, and their deficiency causes diseases. Vitamin B1 deficiency causes beriberi which results in weak muscles and weight loss that when acute can lead to cardiac failure and paralysis. Lack of vitamin B6 causes anaemia and skin disorders such as cracks around the mouth; it can also lead to nervous breakdowns and depression. Vitamin B12 deficiency leads to anaemia and nerve paralysis, muscle paralysis, dementia, fatigue and depression.

- **Vitamin C deficiency:** This can cause scurvy which is characterised by skin spots, bleeding gums (Fig. 7.1) and swelling of the joints; it affects the immune system.

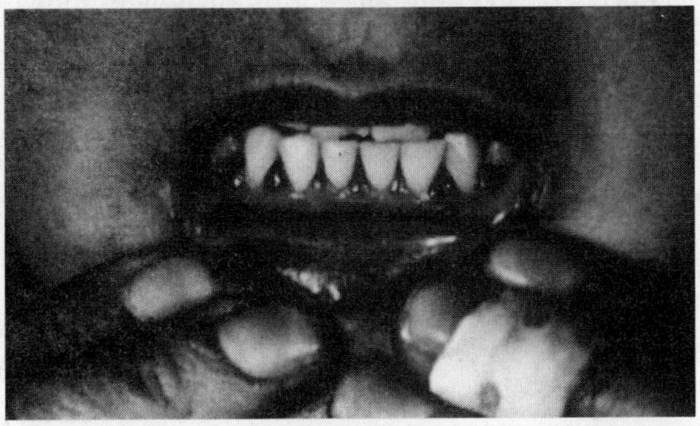

Fig. 7.1 Scurvy
(Scorbutic gums due to a vitamin C deficiency, a symptom of scurvy; Centers for Disease Control and Prevention; 1970; https://commons.wikimedia.org/wiki/File:Scorbutic_gums.jpg. Licensed under: PD US)

- **Vitamin B deficiency:** This causes rickets that leads to weak bones near the joints and tooth decay.
- **Vitamin K deficiency:** This vitamin is important for the coagulation of blood and its deficiency can lead to excessive bleeding.

Minerals are inorganic nutrients that are important for normal functioning; examples are zinc, copper, iron, iodine, potassium, calcium, sodium and magnesium. Deficiency of minerals leads to several dieases:

- **Iodine:** This element is vital for brain development in the foetus and is responsible for the production of hormones such as thyroxine. Salt is an important source of iodine and its deficiency can cause goitre.
- **Iron:** This plays an important role in the formation of haemoglobin and its deficiency causes anaemia which can lead to oxygen deficiency and even death.

- **Non-infectious diseases:** These diseases are caused by pathogens and influenced by other factors like age, gender, lifestyle and nutritional deficiency. For example, diabetes, hypertension, heart disease, asthma, Alzheimer's and cancer. These are non-transmittable.
- **Infectious diseases:** These diseases spread from one human to another; they are also known as communicable diseases. They are caused by microorganisms such as fungi, bacteria, viruses and protozoans. The discharge of an infected person such as bodily fluids through sneezing or coughing contains pathogens that infect a new person. For example, chickenpox, cholera and malaria. The different types of infectious diseases are categorised based on the causative organisms, as shown in Table 7.2.

Table 7.2 Classification of infectious diseases

Type of disease	Description
Bacterial infections	The survival of bacteria in extreme heat and cold and even in radioactive environments makes it one of the most common infectious agents. Bacteria cause cholera, tuberculosis, typhoid and diphtheria. These diseases are treated using antibiotics but increasingly, bacteria are becoming resistant to them.
Viral infections	There are millions of viruses in the world that cause common infections like influenza and cold. A virus enters the cell of the host and attaches itself to it and then releases its genetic material. When the host cell replicates, the virus multiplies. This is followed by lysis of cells and release of the virus. Human papillomavirus causes uncontrolled cell replication, leading to cancer.
Fungal infections	Fungi decompose and absorb organic material using enzymes. There are many fungal infections that appear in the upper layer of the skin and which could penetrate into the deeper layers. When fungal spores are inhaled, fungal infections could result.
Prion diseases	Prion is a type of protein that exists without genetic material; it is folded abnormally and affects the structure of normal proteins, causing deadly diseases that spread rapidly. An example is Creutzfeldt–Jakob disease which replicates in the host and is fatal.

Common Diseases

Let us discuss a few infectious diseases of each type:

- **Cholera:** This is caused by bacteria spread through contaminated water. It causes severe dehydration and diarrhoea. If cholera is left untreated, it could be fatal within hours. Modern water and sewage treatment has almost completely eliminated cholera but it still exists in Africa and Southeast Asia, where poverty is high and sanitation is poor. People who are infected by *Vibrio cholerae* do not become ill but shed this bacteria in their stool for up to 14 days; this is a source of infection for others. The symptoms of cholera are mild diarrhoea which cannot be easily distinguished from normal diarrhoea. The other symptoms of cholera are nausea and vomiting and dehydration. Signs and symptoms of cholera dehydration include fatigue, irritability, dry mouth, sunken eyes, extreme thirst, low blood pressure, irregular heartbeat and shrivelled skin that is very slow to bounce back.
- **AIDS:** Acquired immunodeficiency syndrome (AIDS) is a lethal, chronic condition caused by the Human Immunodeficiency Virus (HIV), which damages the immune system by interfering with the host body's ability to fight diseases and infections (Fig. 7.2). HIV is a sexually transmitted infection (STI) that can also be spread by infected blood and by sharing needles or using illicit injection drugs. HIV can also be transmitted from mother to child during pregnancy, breastfeeding or childbirth. It may take years for the appearance of symptoms. There is no cure for HIV/AIDS, but medications can control progression of infection. The death rate has been reduced by the constant efforts of organisations to increase awareness. The symptoms of AIDS depend on the phase of infection:
    - **For acute infection:** Some reports show that patients are likely to develop a flu-like illness within 2–4 weeks of entry of the virus that may last for a few weeks. The signs and symptoms include:
        - Headache
        - Fever
        - Rash
        - Muscle pain
        - Joint pain
        - Swollen lymph
        - Weight loss
        - Night sweats
        - Cough
        - Diarrhoea
        - Sore throat
    - **For chronic infection:** The symptoms develop as the virus continues to destroy the immune cells of the host and keeps multiplying:
        - Fatigue
        - Fever
        - Diarrhoea
        - Swollen lymph node
        - Weight loss
        - Shingles (herpes)
        - Oral yeast infection
        - Pneumonia
- **Malaria:** This disease is caused by a parasite that spreads to humans through infected mosquitoes.

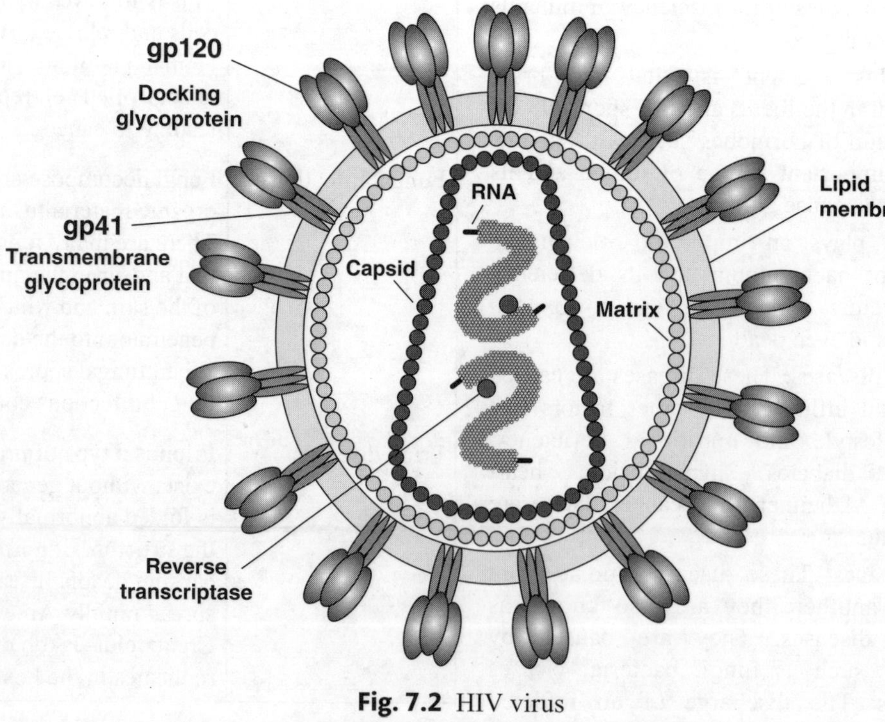

Fig. 7.2 HIV virus

Malaria causes high fever and shaking chills. Tropical and subtropical countries are the most affected; each year nearly 290 million people are infected with malaria, with more than 40,000 people losing their life. In order to reduce the infection, preventative drugs along with insecticide-treated bed nets are distributed by world health programmes to protect people from mosquito bites. People are advised to wear protective clothing and use bed nets and insecticides to protect themselves while travelling in high-risk areas as most malarial parasites have developed resistance against the common medicines. The signs and symptoms of malaria are:

- Chills
- Headache
- Fever
- Nausea and vomiting
- Discomfort
- Muscle and joint pain
- Diarrhoea
- Rapid breathing
- Fatigue
- Rapid heart rate
- Abdominal pain
- Bloody stools

The life cycle of the malarial parasite (Fig. 7.3), which exists in the form of motile sporozoites, involves two hosts. The vector is the female *Anopheles* mosquito that transmits these malarial sporozoites to the host. When these infected mosquitoes bite a human, the sporozoites are injected into the blood through the saliva of the mosquito. These sporozoites initially multiply within the liver, thereby damaging it and rupturing blood cells in the body. The destruction of RBCs is how malaria kills the host as it reproduces asexually in it and releases more parasites by bursting open. This is followed by release of a toxin called haemozoin that causes chills in patients. When the *Anopheles* mosquito bites an infected human, the parasites enter the body of the mosquito along with blood and development and fertilisation occurs to form a sporozoite. The mature parasite reaches the intestine of the mosquito and breaks out of it to migrate to the salivary gland, where the process of infection begins all over again.

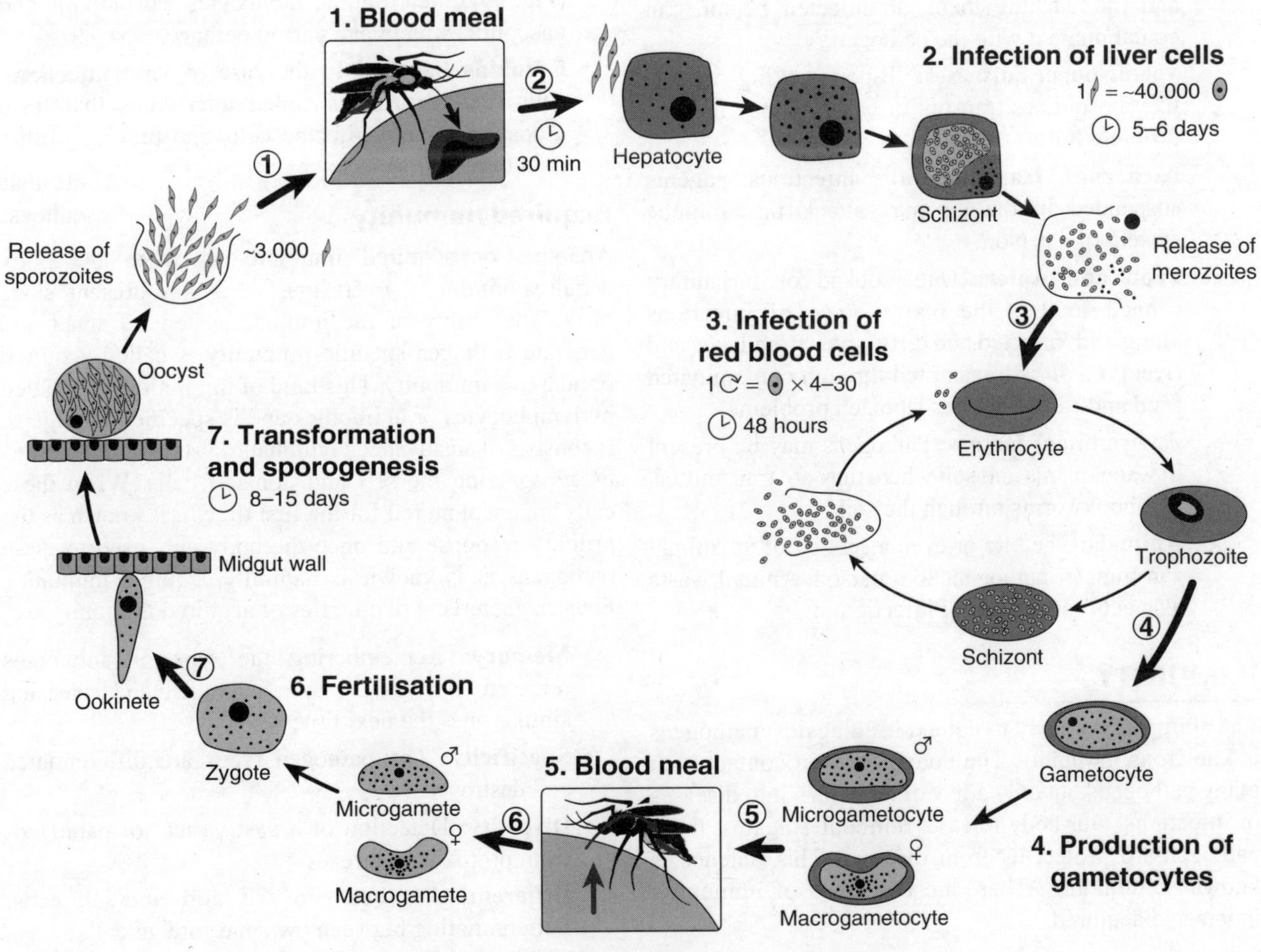

Fig. 7.3 Life cycle of malaria parasite

Transmission of Infectious Diseases

- **Direct transmission:** Here, transmission occurs directly from the infected person. It occurs in different ways:
  - **Person to person:** The exchange of body fluids from infected to non-infected persons.
  - **Droplet transmission:** Droplets can spread when an infected person speaks, sneezes or coughs. They can also be spread by the hands by touching the mouth and nose.
  - **Through body fluids or blood:** The body fluids of an infected person come in contact with the mucous membrane or blood of an uninfected person. For example, AIDS, hepatitis.
  - **By skin:** Infections like chickenpox and ringworm spread when the skin of an infected person comes in contact with the skin of an uninfected person.
- **Indirect transmission:** Here, the infection is transmitted indirectly from one person to another:
  - **Contaminated objects:** Objects carry pathogens and by touching them, an infected person can contaminate it with the pathogen.
  - **Vector-borne diseases:** Blood-sucking insects like mosquitoes transmit diseases as they feed on different hosts.
  - **Airborne transmission:** Infectious agents suspended in the air may attack the immune system of a person.
  - **Food and water:** Undercooked or unsanitary canned food is the main source of infections along with infected and dirty water from lakes and rivers. *E.coli* is transmitted through contaminated food and causes serious stomach problems.
  - **Environment factors:** Pathogens may be present in water, plants and soil where they are transmitted, like hookworms through the soil.
  - **Animals:** The bite or even a scratch of an animal can transfer pathogens to a person. Animal waste also acts as a source of infection.

IMMUNITY

The ability of the body to defend itself against pathogens is known as immunity. The body comes in contact with many pathogens but only a few of them turn into diseases or infections. Our body releases antibodies against these pathogens to protect us from diseases. This defence is known as immunity. There are two types of immunity: innate and acquired.

Innate Immunity

The immunity present since birth and which is activated immediately when there is an attack by a pathogen is known as innate immunity. This type of immunity includes defence mechanisms and barriers for the protection of the body against pathogens. The natural resistance includes natural killer cells, neutrophils, salivary enzymes and intact skin that produces the initial response prior to exposure to the pathogen. There are four **types of barriers**:

- **Physical barrier:** The skin, cilia, hair, eyelash and mucus layer in the gastrointestinal tract and respiratory tract serve as the first line of defence by preventing or trapping the pathogens that enter the body.
- **Physiological barrier:** Hydrochloric acid in the stomach produces an acidic environment that kills most germs. Similarly, saliva and tears have antibiotic properties that prevent the growth of pathogens.
- **Cellular barrier:** Pathogens need to cross other barriers and cells such as leucocytes, lymphocytes (Fig. 7.4), neutrophils, monocytes, eosinophils and basophils, which take part in defence.
- **Cytokine barrier:** In the case of viral infections, cells secrete proteins called interferons that form a coating around infected cells and prevent further infection.

Acquired Immunity

Adaptive or acquired immunity is when the body acquires immunity over time; it is not present since birth. The ability of the immune system to adapt and generate pathogen-specific immunity is called acquired or adaptive immunity. This kind of immunity is mediated by lymphocytes or antibodies and is specific to antigens. It consists of an advanced lymphatic system that functions by recognizing the self and non-self cells. When these cells are encountered for the first time, it is known as the primary response and once the body gets used to these pathogens, it is known as naturally acquired immunity. Four characteristic **properties** of acquired immunity are:

- **Memory:** Remembering the type of antibodies released in response to a pathogen and releasing similar ones the next time
- **Specificity:** The pathogen types are differentiated and destroyed
- **Diversity:** Detection of a vast variety of pathogens from protozoa to viruses
- **Differentiation between self and non-self cells:** Differentiating between own and foreign cells

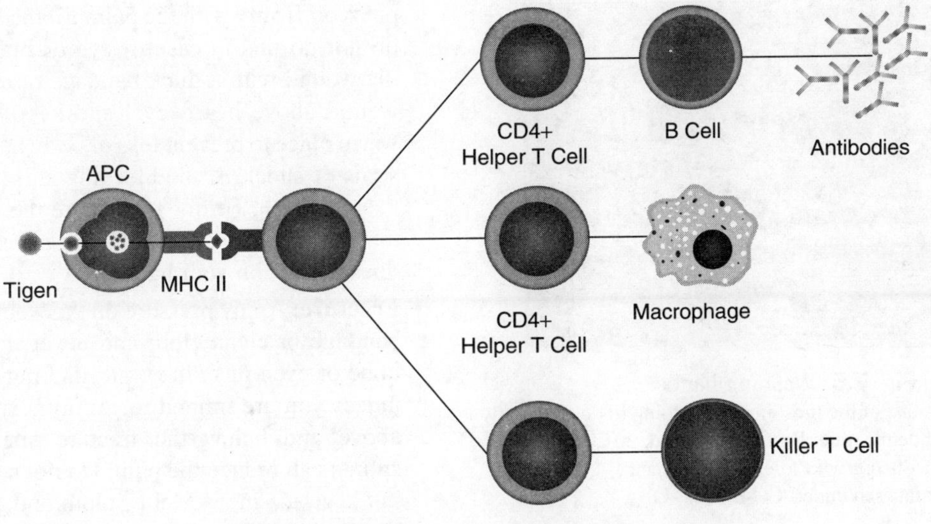

Fig. 7.4 Immune response by lymphocytes

Two types of **responses** are generated by acquired immunity:

- **Humoral response:** Antibodies produced by B-lymphocytes and antigens internalised by them along with those presented by T-cells are included here.
- **Cell-mediated response:** This is initiated by T-cells and promote apoptosis.

TREATMENT PRINCIPLES

The two main treatment principles are as follows:

- Reducing the impact of the disease
  - Treating symptoms of diseases with no cure such as the common cold to alleviate the sickness
  - Using medicines to lessen the impact of the sickness
  - Ensuring personal cleanliness to mitigate the effect of disease
  - Relaxing and giving proper rest to the body
  - Stopping transmission by using sterilised goods and having a nutritious diet
- Eradicating the underlying cause of the disease
  - Consuming specific medication such as antibiotics to disrupt the unique biochemical phase of a microbe to reduce its chances of survival

PREVENTION PRINCIPLES

There are two ways by which disease can be prevented:

- **General:** The hygiene conditions in the surroundings should be maintained with limited exposure to airborne pathogens. Consumption of safe drinking water and clean food is vital and a clean environment prevents breeding of vectors and the spread of vector-borne diseases.
- **Specific:** The immune system fights off microbes and the cells of the system are specialised in killing pathogens. Immunisation is the process where a person is made resistant to a particular disease by a vaccine that stimulates the immune system to prevent certain diseases. Currently, there are vaccines against smallpox, tetanus and measles.

HYGIENE

Hygiene is a set of personal practices known to contribute towards good health and which includes cutting nails and hair periodically, washing hands and bathing. Throwing garbage on the road and defecating in the open leads to poor health and these practices should be stopped. Personal hygiene includes practices such as consuming clean drinking water, keeping the body clean, washing vegetables and fruits before consuming them, washing hands (Fig. 7.5), taking daily baths, etc. Discarding excreta and waste properly like recycling, maintenance of city waste reservoir and waste segregation helps maintain quality hygiene in the home as well as the surroundings. Mosquitoes thrive in stagnant water whereas garbage dumps are home to rats which are vectors that cause diseases.

Sanitation refers to public health conditions that involves keeping the environment clean. It is one of the most important aspect as many of the diseases mentioned here are transmitted through lack of sanitation.

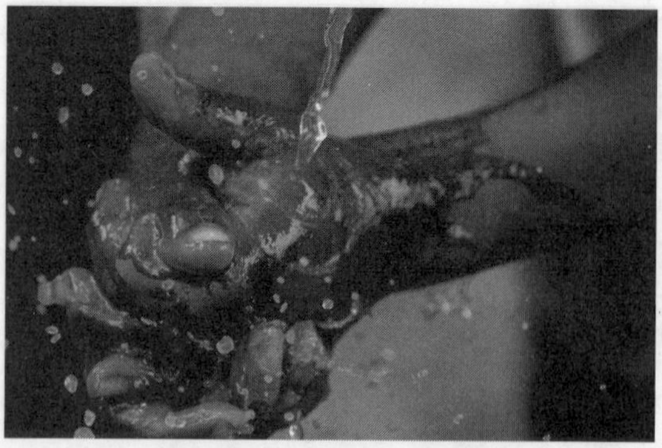

Fig. 7.5 Washing hands
(Washing hands is one of the most important thing to do frequently during this pandemic time; Rasheedhrasheed; 2017; https://commons.wikimedia.org/wiki/File:Washing_hands_Corona.jpg. Licensed under: CC-BY-SA-4.0)

FIRST AID

First aid is the immediate and first response and assistance given to a person with a minor or major injury or illness that can save lives and prevent worsening of the condition or promote recovery. Examples of initial intervention in serious conditions prior to professional medical help are performing CPR (cardiopulmonary resuscitation) while the ambulance is on its way or treatment of minor injuries like burns or snake bites. It is generally performed by a person with basic training in medical procedures, but there are situations when no medical professional is available and so everyone should know basic first aid.

- **Burns:** For minor burns, hold the area under lukewarm running water for 10 minutes; on areas such as the face, apply a cool or wet cloth until the pain eases. In the case of mouth burns from hot drinks or food, put a piece of ice in the mouth for a few minutes. Try to gently remove rings or other tight items worn in the burned area; blisters should not be broken or get infected. Apply lotion with soothing ingredients like aloe vera after the burn area is cool and cover the burn with a clean bandage, but wrap loosely to avoid pressure on the skin.

  In the case of major burns, call the ambulance and meanwhile check that the person is breathing; slowly remove any tight items wore by the patient. Cover the burn loosely with clean cloth or gauze and lift the wound above heart level if possible. Watch for signs of shock such as clammy skin, weak pulse and shallow breathing.
- **Bleeding:** If the bleeding is heavy, call an ambulance and cover the wound with clean gauze or cloth and press on firmly with the palm till the bleeding ceases; do not do this in case of eye or brain injury. Wrap the wound with a thick bandage or cloth and lift the wound above heart level and keep the person in a warm place to prevent loss of body heat and watch for signs of shock. If the bleeding still has not stopped, keep pressing firmly, make sure the person does not move and wash your hands properly so that the blood does not get on your hands.
- **Fracture:** Apply pressure on the wound with a sterile bandage or clean cloth and never try to realign the bone or even push the bone sticking out back inside unless you are trained to. Apply a splint to the area above and below the fracture sites; padding the splints can reduce the pain. Do not apply ice directly but keep ice in a towel or cloth and then apply it on the wound. Check for signs of shock such as weak breathing.
- **Object in the eye:** First wash your hands and make the person sit in a well-lighted area and examine the object in the eye. Pull the lower lid down and ask the person to look up and hold the upper lid while the patient looks down. If the object is floating in the eye as in the film of tears, try using a medicine dropper with clean and warm water to flush it out. Or you can tilt the head back and irrigate the eye surface using clean drinking water. Do not try to remove the object embedded in the eye or rub the eye and do not try to remove any large object in the eye.
- **Unconsciousness:** Open the airways; place one hand on the person's forehead and gently tilt their head back and place the fingertips of the other hand on the person's chin to lift it up. Check the breathing and feel if chest movement is normal. Put them in the recovery position, kneel down on the floor next to them and try to make them lie on their back to shift from lying on their side. Try to find the spinal injury if any and call for help.
- **Swallowing of poison:** Call for help and meanwhile remove anything still present ïn the mouth of the patient as well as any contaminated cloth using gloves. Rinse the skin for 20 minutes in a shower or using a hose if the poison is on the skin; if it is in the eye, then flush the eye out with cool or lukewarm water for 20 minutes until help arrives. If the person has inhaled poison, try to get them into fresh air. If the person starts vomiting, turn their head to their side to avoid choking and give CPR if the person shows no signs of life.
- **Snake bite:** Call for help and while waiting, make the person lie or sit down in a neutral position of comfort

and remove rings or watches before the swelling starts. Wash the bite with soap and water and cover it with a clean cloth or dry dressing. Mark the leading edge of the swelling on the skin and note the time. Try to tie a tight cloth above the bite to slow down poison circulation.

- **Bee sting:** Wash the affected area with soap and water and apply a cold compress. If the sting is on a limb, elevate it and apply hydrocortisone cream to ease redness or itching. Avoid scratching the sting area and go to the doctor if the itching and swelling worsens.

SUMMARY

- Health is defined as the state of physical, biological and social wellbeing.
- Different types of diseases affect the human body: acute, chronic, infectious, non-infectious, degenerative, deficiency and allergic reactions.
- Infectious diseases are communicable and caused by pathogens that harm a specific tissue or organ.
- Pathogens can be single-celled like bacteria or multicelled like worms.
- Infectious agents are of different types and cause different diseases and enter the body through different means.
- To stop the spread of an infectious disease, the various ways of transmission (direct and indirect) should be checked.
- Prevention is better than cure and the prevention of diseases should be preferred over treatment by general or specific means.
- The ability of the body to defend itself against pathogens is known as immunity.
- The immunity present since birth and which is activated immediately when there is an attack by a pathogen is known as innate immunity. This has four basic barriers: physical, physiological, cellular and cytokines.
- Adaptive or acquired immunity is when the body acquires immunity over time; it is not present since birth.
- Humoral response is when antibodies are produced by B-lymphocytes. Cell-mediated response is initiated by T-cells and promotes apoptosis.

PRACTICE QUESTIONS

Analyse Your Concepts (School Exam Based)

Multiple Choice Questions

1. Heart disease is a type of_____ classification.
(a) Anatomical (b) Physiological
(c) Topographical (d) Pathological

2. Meningoencephalitis is caused by:
(a) Virus (b) Amoeba
(c) Fungi (d) Bacteria

3. Fragile X syndrome is a type of:
(a) Deficiency (b) Infection
(c) Degenerative disease (d) Congenital disease

4. Rickets is caused by deficiency of:
(a) Vitamin B (b) Vitamin D
(c) Vitamin C (d) Vitamin K

5. _____ is a type of protein that exists without genetic material.
(a) Prion (b) Bacteria
(c) Fungi (d) Virus

6. Hepatitis transmission occurs through:
(a) Body fluids (b) Skin
(c) Droplets (d) (a) and (b)

7. AIDS is a lethal, chronic condition caused by _______ that damages the immune system by interfering with the host body's ability to fight diseases and infections.
(a) *Vibrio cholera*
(b) Prion
(c) Human Immunodeficiency Virus
(d) None of the above

8. Malarial parasites release a toxin called ______that causes chills in patients.

(a) Hematopoietic
(b) Hemozoin
(c) Reverse transcriptase
(d) Hematoxylin

9. Humoral response produces ______.
(a) B lymphocytes (b) T lymphocytes
(c) Both (a) and (b) (d) None of the above

10. ______ is the process where a person is made resistant to a particular disease by a vaccine that stimulates the immune system to prevent certain diseases.
(a) Health
(b) Immunisation
(c) (a) and (b)
(d) None of the above

Match the Following

Each question contains statements given in two columns which have to be matched. Statements in column I have to be matched with statements in column II.

1. Choose the correct option:

Column I	Column II
i. B lymphocytes	a. Cell-mediated immunity
ii. T lymphocytes	b. Humoral immunity
iii. Skin	c. Cytokines
iv. Interferons	d. Physical

2. Choose the correct option:

Column I	Column II
i. Aspergillosis	a. *Vibrio cholerae*
ii. Meningoencephalitis	b. Hepatitis A
iii. Cholera	c. *Aspergillus*
iv. Hepatitis	d. *Naegleria fowleri*

3. Choose the best match:

Column I	Column II
i. Trisomy 21	a. Vitamin B12 deficiency
ii. Alzheimer's disease	b. Vitamin A deficiency
iii. Xerophthalmia	c. Degenerative
iv. Beriberi	d. Congenital

True or False

Read the following statements and write your answer as true or false.

1. Topographical classification refers to diseases of the affected tissue or organs.
2. A state of complete mental, physical and social well-being is known as health.
3. Cholera is caused by bacteria.
4. Cerebral palsy is an acquired disease.
5. Lack of vitamin B6 causes anaemia and skin disorders.
6. Iron is important for the production of hormones like thyroxine.
7. The vector of malaria is the female *Anopheles* mosquito that transmits these malarial sporozoites to the hosts.
8. The immunity present since birth and which is activated immediately when there is an attack of the pathogen is known as acquired immunity.
9. In viral infections, the cells secrete proteins called interferons.
10. If an object is floating in the eye, as in the film of tears, try using a medicine dropper with clean and warm water to flush it out.

Fill in the Blanks

Complete the following statements with an appropriate word/term to be filled in the blank spaces.

1. Any disturbance in the normal functioning of the body is known as a ______.
2. __________type of classification includes the one that affects either a function or a process like digestion.
3. Malfunction of vital organs due to degeneration of cells worsens to become a ________disease.
4. Vitamin B1 deficiency causes_______.
5. _______is a type of protein that exists without genetic material and is folded abnormally; it affects the structure of the normal protein.
6. AIDS is a lethal, chronic condition caused by _______.
7. The vector of malaria is the female_______mosquito.
8. Hydrochloric acid in the _______ produces an acidic environment that kills most of the germs.

9. ______is a set of personal practices that are known to contribute to good health.

10. Apply ______cream to ease redness in bee stings.

Assertion–Reason Questions

The following questions consist of two statements: Assertion (A) and Reason (R). Answer these questions selecting the appropriate option given below:

(a) Both A and R are true and R is the correct explanation of A
(b) Both A and R are true but R is not the correct explanation of A
(c) A is true but R is false
(d) A is false but R is true

1. Assertion (A): People with malaria experience chills during fever.
   Reason (R): The parasites release a toxin called haemozoin.
2. Assertion (A): There are fewer reports of cholera spreading through contaminated water in the modern world.
   Reason (R): Modern water and sewage treatment has almost eliminated cholera.
3. Assertion (A): Vitamin B1 deficiency causes beriberi.
   Reason (R): Beriberi results in weak muscles and weight loss that when acute leads to cardiac failure and paralysis.
4. Assertion (A): Vitamin K is important for the coagulation of blood.
   Reason (R): Its deficiency can lead to excessive bleeding.
5. Assertion (A): Fragile X syndrome could be diagnosed through genetic testing after pregnancy.
   Reason (R): Trisomy 21 could also be diagnosed by genetic testing.

One-word Answers

1. Which disease is caused by HIV?
2. Which agent is responsible for malaria?
3. Which category does Fragile X syndrome belong to?
4. Which disease is caused by contaminated water?
5. Which micronutrient deficiency could cause disturbance in thyroxine levels?
6. Which toxin is released by malarial *Plasmodium*?
7. Which disease is caused by vitamin C deficiency?
8. Which disease can spread from one person to another?
9. Which barrier does saliva belongs to in innate immunity?
10. Which acquired immunity produces B lymphocytes?

Define the Term

1. Degenerative disease
2. Non-infectious disease
3. Prion diseases
4. Malaria
5. First-aid for fracture

Groups of Terms

Given below are groups of terms. In each group the first pair indicates the relationship between the two terms. Rewrite and complete the second pair on a similar basis.

1. Vitamin C: scurvy :: Vitamin D: _____
2. Cholera : bacteria :: AIDS: _______
3. Skin: physical :: Hydrochloric acid: _____
4. Beriberi : deficiency disease :: Fragile X synbdrome: _____
5. Humoral : B lymphocytes :: Cell mediated: ______

Identify the Odd One Out

1. Cerebral palsy, beriberi, rickets, scurvy
2. Skin, saliva, B lymphocytes, interferons
3. Flu, rickets, fragile X syndrome, goitre
4. Flu, Down syndrome, AIDS, COVID
5. Mosquito, housefly, human, sporozoite

Image-Based Questions

1. What is the type of disease presented in this picture?

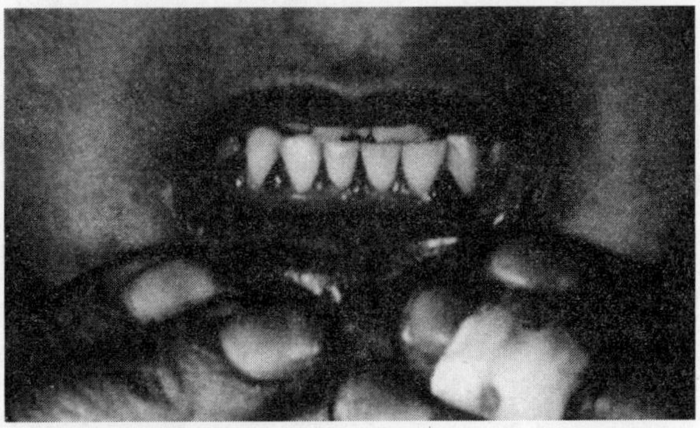

(a) Degenerative (b) Congenital
(c) Infectious (d) Deficiency

2. Which organism's life cycle is depicted here and what disease does it cause?

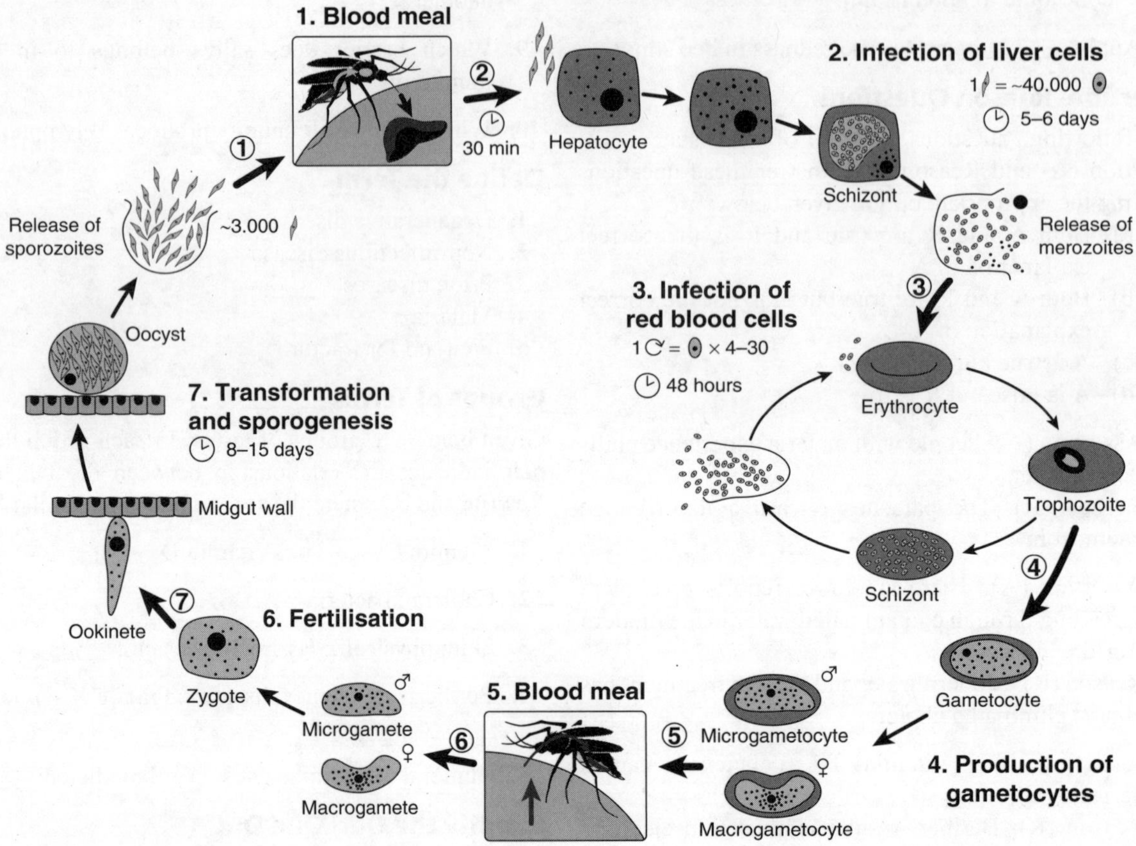

(a) Bacteria, malaria
(b) *Plasmodium*, malaria
(c) Virus, malaria
(d) *Plasmodium*, cholera

3. What are the vectors and hosts of the *Plasmodium* whose life cycle is shown here?

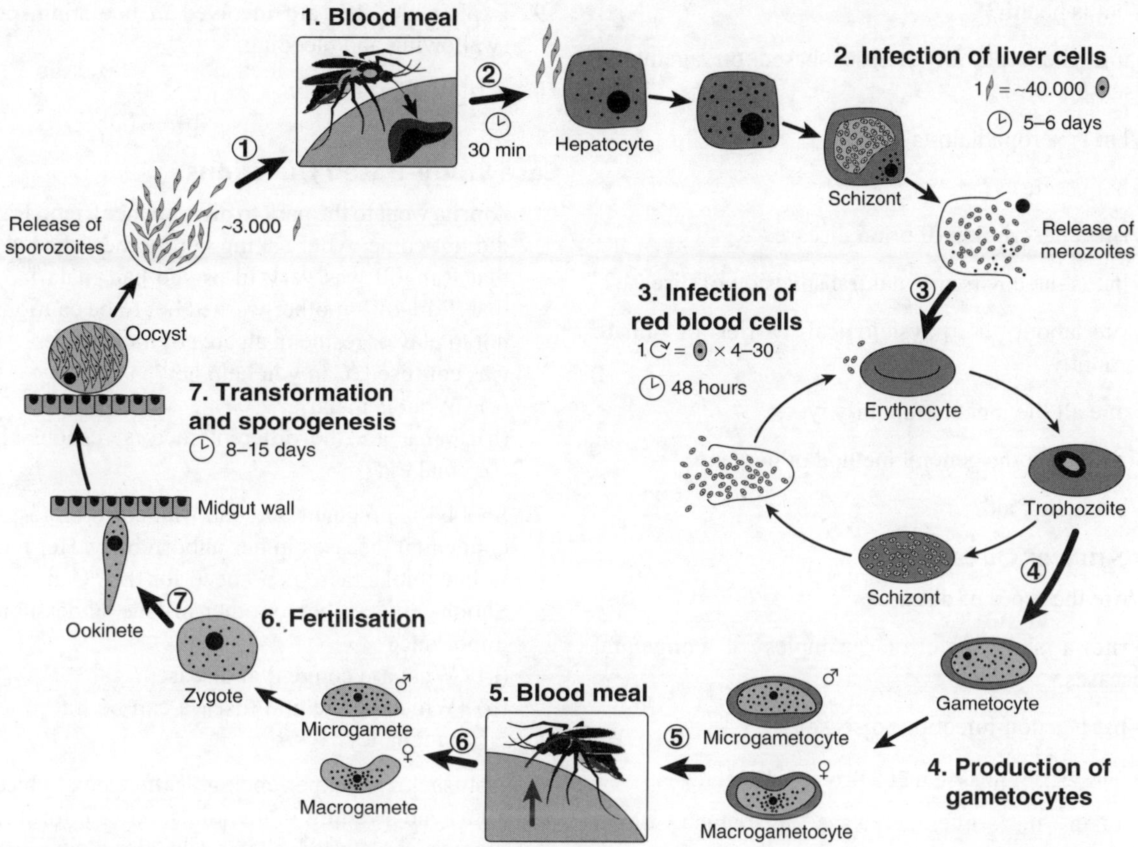

(a) *Anopheles*, pigs (b) *Aedes*, pigs
(c) *Anopheles*, humans (d) *Aedes*, humans

4. What is the name of this agent of disease?

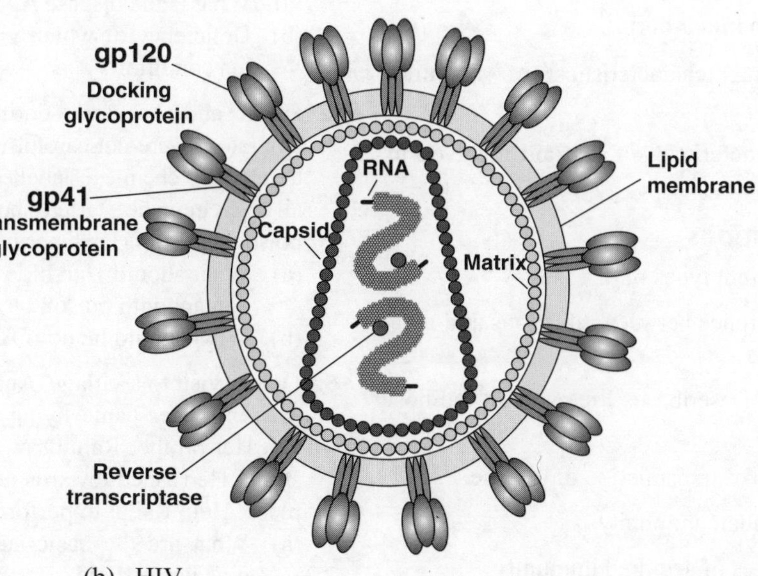

(a) Hepatitis virus (b) HIV
(c) Polio virus (d) Prion

Very Short Answer Questions

1. What is health?
2. Name the types of diseases based on agents of disease.
3. What is xerophthalmia?
4. What is scurvy?
5. Write a description of prion disease.
6. What is the environmental transmission of disease?
7. Write about the physiological barrier of innate immunity.
8. Name all the innate immunity types.
9. Write about the general method of hygiene.
10. What is first aid?

Short Answer Questions

1. Write the types of diseases.
2. Write a short note on examples of congenital diseases.
3. What is a non-infectious disease?
4. What are the factors that cause malaria?
5. Explain the different ways in which direct transmission occurs.
6. Explain the innate immunity barriers.
7. What are the responses shown by acquired immunity?
8. Discuss first aid in minor burn.
9. Write about four characteristics of acquired immunity.
10. Describe the characteristics of first aid involved in unconscious patients.

Long Answer Questions

1. Describe the different types of diseases.
2. Describe the difference between infectious and non-infectious diseases.
3. What is malaria? Describe its life cycle in the host and vector.
4. Explain two types of transmission of disease.
5. Write a note on innate immunity.
6. Explain the features of acquired immunity.
7. Explain the treatment principles.
8. What is prevention and its ways?
9. Explain the first aid involved in bee sting, poison swallowing and bleeding.
10. Write a short note on:
    (a) Cholera (b) AIDS

Case Study-Based Questions

1. Amrita went to the park to play, but her friend Rangoli did not come. After asking her mother, she was told that Rangoli was very ill as she had malaria. After that, Rangoli's mother warned her to be cautious and not to play or reside in an area of mosquitoes. Amrita was confused. Can you help her?
   (a) What is malaria?
   (b) What are the different factors responsible for malaria?
2. Shobha is pregnant and she wanted to be tested for congenital diseases in her unborn baby. Her mother-in-law thinks there is no need for this. Can you help Shobha make her mother-in-law understand its importance?
   (a) What are congenital diseases?
   (b) Which congenital disease can be traced during pregnancy?
3. Abhishek is experiencing skin spots, bleeding gums and swelling in the joints. After informing his mother, she started giving him citrus supplements. This made Abhishek confused about the relationship between his symptoms and citrus fruits. Can you help him understand?
   (a) What is the disease Abhishek is suffering from?
   (b) Deficiency of which vitamin is responsible for his condition?
4. Harsh and Ashutosh during their lecture in the laboratory were not careful with chemicals. Ashutosh touched a chemical labelled poison and spilled it all over him; he started showing symptoms of skin poisoning. Help Harsh perform first aid on him.
   (a) What should Harsh do if Ashutosh has been in contact with poison?
   (b) What should he do if Ahutosh starts vomiting?
5. On her visit to a village, Ananya sat under a tree and suddenly bees came flying out and began to attack her. Her brother Rahul saw her getting attacked and stung. He rushed towards her and took her to a safe place. Help Rahul to perform first aid on her.
   (a) What are the basic steps he should follow to give her first aid?
   (b) What if the swelling and itchiness worsens?

COMPETITION WINDOW (OBJECTIVE TYPE)

1. Humans have acquired immunity that produces antibodies to neutralise pathogens. Still innate immunity is present at the time of birth because it:
   (a) Is very specific and uses different macrophages
   (b) Produces many memory cells for mounting fast secondary response
   (c) Has natural killer cells which can phagocytose and destroy microbes
   (d) Provides passive immunity
2. Which is not an autoimmune disease?
   (a) Psoriasis
   (b) Rheumatoid arthritis
   (c) Alzheimer's
   (d) Vitiligo
3. Which of the following diseases does mosquito transmit causing chronic inflammation of lymphatic vessels?
   (a) Elephantiasis (b) Ascariasis
   (c) Ringworm (d) Amoebiasis
4. Which of the following is caused by bacteria?
   (a) Cholera and tetanus
   (b) Typhoid and smallpox
   (c) Tetanus and mumps
   (d) Herpes and influenza
5. Asthma may be attributed to:
   (a) Allergic reaction of mast cells in the lungs
   (b) Inflammation of the trachea
   (c) Accumulation of fluid in the lungs
   (d) Bacterial infection of the lungs
6. Which of the following diseases is caused by protozoa?
   (a) Syphilis (b) Influenza
   (c) Babesiosis (d) Blastomycosis
7. Which of the following is not transmitted through the semen of an infected male?
   (a) Hepatitis B (b) HIV
   (c) Chikunguniya (d) Ebola
8. The infection by *Ascaris* occurs by:
   (a) Eating half-cooked pork
   (b) Tse tse fly
   (c) Mosquito
   (d) Drinking water containing eggs of *Ascaris*
9. Cirrhosis of liver is caused by chronic intake of:
   (a) Opium
   (b) Alcohol
   (c) Tobacco
   (d) Cocaine
10. Which of the following acts as a physiological barrier to the entry of microbes in the human body?
    (a) Tears
    (b) Monocytes
    (c) Skin
    (d) Epithelium of urogenital tract

Answer Keys

Multiple Choice Questions

1. (a) **2.** (b) **3.** (d) **4.** (b) **5.** (a)
6. (a) **7.** (c) **8.** (b) **9.** (a) **10.** (b)

Match the Following

1. i (b), ii (a), iii (d), iv (c)
2. i (c), ii (d), iii (a), iv (b)
3. i (d), ii (c), iii (b), iv (a)

True or False

1. F **2.** T **3.** T **4.** F **5.** T
6. F **7.** T **8.** F **9.** T **10.** T

Fill in the Blanks

1. Disease
2. Physiological
3. Degenerative
4. Beriberi
5. Prion
6. Human Immunodeficiency Virus
7. *Anopheles*
8. Stomach
9. Hygiene
10. Hydrocortisone

Assertion–Reason Questions

1. (a) **2.** (a) **3.** (b) **4.** (a) **5.** (d)

One-word Answers

1. AIDS
2. *Plasmodium*
3. Congenital
4. Cholera or typhoid
5. Iodine
6. Haemazoin
7. Scurvy
8. Infectious disease
9. Physiological barrier
10. Humoral immunity

Define the Term

1. Malfunction of vital organs due to degeneration of cells gives rise to degenerative diseases like osteoporosis, which shows increasing bone weakness, leading to an increase in the risk of bone fractures. Neurons also undergo degeneration, leading to neurodegenerative disorders like Alzheimer's disease which is caused by ageing, hereditary or even lifestyle choices.
2. The diseases caused by pathogens along with the influence of other factors such as age, gender, lifestyle and nutritional deficiency. For example, diabetes, hypertension and cancer. They are non-transmittable.
3. Prion is a type of protein that exists without genetic material and is folded abnormally. It affects the structure of normal proteins, causing deadly disease that spreads rapidly. An example is Creutzfeldt-Jakob disease which is fatal and replicates in the host.
4. The diseases caused by parasites that spread to humans through infected mosquitoes. Malaria causes high fever and shaking chills. Tropical and subtropical countries are the most affected; ever year, nearly 290 million people are infected by malaria with more than 40,000 people losing their life. Preventative drugs along with insecticide-treated bed nets are distributed by world health programmes to protect people from mosquito bites.
5. Apply pressure on the wound with a sterile bandage or clean cloth and never try to realign the bone or even push the bone sticking out back inside unless you are trained to do so. Apply a splint to the area above and below the fracture sites and pad the splints to reduce the pain. Do not apply ice directly but wrap ice in a towel or cloth and then apply on the wound. Check for shock in the patient like weak breathing.

Groups of Terms

1. Rickets
2. Virus
3. Physiological
4. Congenital
5. T lymphocytes

Identify the Odd One Out

1. Cerebral palsy
2. B lymphocytes
3. Flu
4. Down syndrome
5. Housefly

Image-Based Questions

1. (d) **2.** (b) **3.** (c) **4.** (b)

Very Short Answer Questions

1. A state of complete mental, physical and social well-being is known as health and not merely the absence of infirmity and disease.
2. A disease is caused by different microorganisms that can be classified on the basis of agents as bacteria, fungi, viruses, etc. Some diseases are caused by multicellular organisms like worms.
3. Some children below the age of 5 years suffer from a serious eye disorder called xerophthalmia due to vitamin A deficiency. This places them at risk of becoming blind; it also caused night blindness.
4. Vitamin C deficiency can cause scurvy and is characterised by skin spots, bleeding gums and swelling in the joints that affects the immune system.
5. Prion is a type of protein that exists without genetic material and is folded abnormally. It affects the structure of the normal protein, causing deadly diseases that spread rapidly. One example is Creutzfeldt-Jakob disease which is fatal and replicates in the host.
6. The pathogens are present in water, plants and soil; they may be transmitted like hookworm through the soil.
7. Hydrochloric acid in the stomach produces an acidic environment that kills most of the germs. Similarly, saliva and tears have an antibiotic property that prevents the growth of the pathogen.
8. The innate immunity barriers are physical, physiological, cellular and cytokine.

9. The hygiene conditions in the surroundings should be maintained with limited exposure to airborne pathogens. Consumption of safe drinking water and clean food is important. A clean environment prevents mosquito breeding to prevent the spread of vector-borne diseases.
10. First aid is the immediate and first response and assistance given to a person with minor or major injury or illness that could be carried out to save their life and prevent worsening of the condition or to promote recovery.

Short Answer Questions

1. Anatomical classification: The type of classification that refers to the disease of the affected tissue or organs. For example, heart disease is a type of anatomical classification.

   Topographical classification: The type of classification that is handled by specialisations in medicine to follow the topographical categories. These types of diseases cover chest disease, vascular disease, abdominal disease and gastrointestinal disease. For example, ENT specialists only treat diseases specific to the ear, nose and throat.

   Physiological classification: This type of classification includes diseases that affect a function or a process like digestion, respiration and metabolism. For example, diabetes.

   Pathological classification: This type of classification considers the disease's nature; cancer is associated with uncontrolled cell proliferation and there are variations in the disease. For example, neoplastic disease involves the same uncontrolled proliferation of cells as in cancer, and inflammatory diseases like autoimmunity.

   Epidemiological classification: This type of classification refers to the rate of distribution, occurrence and control of the disease in a population.
2. The congenital disorders are congenital heart disease which includes a hole in the heart and problems with blood vessels or valves. Diagnosis is possible during routine scans during pregnancy or during screening of the newborn. Cerebral palsy is also a congenital disease, and Fragile X syndrome could be diagnosed through genetic testing during pregnancy. Trisomy 21 or Down syndrome could also be diagnosed by genetic testing. The sweat glands secrete sweat. The sebaceous glands secrete oil and many young girls and boys develop acne and pimples due to this.
3. The disease that is also caused by pathogens and is affected by other factors like age, gender, lifestyle, and nutritional deficiency of an individual. For example, diabetes, hypertension and cancer. They are non-transmittable; more examples of such diseases are heart disease, asthma and Alzheimer's disease.
4. There are different factors that cause malaria:
   - Use of infected or share syringes
   - Bitten by a malaria vector (*Anopheles stephensi*)
   - Transfusion
   - Organ transplantation
   - From infected mother to baby during birth
   - Person to person: The exchange of body fluids from an infected to a non-infected person
   - Droplet transmission: Droplets during speaking, sneezing and coughing by an infected person can spread from one person to another. The droplets could also be spread by touching the mouth and nose which travel only a short distance.
   - Through body fluids or blood: The infection can spread when it comes in contact with the mucous membrane or blood of an uninfected person. For example, AIDS and hepatitis.
   - By skin: infections like chickenpox and ringworm that spread when the skin of infected person comes in contact with skin of non-infected
5. There are four types of barriers:
   - Physical barrier: Skin, cilia, hair, eyelash and mucus layer in the gastrointestinal tract and respiratory tract serve as the first line of defence by preventing or trapping the pathogens that could enter the body.
   - Physiological barrier: Hydrochloric acid in the stomach produces an acidic environment that kills most of the germs. Similarly, saliva and tears have an antibiotic property that prevents the growth of the pathogen.
   - Cellular barriers: There are pathogens present in our body crossing other barriers and cells like leucocytes, lymphocytes, neutrophils, monocytes, eosinophils and basophils take part in this defence.
   - Cytokine barriers: In case of viral infections, the cells secrete proteins called interferons that form a coating around the infected cells and prevent further infection.
6. There are two types of responses generated by acquired immunity:
   - Humoral response: antibodies produced by B-lymphocytes and antigens internalised by them along with those presented by T-cells are included under this.

- Cell-mediated response: Initiated by T-cells and promoting apoptosis.

7. For minor burns, hold the area under not-very-cold running water for 10 minutes and on areas the like face, apply cool or wet cloth until the pain eases. In case of mouth burns from hot drinks or food, insert a piece of ice for a few minutes. Try to gently remove rings or other tight items worn in the burned area; blisters should not be broken or get infected. Apply lotion after the burn area becomes cool, with soothing ingredients like aloe vera and cover the burn with a clean bandage but wrap loosely to avoid pressure on the skin.
8. Four characteristic properties of acquired immunity are:
   - Memory: Remembering the type of antibodies released in response to a pathogen and releasing similar ones the next time
   - Specificity: Type of pathogens could be differentiated and destroyed
   - Diversity: Detection of a vast variety of pathogens from protozoa to virus
   - Differentiate between self and non-self cells: Differentiating between own and foreign cells
9. Open the airways of breathing by placing one hand on the person's forehead and gently tilt their head back. Use your fingertips to lift their chin up. Check the breathing and feel if chest movement is normal and put them in the recovery position. Kneel down next to them on the floor and try to make them lie on their back to shift from lying on their side. Try to find a spinal injury, if any, and call for help.

Long Answer Questions

1. Hint: Explain all the types of diseases mentioned in this chapter, with examples.
2. Hint: Explain non-infectious and infectious diseases, with examples.
3. Hint: *Plasmodium*-mediated malaria along with its factors, agents, vectors and host should be explained with a life cycle diagram.
4. Hint: Direct and indirect transmission should be explained.
5. Hint: Innate immunity and its components should be explained.
6. Hint: Explain in detail acquired immunity and its components.
7. Hint: Explain the treatment principles mentioned in this chapter.
8. Hint: Explain the ways of prevention of disease mentioned in this chapter.
9. Hint: Explain the first aid steps involved in bee sting, poison swallowing and bleeding.
10. Hint: Explain cholera and AIDS with their symptoms and agents along with the factors affecting them.

Case Study-Based Questions

1. (a) Malaria is caused by a parasite that is spread to humans through infected mosquitoes that causes high fever and shaking chills.

   (b) There are different factors that cause malaria:
   - Use of infected or share syringes
   - Bitten by a malaria vector (*Anopheles stephensi*)
   - Transfusion
   - Organ transplantation
   - From infected mother to baby during birth

   The vector of malaria is the female *Anopheles* mosquito that transmits these malarial sporozoites to the hosts. When these infected mosquitoes bite a human, the sporozoites are injected into the blood through the saliva of the mosquito.
2. (a) Congenital condition is one that is present at birth. The reasons for this could either be environmental factors like chemicals, smoking and infection during pregnancy or genes. The impact of this on the development and health of the child is not always severe and instead could be quite mild. Congenital heart disease, cerebral palsy, trisomy 21, and Fragile X syndrome are examples.
3. (a) Scurvy

   (b) Vitamin C
4. (a) Call for help and remove anything still present in mouth of the patient. Remove any contaminated cloth using gloves and rinse the skin for 20 minutes in a shower or hose. If the poison is in the eye, flush the eye with cool or lukewarm water for 20 minutes until help arrives. If the person has inhaled poison, try to get them to fresh air.

   (b) If the person starts vomiting, turn their head to their side to avoid choking and give CPR if they show no signs of life.
5. (a) Wash the affected area with soap and water and apply a cold compress. If the sting is on the leg or arm, elevate it and apply hydrocortisone cream to ease redness or itching and avoid scratching the sting area.

   (b) Go to a doctor if itching and swelling worsens.

Competition Window

1. (c)	**2.** (c)	**3.** (a)	**4.** (a)	**5.** (a)
6. (c)	**7.** (c)	**8.** (d)	**9.** (b)	**10.** (a)

CHAPTER **8**

Transportation and Reproduction in Plants

PREREQUISITES

You must have knowledge of:

- Basic structure of plants
- Cells and tissues
- Ecosystem

LEARNING OBJECTIVES

At the end of this chapter, you will be able to:

- Describe the tissues involved in the transportation of substances in plants
- Explain the modes of transportation in plants
- List the factors in the transportation of water in plants
- List the modes of reproduction in plants
- Compare and contrast the asexual and sexual modes of reproduction
- Describe sexual reproduction in plants

INTRODUCTION

All organisms require water, food, minerals and oxygen to be transported to the different parts of the body. This is necessary for the growth and respiration of cells. Plants and animals have different organs and processes for the transport of various substances.

Plants need raw materials for photosynthesis. They take in CO_2 through the stomatal pores and it diffuses into their cells. They also take in nitrogen, phosphorus and minerals through the roots. The distance between the soil and the chlorophyll-containing organs is large; hence, diffusion is not sufficient to send the raw materials to the leaves and energy to the roots. Thus, a proper transportation system is required.

In plants, the energy requirement is low because they do not move. They have a large proportion of dead cells in many tissues. So, the transport system in plants is a very slow process, though the distances covered are vast.

WATER

The root system of plants consists of a main root that has lateral roots bearing many fine outgrowths known as **root hairs**.

These are long protuberances of cells. They contain sap or fluid that is more concentrated than the surrounding soil water. Root hairs are covered by a thin cell membrane that allows water molecules to pass through; however larger molecules cannot. This type of membrane is known as a **semipermeable membrane**.

The main **features of root hairs** are as follows:

- Root hairs have a large surface area, which increases absorption
- Each root hair has a semipermeable membrane
- It contains cell sap of higher concentration than water

SCIENCE IN DAILY LIFE

1. Mark four test tubes as A, B, C and D. Fill up to three-fourths of A, B and C with water. Pour only a little amount of water in D. Fix a cork over the mouth of A.
2. Take three small plants with intact roots. Wash their roots and insert the plants in test tubes B, C and D in such way that the roots are dipped in water for B and C, but not in D.
3. In test tube C, add a colour dye. Pour oil in B and C. This will prevent loss of water by evaporation.
4. Mark the water level with a pencil and leave the set-up for 24 hours. Observe the change in water levels. You will see that the water level in test tube A remains the same, B is reduced as some is absorbed by the plant, C has a higher water level than B and the plant has leaves with red-coloured veins, while D contains a dried plant with lower water level due to evaporation.

MINERALS

Minerals are also absorbed by plants with water serving as the medium. Minerals are important for plants to grow stems, leaves and roots. Proper nutrition is important for flowers, seeds and the production of food. Plants need minerals in order to produce chlorophyll and also for circulation of water, reproduction and general survival.

Thus, minerals are **important** to plants for the following reasons:

- For development and reproduction, phosphorous and magnesium are essential.
- The life cycle of plants along with seed production are supported by the absorption of minerals by the roots.
- Deficiency of minerals such as nitrogen, copper, iron and sulphur will result in cell senescence, stunted growth and chlorosis or necrosis.
- The quality of fruits and vegetables is increased by potassium.
- Sodium, magnesium and potassium play a direct role in plant metabolism.
- The cell wall of plants is made up of calcium.
- Proteins are made up of nitrogen, carbon and hydrogen along with nucleotides.
- Osmotic potential along with turbidity of cells are maintained by minerals.
- Copper, iron and zinc show catalytic effects.

Based on their quantitative requirements, minerals can be divided in two **categories**:

- **Macronutrients:** The minerals required by plants in large quantities are sodium, phosphorous, carbon, silicon, magnesium, nitrogen, calcium, hydrogen, sulphur, selenium and oxygen.
- **Micronutrients:** The minerals required by plants in small quantities are zinc, cobalt, molybdenum, iron, nickel, manganese, boron and chlorine.

There are two ways in which **minerals are absorbed** by plants:

- **Passive mineral absorption:** Here, no energy is required as it takes place along the concentration gradient by simple diffusion.
- **Active mineral absorption:** This process requires energy and occurs against the concentration gradient.

Deficiency of minerals will result in different symptoms in plants:

- **Nitrogen deficiency:** Chlorosis occurs in the leaf and the green colour becomes yellow. This is followed by necrosis; the leaf then turns brown and dies. Plant growth is stunted and they mature early.
- **Phosphorous deficiency:** This results in stunted growth; the underside of older leaves becomes purple.
- **Potassium deficiency:** The edges of the leaves look burnt; this is known as scorching of leaves. Plants will lodge and the production of seeds with fruit will be of poor quality.
- **Iron deficiency:** Interveinal chlorosis occurs in young leaves with bleaching of new growth.

TRANSPORTATION USING XYLEM AND PHLOEM

Plants have specialised vascular tissues for the transportation of substances. There are two types of vascular tissues (Fig. 8.1):

- **Xylem:** Xylem is used for the transportation of water and minerals. It is composed of tracheids, xylem vessels, xylem parenchyma and xylem fibres. Tracheids and xylem vessels are the conducting elements. The xylem makes a continuous tube in plants which runs from the roots to the stem, right up to the veins of leaves. It carries water and minerals from the roots to the leaves of the plant.
- **Phloem:** Phloem is responsible for the transportation of food. Phloem consists of sieve tubes, companion

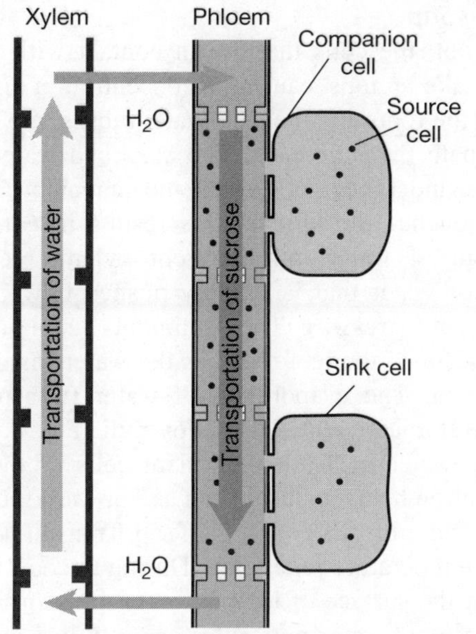

Fig. 8.1 Transport in xylem and phloem

cells, phloem parenchyma and bast fibres. Sieve tubes are the conducting elements in phloem. Phloem carries the product of photosynthesis from the leaves to other parts of the plant.

Like the circulatory system in humans, the xylem and phloem tissues extend throughout the plant body. Gaseous exchange takes place through the stomata. Roots exchange gases within the air spaces in the soil.

We will study the xylem and phloem in greater detail later in this chapter.

Means of Transportation in Plants

- **Diffusion:** This is a transportation process (Fig. 8.2) that involves the movement of a substance from cell to cell. It does not require the expenditure of energy. Molecules move in a random manner. It is a slow process.
- **Osmosis:** This is the movement of water molecules from a region of higher concentration to one of lower concentration through a semipermeable membrane.

SCIENCE IN DAILY LIFE

1. Place some raisins in a bowl of water.
2. Place some soft-skinned grapes in another bowl filled with sugar syrup.
3. Leave the bowls for 10–12 hours. You will observe that the raisins have swelled up but the grapes have shrunk; the raisins absorbed water while the grapes lost it. This transport occurs across a semipermeable membrane.

- **Facilitated Diffusion:** The gradient is a vital component in the process of diffusion. Hence, a smaller substance can be diffused faster when compared to large one. Facilitated diffusion is a passive diffusion process that comprises antiport, uniport and symport proteins (Fig. 8.3).

  Antiport proteins exchange solutes by transporting them in and out of the cell. The main function of the

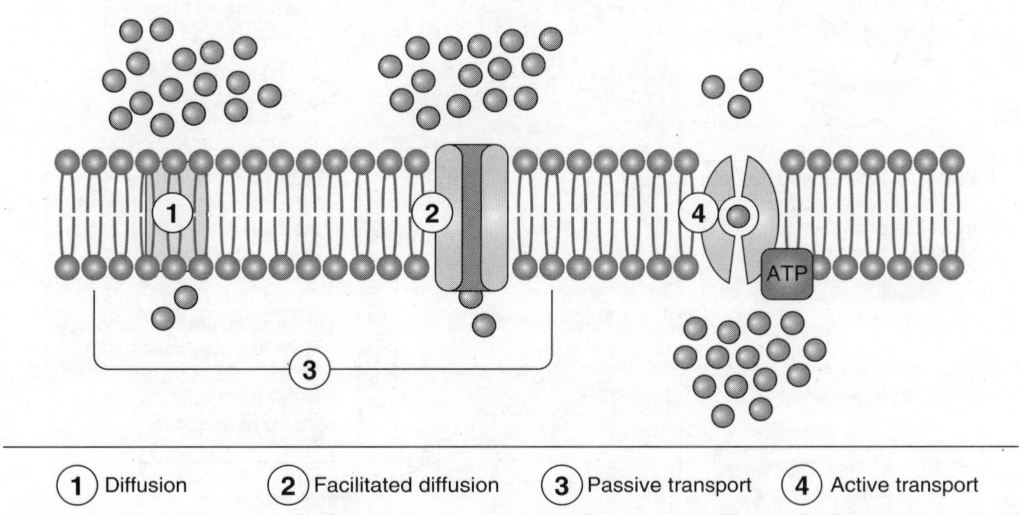

Fig. 8.2 Different types of transport

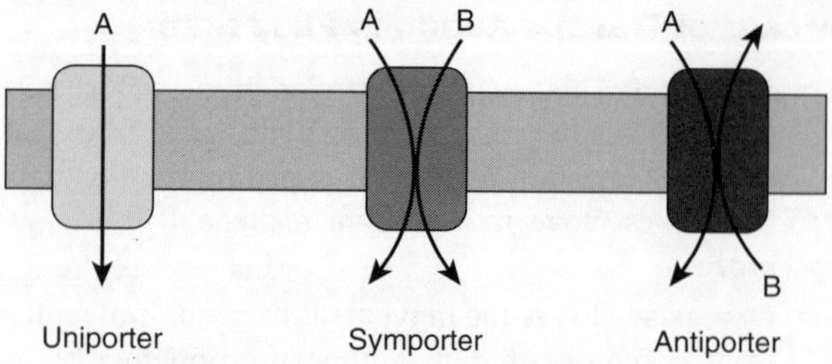

Fig. 8.3 Types of facilitated diffusion

uniport protein is to carry a single solute across the membrane. Symport proteins transfer two different solutes simultaneously in the same direction.

- **Active Transport:** In active transport, molecules move against the concentration gradient. Energy provided as ATP is used to drive the pump. ATP gives a phosphate which then pumps the molecule across the membrane.

Transport of Water through Xylem

Xylem is a specialised vascular tissue that transports water and nutrients from the soil to the leaves. The flow of water is unidirectional and is in the upward direction. It is responsible for replacing water lost through transpiration and photosynthesis. The water is absorbed by the root hairs and undergoes cell-to-cell movement by osmosis. The water then travels through the plant to reach the leaves. This is called ascent of sap.

Ascent of Sap

At the root, the cells that are in contact with the soil actively take up ions, causing a concentration difference between the roots and the soil. Water moves into the root to eliminate the difference. The absorbed water at the root hair undergoes cell-to-cell movement by osmosis until it reaches the xylem. This results in continuous movement of water into the root xylem, creating a column of water that is steadily pushed upwards; this is called **root pressure**. The continuous supply of water to the xylem of the leaf replaces the water lost through the stomata. The evaporation of water (transpiration) from the stomata creates a suction pull (Fig. 8.4). This pulls the water up from the xylem cells of the roots. Transpiration helps in absorption and upward movement of water and minerals or ascent of sap. Transpiration also helps in temperature regulation. During the day, water is lost from the surface of leaves by transpiration and this results in more water molecules being pulled up due to their tendency to cohesion. This pulling force created by leaves helps in the upward conduction of water in tall trees.

Factors affecting Transpiration

- **Sunlight:** During the day, the rate of transpiration is higher as the stomata are open and this allows inward diffusion of carbon dioxide for photosynthesis. In the dark, the stomata are closed and transpiration hardly occurs.

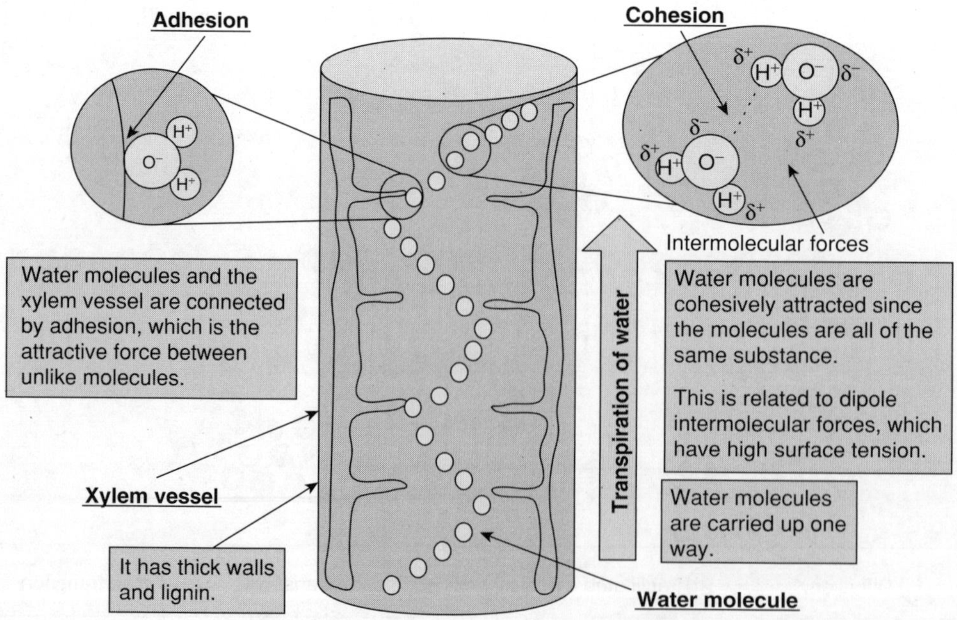

Fig. 8.4 Transpiration of water in xylem

- **Wind:** Transpiration is greater when the wind blows at speed as water vapour is removed faster from the leaves.
- **Temperature:** Transpiration is higher in the hot days of summer than in the cold days of winter.
- **Humidity:** Transpiration decreases if the air is humid as air cannot hold more water molecules (already present as moisture).

Practical Learning

1. Take a well-watered, small, potted plant and cover it with a transparent polythene bag.
2. Tie the mouth around the stem base.
3. Leave the plant in sunlight and mark the changes after a few hours. You will observe drops of water inside the bag. This is due to water vapour saturation.

Importance of Transpiration

- **Cooling effect:** When water evaporates from a plant, the heat used for transpiration is taken from the plant, enabling it to cool itself.
- **Concentration of sap:** It maintains the concentration of sap inside the plant body. As the roots continue to absorb water, the excess water that is not evaporated out results in dilution of sap. This prevents further absorption of water from the roots.

Factors affecting Ascent of Sap

- **Root pressure:** This is a positive pressure created by the movement of water from the soil to the roots, for upward movement of water. The effect of root pressure in the transport of water is more important at night.
- **Transpiration pull:** This is the pull of water as a result of tension created by transpiration in the aerial parts of the plant. It is the main driving force of water movement upwards in a plant during the day. High temperature, wind velocity and low humidity influence transpiration pull (Fig. 8.5).
- **Capillary action:** A capillary is a very fine tube. Water or liquid rises in the capillary because of physical forces and this phenomenon is called capillary action. Water in the stem rises to a certain height because of capillary action.
- **Adhesion-cohesion of water molecules:** Water molecules make a continuous column in the xylem because of the forces of adhesion and cohesion among them.

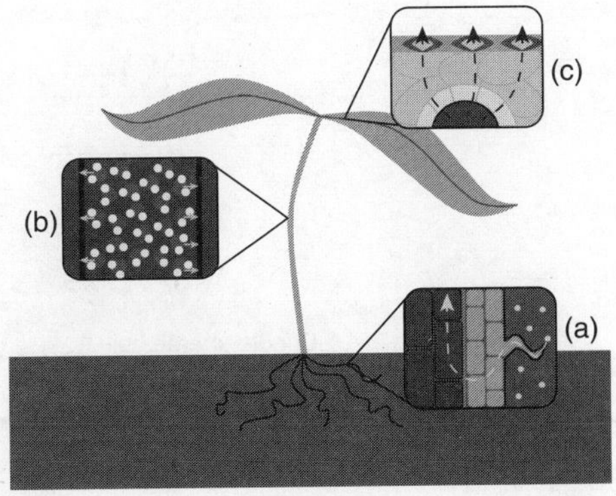

Fig. 8.5 Ascent of sap: (a) Water molecules enter the roots passively. A concentration gradient of solutes within the xylem leads to further passive transport of water into the tracheids. (b) Adhesion causes the molecules to adhere to the sides of the vessels, and cohesion causes the molecules to adhere to each other. (c) Water from the xylem enters the spongy mesophyll, where it can come into contact with stomata. When the stomata are open, evaporation occurs. When water molecules evaporate out of the stomata, they create tension in the water column, creating a pulling effect in the xylem.

Transport of Food through Phloem

The transport of food/glucose and amino acids as well as other substances from the leaves to the different parts of the plant body is called **translocation**. It takes place in the phloem (Fig. 8.6) and delivers substances to the storage organs such as roots and fruits and to the growth organs. Phloem is a complex tissue that is composed of four types of cells (sieve tubes, companion cells, phloem fibre and phloem parenchyma). The only dead component of phloem is the phloem fibre. Translocation is bidirectional; that is, it takes place in the upward and downward directions. It occurs in the sieve tubes with the help of companion cells. Ascent of sap takes place by simple physical forces but translocation in phloem is achieved by utilising energy.

Materials like sucrose formed in the leaf are transferred into the phloem using energy from ATP. This increases the osmotic pressure of the tissue. Water from the neighbouring xylem now moves into the sieve tube through the companion cells. The bulk pressure increases in the sieve tubes and the material moves in the phloem to tissues which have less pressure. This allows the phloem to move materials according to the plant's needs in both directions, up and down. The materials may be stored in the roots, fruits, seeds, etc. So, sugars stored in the root or stem tissues may move to the growing buds in spring as they need energy.

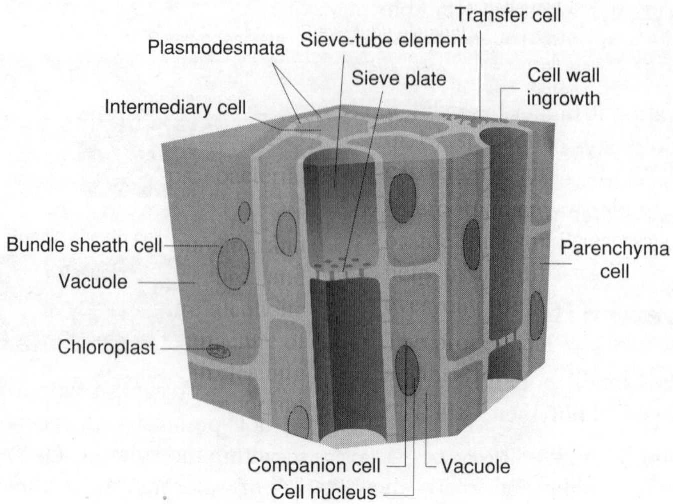

Fig. 8.6 Phloem of plants

REPRODUCTION

Reproduction is a characteristic aspect of every living organism. It is the biological process that ensures continuity of the species.

Reproduction can be asexual or sexual. Single individuals can produce bacterial and protozoan progeny. Sexual reproduction involves two parents and the fusion of reproductive cells. When a male gamete fuses with a female gamete, a new individual is born. In plants and animals, this is a common way of reproduction.

Let us study each of these modes in detail.

Asexual Reproduction

Asexual reproduction is one where new organisms are produced from a single parent without the involvement of sex cells. Sponges and *Hydra* are examples of animals that reproduce asexually and sexually.

The different **modes of asexual reproduction** are as follows:

- **Binary fission:** Here, the parent cell divides into daughter cells. When fission results in two daughter cells, it is binary fission. For example, *Amoeba* and *Paramecium*.
- **Budding:** In some organisms, a small bud is formed which develops into a tiny individual. It detaches from the parent body after maturation and grows into a new individual. For example, *Hydra* and yeast.
- **Fragmentation:** *Spirogyra* and similar organisms are made of long, ribbon-like filaments that reproduce by fragmentation. The division and growth of filaments into two or equal parts takes place and each part grows into an individual.
- **Spore formation:** This is common in ferns as these ornamental plants bear spores on the underside of leaves. The spores float in the air as they are light and are carried to different places.
- **Vegetative propagation:** New plants are produced by the vegetative parts such as leaf, root and stem. Ginger, potato and mint reproduce by this method. Humans have adapted natural vegetative propagation methods (Table 8.1) to artificially propagate a large number of plants:
    - **Cutting:** A branch is cut and planted in the ground. A new plant emerges as a result of the formation of roots.
    - **Layering:** A branch is placed in the ground, with a section of it buried in the dirt. The plant's shoot that will be utilised as stock is chopped off. The stem is sliced vertically, and the scion is placed and taped into the stock. This component takes hold and a new plant sprouts (Fig. 8.7).

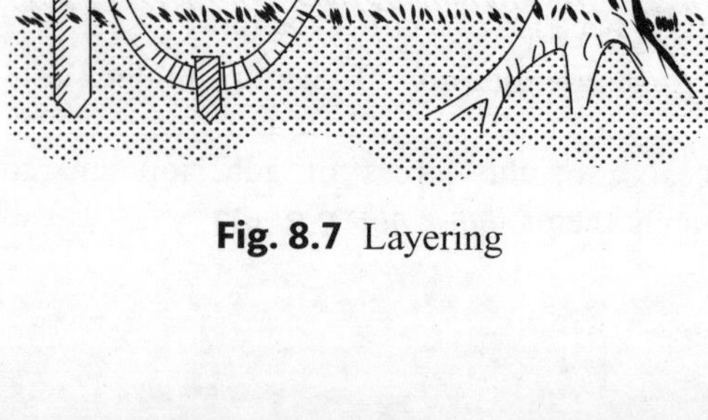

Fig. 8.7 Layering

- Grafting: The joining of two plants where the upper part of the graft (known as scion) becomes the top of plant and the lower portion (known as understock) becomes the trunk or the root system. Several plants could also be combined to form a graft (Fig. 8.8).

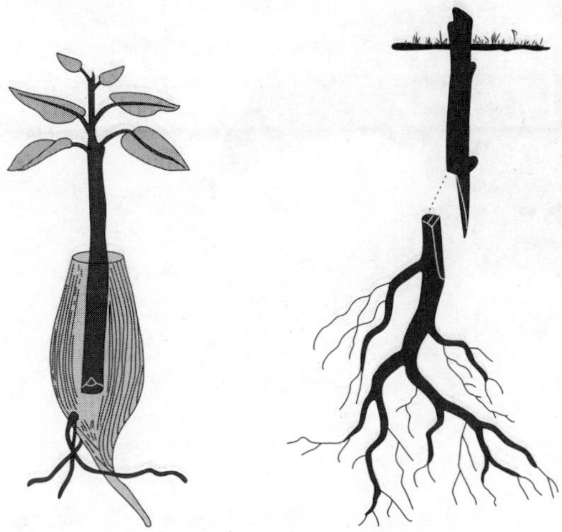

Fig. 8.8 Grafting

Table 8.1 Natural and artificial vegetative propagation

Mode of reproduction	Plant part involved	Example
Natural		
(a) Root	Adventitious root	Sweet potato
(b) Stem	Runner	Lawn grass
	Sucker	*Chrysanthemum*
	Bulb	Onion
	Tuber	Potato
	Rhizome	Ginger
	Adventitious buds	*Bryophyllum*
Artificial	Cutting	Rose, money plant
	Layering	Jasmine, mint
	Grafting	Apple, mango

SCIENCE IN DAILY LIFE

1. Place a piece of lawn grass in a pot and cover a portion of it with moist soil.
2. Leave it for 10–12 days.
3. Sprinkle some water every day and keep the soil moist. After few days, new grass shoots grow from buds on nodes. Lawn internodes can be seen creeping on the surface of the ground, and from each node, roots will grow out. The axillary buds at the nodes will produce new shoots.

- **Tissue culture:** In tissue culture, new plants are grown by removing tissue from the growing tip of a plant. The cells are placed in an artificial medium (growth medium) where they divide rapidly to form a small group of cells known as callus. The callus is transferred to another medium containing hormones which are responsible for growth and differentiation. The plantlets are then placed in the soil so that they can grow into developed and mature plants.
- The **importance of vegetative propagation** is as follows:
  - Plants can bear flowers and fruits quickly.
  - Plants that have lost the ability to produce viable and healthy seeds can reproduce by vegetative propagation.
  - All plants are genetically similar to the parent plant.
  - Seedless varieties like banana can be obtained.

Sexual Reproduction

Sexual reproduction necessitates the joining of male and female gametes. It involves two parents: male and female. A male gamete fuses with a female gamete to form a new cell called the zygote. The zygote then grows and develops into a new organism.

The flower is the main reproductive organ of plants (Fig. 8.9). It consists of four parts: sepals, petals, stamen and pistil. The **stamen** is the male reproductive part and produces pollen grains that contain the male gametes. Each stamen has two parts: filament and anther. The **carpel** is the female reproductive part and produces ovules that contain female gametes. It has three parts: stigma, style and ovary.

Flowers can be of two types:

- **Unisexual:** This flower contains either male or female reproductive parts. For example, papaya and watermelon.
- **Bisexual:** This flower contains both male and female reproductive parts. For example, *Hibiscus* and mustard.

Pollination

The pollen grains and nuclei of the ovules are brought together by a variety of forces, including wind, water and insects. Pollination is the transport of pollen grains from the anther to the stigma of a flower. This can take place in three ways:

- **Autogamy:** The pollen grains of a flower fall on the stigma of the same flower.

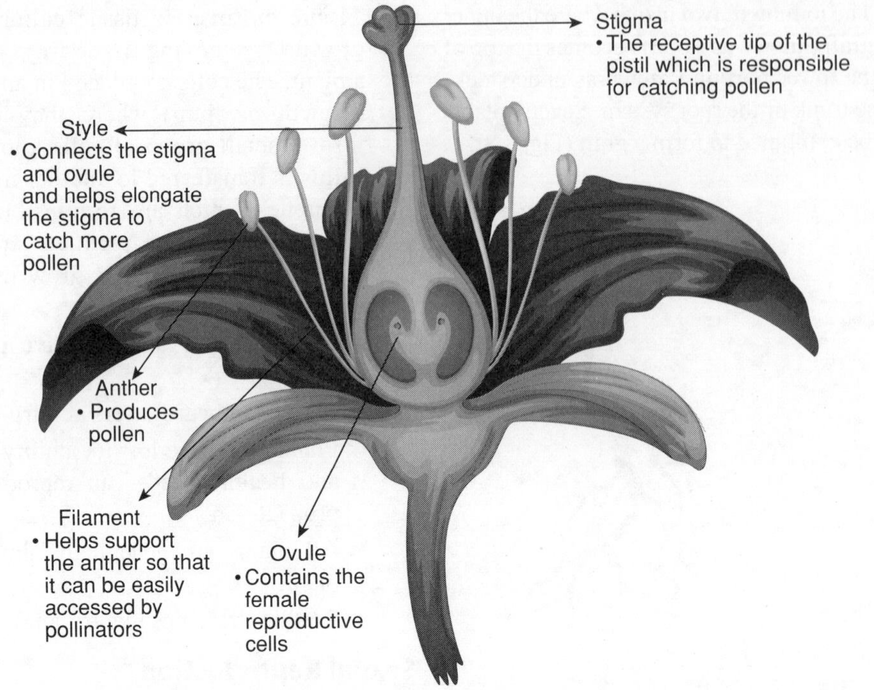

Fig. 8.9 Longitudinal section of a flower

- **Geitonogamy:** The pollen grains of a flower of a plant fall on the stigma of a different flower of the same plant.
- **Allogamy:** The pollen grains of a flower of a plant fall on the stigma of a flower of another plant of the same species.

When pollination occurs through insects, wind or other agents, it is known as **cross-pollination** (Fig. 8.10).

When the pollen from a flower falls on its own stigma and fertilises the ovule, this is known as **self-pollination**.

Process: Pollination begins when the pollen grains of a flower of a species lands on the stigma of a flower of the same species to form a pollen tube that connects the ovary and stigma. When the pollen tube is complete, the pollen grain transmits sperm cells to the ovary. Fertilisation takes place when the sperm cells reach the ovary and egg cell.

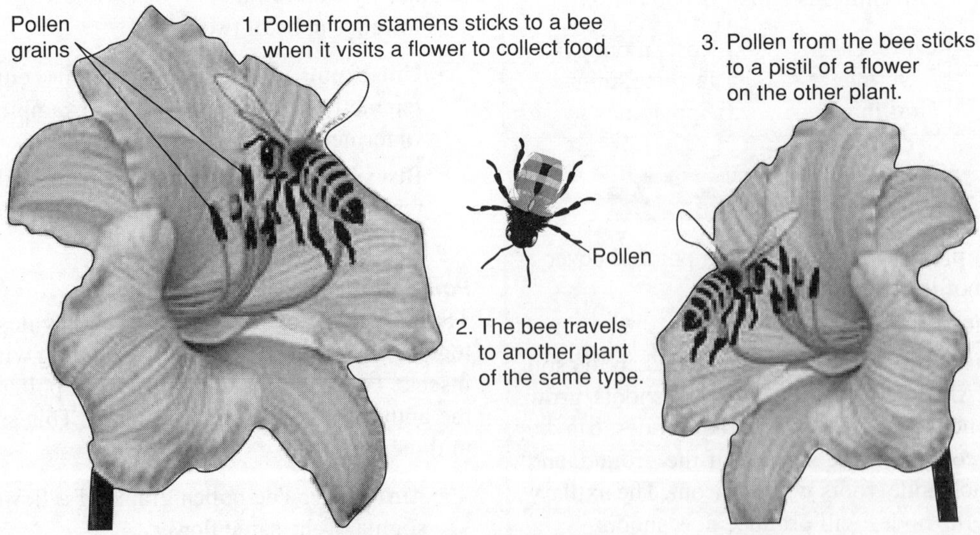

Fig. 8.10 Cross-pollination

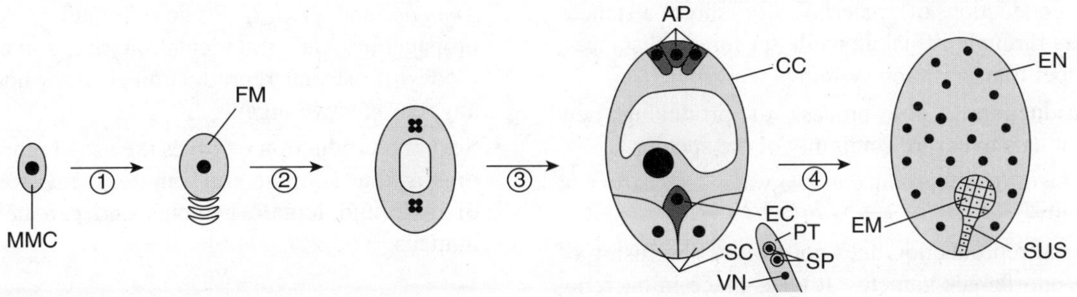

Fig. 8.11 Double fertilisation: The diploid megaspore mother cell (MMC) after two meiotic divisions (1) develops into a haploid functional megaspore (FM). Eight nuclear cells are formed after three syncytial mitotic divisions (2) in the FM. A seven-celled embryo sac is formed after nuclear migration, fusion and differentiation (3). This embryo sac contains an egg cell (EC), two synergid cells (SC) at the micropylar pole, three antipodal cells (AP) at the chalazal end and one homodiploid central cell (CC) in the middle. AP cells degenerate leading to the entry of one pollen tube (PT) and two sperm cells (SP) that fertilise the egg and CC (4). This leads to the development of a diploid embryo (EM) attached by a suspender (SUS) and a triploid endosperm (EN).

When a person transfers pollen to the stigma of a flower, it is known as **artificial pollination**. In the Babylonian civilisation, sprinkling of male flowers of palms on the female flowers was a common practice, though they had no idea of the concept of pollination at that time. In modern times, artificial pollination is a standard practice adopted by breeders of plants or scientists in their effort to evolve new varieties. **Emasculation** is the removal of anthers in young flowers; these flowers are covered by plastic bags. Later, pollination of these flowers with pollen from the desired plants is performed.

Fertilisation

Germination of pollen grains takes place when the stigma receives pollen of the same species and allows it to disintegrate. Germination is stimulated by secretion of sugar by the stigma. Exine in the pollen grows to form a tube-like structure that has a generative nucleus at the tip. The tube nucleus divides into two parts, known as **male/sperm nuclei.** These three nuclei (two sperm nuclei and one generative nucleus), which are not separated by a cell wall, share the same cytoplasm and travel to the ovule.

The pollen tube grows through the style and stigma by enzymatically dissolving the tissues and reaches the ovary. It pushes through the micropyle and reaches the embryo sac. The tube nucleus that directed the growth of the pollen tube now disintegrates. Two sperm nuclei are released when the pollen tube enters one of the synergids. One of these sperm nuclei fuses with the egg nucleus to form the **zygote** while the other sperm moves towards the two polar nuclei in the central cell and fuses with it. These two fertilisation processes that take place are together known as **double fertilisation** (Fig. 8.11).

Seeds are disseminated or moved away from the parent plant once they are created, and they germinate under favourable conditions (Fig. 8.12).

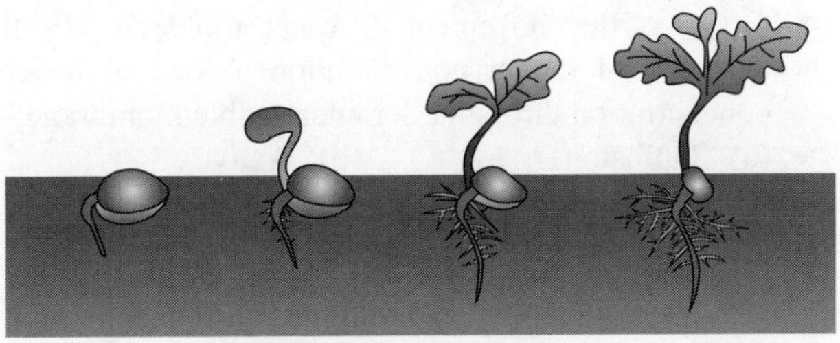

Fig. 8.12 Germination

SUMMARY

- Plants prepare food from the carbon dioxide and water in their leaves.
- To supply water and minerals to the plant and to transport food prepared in the leaves to the various parts of the plant, a transport system is required.
- Transportation in plants takes place in many ways such as diffusion, osmosis and facultative diffusion through transporters like uniport, symport and antiport.
- Phloem helps in ascent of sap whereas xylem is responsible for transpiration.

- The conduction of material over short distances occurs through diffusion, while for longer distances, a proper transportation system is required.
- Reproduction is the process of producing new individuals to ensure continuity of the species.
- Organisms can reproduce in two ways: asexually and sexually.
- Asexual reproduction does not involve the fusion of male and female gametes. It takes place in bacteria, *Amoeba* and *Hydra*. Fission, budding, vegetative propagation and fragmentation are some of the modes of asexual reproduction. It does not require any reproductive organs.
- Sexual reproduction involves the use of reproductive organs, that is, male and female. It involves fusion of male and female gametes and can be seen in humans.

PRACTICE QUESTIONS

Analyse Your Concepts (School Exam Based)

Multiple Choice Questions

1. The main plant root has lateral roots bearing many fine outgrowths known as:
   (a) Root hairs (b) Adventitious root
   (c) Lateral root (d) Roots
2. The quality of fruits and vegetables is increased by:
   (a) Sodium (b) Calcium
   (c) Potassium (d) Sulphur
3. _______ carries the product of photosynthesis from the leaves to other parts of the plant.
   (a) Xylem (b) Phloem
   (c) Stomata (d) Collenchyma
4. _____ is the movement of water molecules from a region of higher concentration to one of lower concentration through a semipermeable membrane.
   (a) Diffusion (b) Active
   (c) Osmosis (d) Reverse osmosis
5. _________ is the pull of the water as a result of tension created by transpiration in the aerial parts of the plant.
   (a) Transpiration pull (b) Osmotic pull
   (c) Diffusion pull (d) Ascent of sap
6. *Spirogyra* reproduces by:
   (a) Binary fission
   (b) Vegetative propagation
   (c) Budding
   (d) Fragmentation
7. The ______ is the female reproductive part.
   (a) Stamen (b) Petal
   (c) Carpel (d) Sepal
8. When pollination occurs through insects, wind or other agents, it is known as:
   (a) Self-pollination (b) Cross-pollination
   (c) Both (a) and (b) (d) None of the above
9. ______ is the removal of anthers in young flowers.
   (a) Pollination (b) Cutting
   (c) Layering (d) Emasculation
10. Two fertilisation processes that take place are together known as:
    (a) Fertilisation (b) Pollination
    (c) Double fertilisation (d) None of the above

Match the Following

Each question contains statements given in two columns which have to be matched. Statements in column I have to be matched with statements in column II.

1. Choose the correct option:

Column I	Column II
i. Tracheids	a. Phloem
ii. Sieve tubes	b. Xylem
iii. Carpel	c. Yeast
iv. Budding	d. Flower

2. Choose the correct option:

Column I	Column II
i. Layering	a. Artificial propagation
ii. Fragmentation	b. Sexual reproduction
iii. Fertilisation	c. Vegetative propagation
iv. Tissue culture	d. Asexual reproduction

3. Choose the best match:

Column I	Column II
i. Xylem	a. Osmosis
ii. Phloem	b. Transport of water
iii. Semi-permeable membrane	c. Food transport

True or False

Read the following statements and write your answer as true or false.

1. Facilitated diffusion is a passive diffusion process that comprises antiport, uniport and symport.
2. Xylem is used for transportation of water and minerals.
3. Root hairs have a large surface area which decreases absorption.
4. Transpiration helps in absorption and upward movement of water and minerals, also called ascent of sap.
5. Diffusion is the movement of water molecules from higher concentration to lower concentration through a semi-permeable membrane.
6. Transpiration pull is the positive pressure created by the movement of water from the soil to the roots, for upward movement of water.
7. Transpiration is faster on hot days than cold days.
8. *Spirogyra* and other organisms are made of long ribbon-like filaments that reproduce by spore formation.
9. Stamen is the male reproductive part of the flower.
10. Geitonogamy is when the pollen of a flower falls on its own stigma.

Fill in the Blanks

Complete the following statements with an appropriate word/term to be filled in the blank spaces.

1. ________ contains cell sap of higher concentration than water.
2. _______ is used for transportation of water and minerals.
3. Phloem consists of ______, _______, _______ and ________.
4. ________ is a transportation process that involves the movement of a substance from cell to cell.
5. _______ molecules move against the concentration gradient.
6. _______ helps in absorption and upward movement of water and minerals, also called ______.
7. Transpiration is _____when wind is blowing faster as water vapour is removed ____from leaves.
8. Water molecules make a continuous column in the xylem because of forces of ______ and ______ among the molecules.
9. ________ is the biological process by which organisms produce offspring of their own species.
10. In _______, a small bud is formed which develops into a tiny individual.
11. ______ is a plant branch is placed on the ground, with a section of it buried in the soil.
12. ______ is the female reproductive part of the flower and produces _____that contain female gametes.

Assertion-Reason Questions

The following questions consist of two statements: Assertion (A) and Reason (R). Answer these questions selecting the appropriate option given below:

(a) Both A and R are true and R is the correct explanation of A
(b) Both A and R are true but R is not the correct explanation of A
(c) A is true but R is false
(d) A is false but R is true

1. Assertion (A): A male gamete fuses with a female gamete to form a new cell called the zygote.
   Reason (R): The zygote grows and develops into a new organism.
2. Assertion (A): Transpiration is faster in the hot days of summer.
   Reason (R): Transpiration is faster in the cold days of winter.
3. Assertion (A): Passive mineral absorption requires no energy as it takes place along the concentration gradient by simple diffusion.
   Reason (R): Passive mineral absorption takes place along the concentration gradient by simple diffusion.
4. Assertion (A): Sodium plays a direct role in plant metabolism.

Reason (R): The cell wall of plants is made up of calcium.

5. Assertion (A): The quality of fruits and vegetables is increased by sodium.
Reason (R): Sodium also plays a direct role in plant metabolism.

One-word Answers

1. Name the phenomenon that occurs in the leaf where the green colour becomes yellow.
2. Which vascular tissue is used for the transportation of water and minerals?
3. The phenomenon of physical forces that raises the water in the capillary is called?
4. Which asexual reproduction involves a small bud is formed which develops into a tiny individual?
5. What is the male reproductive part of a flower?

Define the Term

1. Ascent of sap
2. Root pressure
3. Binary fission
4. Layering
5. Process of pollination

Groups of Terms

Given below are groups of terms. In each group the first pair indicates the relationship between the two terms. Rewrite and complete the second pair on a similar basis.

1. Xylem: water :: Phloem: _____
2. Male: stamen :: Female: ______
3. Tracheids: xylem :: Companion cell: _____
4. Carbon : macronutrient :: Boron: _____
5. *Hydra*: budding :: *Spirogyra*: ______

Identify the Odd One Out

1. Tracheids, vessels, xylem fibre, companion cells
2. Phosphorous, carbon, boron, magnesium
3. Binary fission, fertilisation, fragmentation, budding
4. Sepal, stomata, petal, pistil
5. Autogamy, geitonogamy, allogamy, monogamy

Image-Based Questions

1. Which is the uniporter?

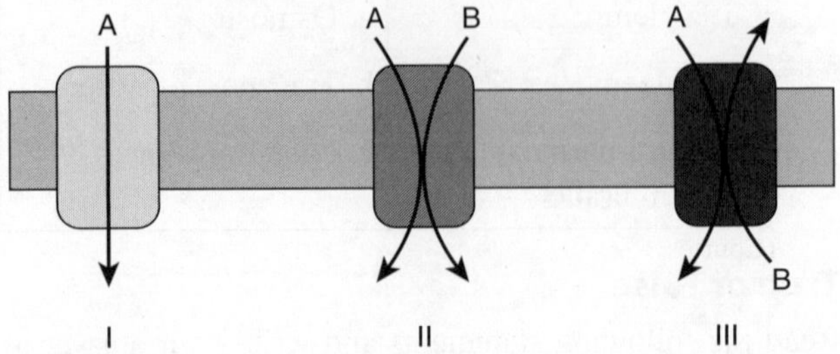

(a) II (b) III
(c) I, II (d) I

2. Which type of asexual reproduction is shown here?

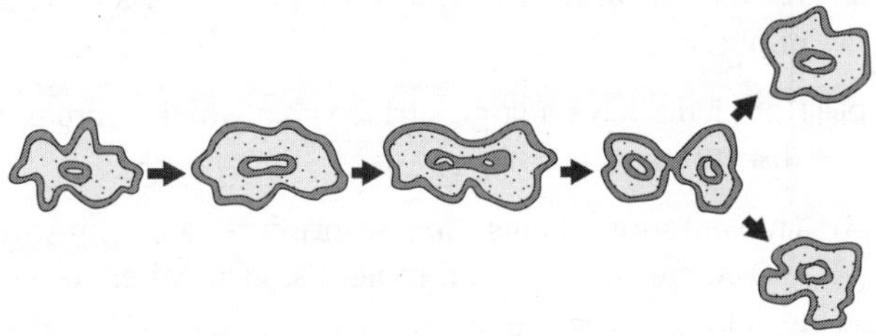

(a) Fragmentation
(b) Fission
(c) Budding
(d) Spore formation

3. What is the female part of this called?

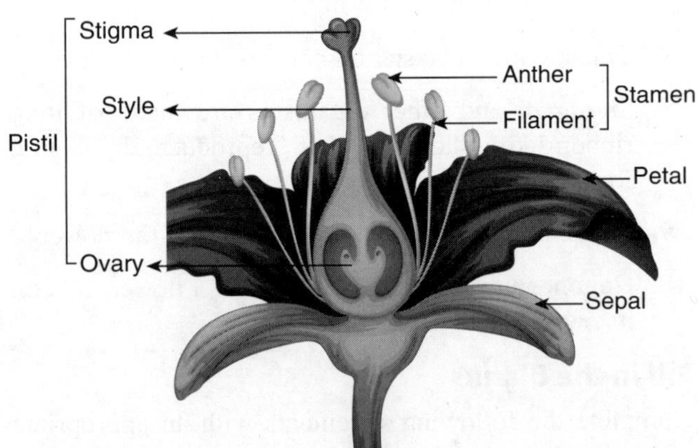

(a) Carpel
(b) Stamen
(c) Petal
(d) Sepal

4. Which component performs food transport?

(Courtesy: Stefan Schweihofer; https://pixabay.com/photos/wheat-crops-barley-crops-cereals-143373/)

(a) Xylem (b) Root
(c) Phloem (d) Chlorophyll

5. Choose the male part of the flower:

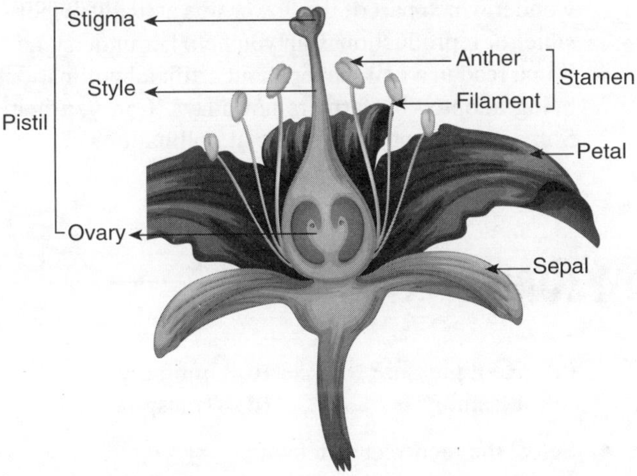

(a) Stamen (b) Carpel
(c) Sepal (d) Petal

6. Name the organisms that reproduces by this method:

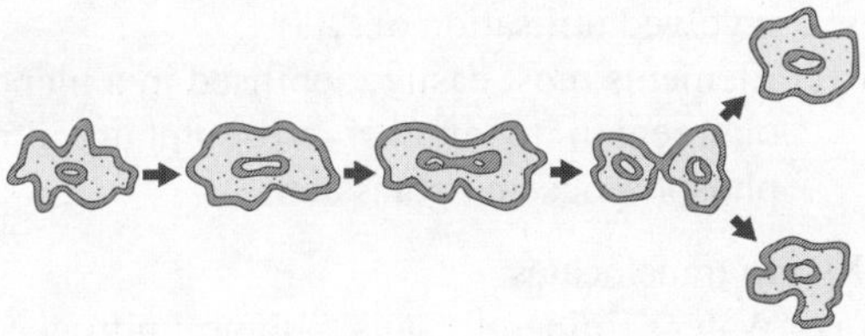

(a) Animal (b) *Amoeba*
(c) Fungus (d) None

7. Which of the following is antiporter?

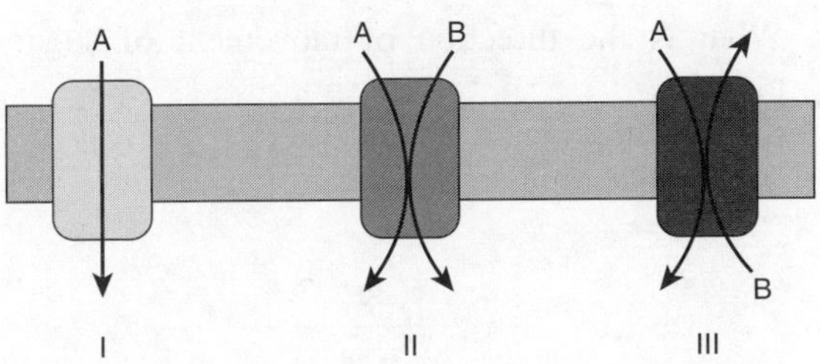

(a) I (b) II
(c) III, II (d) III

8. Choose the plant part that participates in water transport:

(a) Phloem (b) Xylem
(c) Chlorophyll (d) Thorns

Very Short Answer Questions

1. Define transportation in plants.
2. What are the components of the plant vascular system?
3. Give the types of factors influencing transpiration.
4. What is ascent of sap?
5. What is root pressure in plants?
6. What are the types of transportation in plants?
7. What is a flower?
8. Name the parts of a flower.
9. What is pollination?
10. Write a few lines about fertilisation.

Short Answer Questions

1. Explain the types of reproduction in plants.
2. Explain the role of the flower in reproduction.
3. What is pollination? What are the different types of pollination?
4. Write about
   (a) Active transport
   (b) Diffusion
   (c) Xylem
   (d) Phloem
5. Write about the relation between pollination and fertilisation.
6. Write about xylem and phloem.
7. What are the features of ascent of sap?
8. Write a short note on vegetative propagation.

9. Write a short note on transpiration.

10. What is fragmentation?

Long Answer Questions

1. Explain the structure of a flower.
2. Explain pollination and fertilisation in detail.
3. Write an essay on reproduction in plants.
4. Explain transpiration.
5. Define the types of transport in plants.
6. Write a brief account on the parts of plants that participate in transportation.
7. Write an essay on tissue culture.
8. Explain the types of vegetative propagation.
9. Write notes on fertilisation and double fertilisation.
10. Write an essay on active and passive transport.

Case Study-Based Questions

1. Disha was watering her plants when it struck her that if she was watering the roots, how did the leaves get water to perform essential functions like photosynthesis? Can you help her understand which part of the plant is responsible for water transport?
2. Sandeep was reading about different types of vegetative propagation being adopted all over the world. This made him think about why this technique is important in agriculture. Can you help him understand?
3. While observing a flower, Mansi saw that it had various complex structures. In another flower, a butterfly was sitting on the anther. This made Mansi wonder which part of the flower was actually responsible for reproduction. Can you help her understand?
4. Soma read in a newspaper about artificial pollination being adopted by farmers nowadays. Can you help Soma understand what artificial pollination is?

COMPETITION WINDOW (OBJECTIVE TYPE)

1. Root pressure develops due to:
   (a) Active absorption
   (b) Low osmotic potential in soil
   (c) Passive absorption
   (d) Increase in transpiration

2. Which of the following is the correct match?

Column I	Column II
A. Cohesion	i. More attraction in liquid phase
B. Adhesion	ii. Mutual attraction among water molecules
C. Surface tension	iii. Water loss in liquid phase
D. Guttation	iv. Attraction towards polar surfaces

   (a) A iii, b i, c iv, d ii (b) A ii, b i, c iv, d iii
   (c) A ii, b iv, c i, d iii (d) A iv, b iii, c ii, d i

3. The process responsible for facilitating loss of water in liquid form from the tip of grass blades at night and early morning is:
   (a) Root pressure (b) Imbibition
   (c) Plasmolysis (d) Transpiration

4. Select the incorrect statement:
   (a) Transport of molecules in phloem can be bidirectional
   (b) Movement of minerals in xylem is unidirectional
   (c) Unloading of sucrose at the sink does not involved utilisation of ATP
   (d) Elements most easily mobilised in a plant from one region to another are sulphur, nitrogen, phosphorous and potassium

5. Xylem translocates:
   (a) Water, mineral salts, some nitrogen and hormones
   (b) Water only
   (c) Water and mineral salts only
   (d) Water, mineral and some organic nitrogen only

6. What is the direction of movement of sugars in phloem?
   (a) Bi-directional
   (b) Non-multidirectional

(c) Upward
(d) Downwards

7. The main difference between active and passive transport across the cell membrane is:
(a) Passive transport is non-selective whereas active transport is selective
(b) Passive transport requires a concentration gradient across a membrane whereas active transport requires energy to move solutes
(c) Passive transport is confined to anionic carrier proteins whereas transport is confined to the cationic channel

8. Which of the following is not a feature of active transport of solutes in plants
(a) Occurs against gradient concentration
(b) Non-selective
(c) Occurs through membrane
(d) Requires ATP

9. What will be the direction of flow of water when a plant cell is placed in hypotonic solution?
(a) Water will flow in both directions
(b) Water will flow out of cell
(c) Water will flow into cell
(d) No flow of water in any direction

10. Stomatal movement is not affected by:
(a) Temperature
(b) Light
(c) Oxygen concentration
(d) Carbon dioxide concentration

Answer Keys

Multiple Choice Questions

1. (a)	2. (c)	3. (b)	4. (c)	5. (a)
6. (d)	7. (c)	8. (b)	9. (d)	10. (c)

Match the Following

1. i (b), ii (a), iii (d), iv (c)
2. i (c), ii (d), iii (b), iv (a)
3. i (b), ii (c), iii (a)

True or False

1. T	2. T	3. F	4. T	5. F
6. F	7. T	8. F	9. T	10. F

Fill in the Blanks

1. Root hairs
2. Xylem
3. Sieve tubes, companion cells, phloem parenchyma and bast fibres
4. Diffusion
5. Active transport
6. Transpiration, ascent of sap
7. More, faster
8. Adhesion and cohesion
9. Reproduction
10. Layering
11. Pancreas
12. Carpel, ovules

Assertion–Reason Questions

1. (b)	2. (c)	3. (a)	4. (b)	5. (d)

One-word Answers

1. Chlorosis
2. Xylem
3. Capillary action
4. Budding
5. Stamen

Define the Term

1. The flow of water is unidirectional and is in the upward direction. It is responsible for replacing water lost through transpiration and photosynthesis. The water is absorbed by the root hair and undergoes cell-to-cell movement by osmosis. The water then travels through the plant to reach the leaves. This is called ascent of sap.
2. Root pressure is the positive pressure created by the movement of water from the soil to the roots, for upward movement of water. The effect of root pressure in the transport of water is more important at night.
3. Fission is a mode of reproduction in which the parent cell divides into daughter cells. When fission results in two daughter cells, it is binary fission. For example, Amoeba and Paramecium.
4. A branch of a plant is placed on the ground, with a section of it buried in the soil. The plant's shoot

that will be utilised as stock is chopped off. The stem is sliced vertically, and the scion is placed and taped into the stock. This component takes hold and sprouts a new plant.

5. The pollination process begins when pollen grains of the flower of the same species lands on the stigma to form a pollen tube that connects the ovary and stigma. When the pollen tube is completed, the pollen grain starts to transmit sperm cells from the pollen to the ovary. Fertilisation takes place when the sperm cells reach the ovary and egg cell.

Groups of Terms

1. Food
2. Carpel
3. Phloem
4. Micronutrient
5. Fragmentation

Identify the Odd One Out

1. Companion cells
2. Boron
3. Fertilisation
4. Stomata
5. Monogamy

Image-Based Questions

1. (d) 2. (c) 3. (a) 4. (c) 5. (a)
6. (b) 7. (d) 8. (b)

Very Short Answer Questions

1. All organisms require water, food, minerals and oxygen to be transported to the different parts of the body for the growth and respiration of cells. Plants need raw materials for photosynthesis as they take in CO_2 through the stomatal pores and it diffuses into their cells while nitrogen, phosphorus and minerals uptake happens through the roots. The distance between the soil and the chlorophyll-containing organs are large; hence, a proper transportation system is required.
2. The two main components of the plant vascular system are xylem and phloem.
3. The four factors that influence transpiration are sunlight, wind, temperature and humidity.
4. Water is absorbed by the root hairs and undergoes cell-to-cell movement by osmosis. The water then travels through the plant to reach the leaves, known as the ascent of sap.
5. Root pressure is a positive pressure created by the movement of water from the soil to the roots, for the upward movement of water which is more important for water transportation at night.
6. Transport in plants takes place in many ways like diffusion, osmosis and facultative diffusion through transporters like uniport, symport and antiport.
7. The flower is the main reproductive organ of plants.
8. The flower consists of four parts: sepals, petals, stamen and pistil.
9. Pollination is the transport of pollen grains from the anther to the stigma of a flower.
10. Germination of pollen grains takes place when the stigma receives pollen of the same species and allows it to disintegrate. Two sperm nuclei are released when the pollen tube enters one of the synergids. One of these sperm nuclei fuses with the egg nucleus to form the zygote while the other sperm moves towards the two polar nuclei in the central cell and fuses with it.

Short Answer Questions

1. Reproduction can be asexual or sexual. Asexual reproduction is one where new organisms are produced from a single parent without the involvement of sex cells. It is of different types: binary fission, budding, spore formation, fragmentation and vegetative propagation (cutting, layering and grafting). Sexual reproduction necessitates the joining of male and female gametes. It involves two parents: male and female. A male gamete fuses with a female gamete to form a new cell called the zygote. The zygote then grows and develops into a new organism.
2. Flower is the reproductive part of the plant which has four components. The petals and carpels are necessary for pollination which is the transport of pollen grains from the anther to the stigma of a flower. The other two components, stamen, and pistil, are necessary for sexual reproduction involving fertilisation of the male and female gamete.
3. The pollen grains and nuclei of the ovules are brought together by a variety of forces, including wind, water and insects. Pollination is the transport of pollen grains from the anther to the stigma of a flower. This can happen in three ways: autogamy, where pollen grains of a flower fall on the stigma of the same flower; geitonogamy, where the pollen grains of a flower of a plant fall on the stigma of a different flower of the same plant; and allogamy, where the pollen grains of a flower of a plant fall on the stigma of a flower of

another plant of the same species. When pollination occurs through insects, wind or other agents, it is known as cross-pollination. When the pollen from a flower falls on its own stigma and fertilises the ovule, this is known as self-pollination.

4. (a) In active transport, molecules move against the concentration gradient. The energy provided as ATP is used to drive the pump. ATP gives a phosphate which then pumps the molecule across the membrane.
   (b) Diffusion is a transportation process that involves the movement of a substance from cell to cell. It does not require the expenditure of energy. Molecules move in a random manner. It is a slow process.
   (c) Xylem is a specialised vascular tissue that transports water and nutrients from the soil to the leaves. The flow of water is unidirectional and is in the upward direction. It is responsible for replacing water lost through transpiration and photosynthesis.
   (d) The transport of food/glucose and amino acids , as well as other substances from the leaves to the different parts of the plant body, is called translocation. It takes place in the phloem and delivers substances to the storage organs such as roots and fruits and to the growth organs. Phloem is a complex tissue that is composed of four types of cells (sieve tubes, companion cells, phloem fibre and phloem parenchyma).
5. Pollination begins when the pollen grains of a flower of a species land on the stigma of a flower of the same species to form a pollen tube that connects the ovary and stigma. When the pollen tube is complete, the pollen grain transmits sperm cells to the ovary. Fertilisation takes place when the sperm cells reach the ovary and egg cells.
6. Xylem: The xylem is used for the transportation of water and minerals. It is composed of tracheids, xylem vessels, xylem parenchyma and xylem fibre. Tracheids and xylem vessels are the conducting elements. The xylem makes a continuous tube in plants that runs from the roots to the stem, right up to the veins of leaves. It carries water and minerals from the roots to the leaves of the plant.

   Phloem: Phloem is responsible for the transportation of food. Phloem consists of sieve tubes, companion cells, phloem parenchyma and bast fibres. Sieve tubes are the conducting elements in phloem. Phloem carries the product of photosynthesis from the leaves to other parts of the plant.
7. At the root, the cells that are in contact with the soil actively take up ions, causing a concentration difference between the roots and the soil. Water moves into the root to eliminate the difference. The flow of water is unidirectional and is in the upward direction. The absorbed water at the root hair undergoes cell-to-cell movement by osmosis until it reaches the xylem, resulting in a continuous movement of water into the root xylem, creating a column of water that is steadily pushed upwards. The continuous supply of water to the xylem of the leaf replaces the water lost through the stomata. Xylem is responsible for replacing water lost through transpiration and photosynthesis. The water then travels through the plant to reach the leaves. This is called the ascent of sap.
8. Vegetative propagation is a type of asexual reproduction in plants. New plants are produced by vegetative parts such as leaf, root and stem. Ginger, potato and mint reproduce by this method. Humans have adapted natural vegetative propagation methods to artificially propagate a large number of plants:

   Cutting: A branch is cut and planted in the ground. A new plant emerges as a result of the formation of roots; Layering: A branch is placed in the ground, with a section of it buried in the soil. The plant's shoot will be utilised as stock is chopped off. The stem is sliced vertically, and the scion is placed and taped into the stock. This component takes hold and a new plant sprouts; Grafting: The act of joining two plants where the upper part of the graft (known as scion) becomes the top of plant whereas the lower portion (known as understock) becomes the trunk or the root system. Several plants could also be combined to form a graft.
9. Transpiration is the evaporation of water from the stomata of leaves, creating a suction pull. This pulls the water up from the xylem cells of the roots. Transpiration helps in absorption and upward movement of water and minerals or the ascent of sap. Transpiration also helps in temperature regulation. During the day, water is lost from the surface of leaves by transpiration and this results in more water molecules being pulled up due to their tendency to cohesion. This pulling force created by leaves helps in the upward conduction of water in tall trees.
10. *Spirogyra* and similar organisms are made of long, ribbon-like filaments that reproduce by fragmentation. The division and growth of filaments into two or equal parts take place and each part grows into an individual.

Long Answer Questions

1. Hint: Explain the structure of the flower with a well-labelled diagram.
2. Hint: Explain pollination, its process, significance and types and its relation to fertilisation.
3. Hint: Explain the two types of reproduction and their processes.
4. Hint: Explain transpiration and its role in the ascent of sap, along with a well-labelled diagram.
5. Hint: Describe the types of transport in plants with a diagram.
6. Hint: Write about the role of roots, leaves, xylem, phloem and stomata in the transportation of water and food.
7. Hint: Explain tissue culture, its steps and significance with the help of a diagram.
8. Hint: Explain vegetative propagation and its types.
9. Hint: Explain the process of fertilisation and double fertilisation.
10. Hint: Explain in detail active and passive transport with a diagram.

Case Study-Based Questions

1. Xylem is used for transportation of water and minerals. It is composed of trachieds, xylem vessels, xylem parenchyma and xylem fibre. Trachieds and xylem vessels are the conducting elements. The xylem makes a continuous tube in plants which runs from the roots to the stem and up to the veins of leaves. It carries water and minerals from the roots to leaves of the plant.
2. Importance of vegetative propagation: Plants can bear flowers and fruits faster. Plants which have lost the ability to produce viable and healthy seeds can reproduce by vegetative propagation. All plants are genetically similar to the parent plant. Seedless varieties like banana can be obtained by this method.
3. A flower has four parts: petal, sepal, stamen and carpel. The stamen is the male reproductive part and its produces pollen grains that contain the male gametes. Each stamen has two parts: filament and anther. The carpel is the female reproductive part and it produces ovules that contain female gametes. It has three parts: stigma, style and ovary.
4. When a person transfers pollen to the stigma, it is known as artificial pollination. In the Babylonia civilisation, sprinkling of male palm flowers on the female flowers was a common practice, though they had no idea of pollination at that time. In modern times, artificial pollination is a standard practice adopted by breeders of plants or scientists in their effort to evolve new varieties.

Competition Window

1. (a)	**2.** (c)	**3.** (a)	**4.** (c)	**5.** (a)
6. (a)	**7.** (b)	**8.** (b)	**9.** (c)	**10.** (c)

CHAPTER 9

Crop Production and Management

PREREQUISITES

You must have knowledge of:

- Crops
- Plants and food
- Nature and the ecosystem

LEARNING OBJECTIVES

At the end of this chapter, you will be able to:

- Define agriculture
- Describe the seasons for growing different crops
- Describe the process of preparation of soil for growing crops
- Clarify the use of fertilisers
- Explain the importance of irrigation and the different types of irrigation
- Elucidate the importance of crop rotation
- Give details on how to protect crops from weeds
- List the different foods we obtain from animals

INTRODUCTION

Food is the most vital source of energy for all organisms. We know that among living beings, plants prepare their own food while all other animals, including human beings, depend on plants for food. Agriculture began about 12,000 years ago. At that time, humans were hunter-gatherers. They ate raw fruits and vegetables and hunted animals. They then learnt how to grow crops and this was the beginning of agriculture. They settled in farming communities and began domesticating animals for food.

AGRICULTURE

The word agriculture comes from the Latin word '*ager*' meaning 'field' and '*cultura*' meaning 'care'. Agriculture refers to the growing of plants or raising of animals for human use. Quality of agriculture is mostly dependent on the composition of those activities that are essential for improving the environment for the growth or rearing of plants and animals, thereby improving their usefulness to humans. The practice or business of agriculture is called **farming** and the place where the work is carried out is called a **farm**.

Plants of the same type grown and cultivated at one place in large quantities in a field are known as **crops**. For example, a crop of rice indicates that all the plants grown in that field will be paddy plants. The products obtained from cultivated crops is known as **crop produce**. Based on the type of products obtained, crops can be classified as cereals, pulses, oil crops and fibre crops.

The word **horticulture** is derived from the Latin word '*hortus*' meaning 'garden' and '*cultura*' which means 'care'. Thus, horticulture is the branch of science that deals with growing and producing vegetables, fruits, flowers and ornamental plants.

CROP SEASONS IN INDIA

In India, there are two main crop varieties, depending on the season. They are:

- **Kharif crops:** These are sown in the rainy season, starting in the month of June, and harvested around

October. For example, paddy, pigeon pea, maize (Fig. 9.1), cotton, soybean and groundnut.

Fig. 9.1 Maize

(Alex Lomas; 2007; https://commons.wikimedia.org/wiki/File:Zea_mays_(2943636425).jpg. Licensed under: CC-BY-2.0)

- **Rabi crops:** These are sown in winter, starting in the month of October, and harvested around April. For example, wheat, gram, pea (Fig. 9.2), mustard, linseed, oilseed and barley.

Fig. 9.2 Pea

(Photo Garden Pea; Zenon Sych; 2010; https://commons.wikimedia.org/wiki/File:Pisum_sativum_%D0%93%D0%BE%D1%80%D0%BE%D1%85_%D0%BB%D1%83%D1%89%D0%B8%D0%BB%D1%8C% D0%BD%D0%B8%D0%B9_%D1%83_% D1%82%D0%B5%D1%85%D0%BD%D1%96% D1%87%D0%BD%D1%96%D0%B9_%D1%81%D1% 82%D0%B8%D0%B3%D0%BB%D0%BE%D1%81% D1%82%D1%96.jpg. Licensed under: CC-BY-SA-4.0)

AGRICULTURE PRACTICES

The steps involved in agriculture, which are followed by farmers for good crop yield, are known as agriculture practices. These steps (Fig. 9.3) are as follows:

1. Soil preparation
2. Selection and sowing of seeds
3. Manuring or adding fertilisers
4. Irrigation
5. Weeding
6. Harvesting
7. Threshing and winnowing
8. Storage

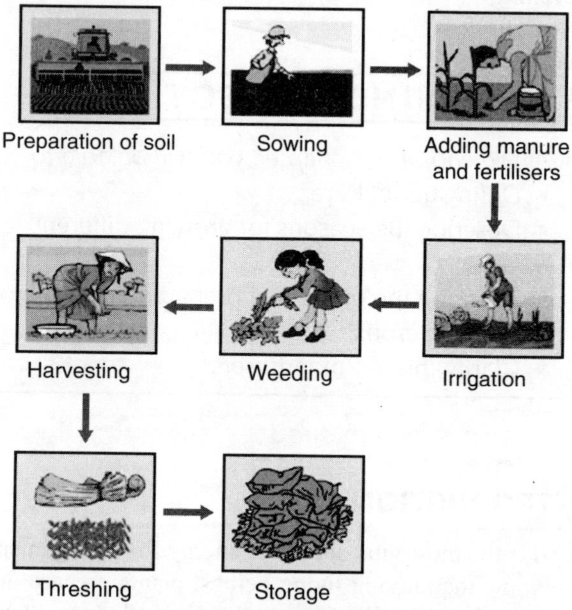

Fig. 9.3 Agriculture practices

Let us discuss each of these steps in detail.

Soil Preparation

First, the soil needs to be prepared and made ready for cultivation. This step involves basic tasks such as ploughing or tilling and levelling.

Ploughing

The loosening and turning of a few inches of topsoil is known as ploughing or tilling. If the soil is dry, it may need to be watered before ploughing. Ploughing is important because:

- It helps in loosening the soil so that it can hold more water and air. Air helps the roots of crops to breathe, even though they penetrate deep into the soil.

- The loose soil also helps in the growth of earthworms and microbes, which make the soil rich in nutrients. They are known as 'friends of the farmer' as they help in further turning and loosening the soil. They also add humus to the soil.
- It is easy to mix fertilisers and manure in loose soil.
- Weeds can be easily removed by uprooting them.
- Loose soil facilitates the growth of roots deep into the soil for firm binding.
- Soil contains minerals, water, air and some living organisms. Dead plants and animals are decomposed by soil organisms and the nutrients from them are again absorbed by plants.

Levelling

Even after ploughing, big lumps of soil (called crumbs) may remain in the field. These are crushed using wooden or iron planks called levellers, and this process is called levelling. A leveller is driven by a machine or an animal. Levelling is important because:

- It helps in preventing loss of moisture from the ploughed soil.
- It enables the soil to settle so that it is not eroded by wind or water.
- It brings uniformity to the field so that water and manure can be distributed evenly to all the parts of the field.

Agricultural Implements

The main implements used for ploughing and levelling are the plough, hoe and cultivator.

- **Plough:** This tool has been used since ancient times for tilling (or ploughing) the soil, adding fertilisers and removing weeds. It comprises a wedge-shaped block of wood or iron, which is driven by oxen, horses, camels or even tractors (Fig. 9.4). It has a strong triangular iron strip known as the ploughshare that penetrates the soil to make furrows. The ploughshare is attached to a long log of wood, which is known as the ploughshaft. This has a handle at one end. The other end of the ploughshaft is attached to a beam which is placed on the animal's neck while ploughing.

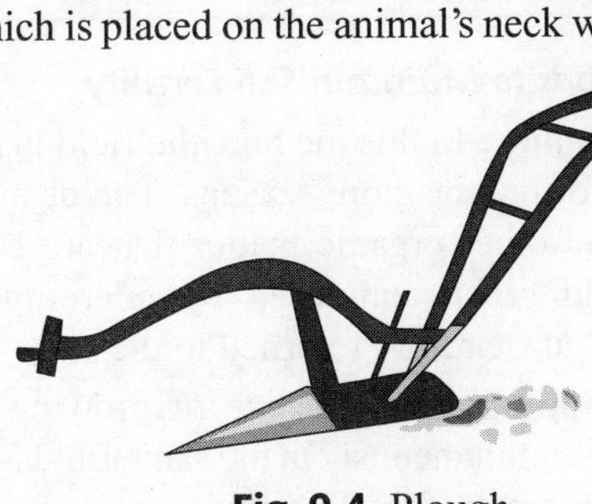

Fig. 9.4 Plough

- **Hoe:** This tool is used for removing weeds and for loosening the soil. It consists of a long rod of wood or iron. There is a handle at one end of the hoe. A strong, broad and bent plate of iron is fixed below the handle and it works like a blade (Fig. 9.5). The other end is attached to a beam which is placed on the neck of the animal. Thus, the hoe is also pulled by animals.

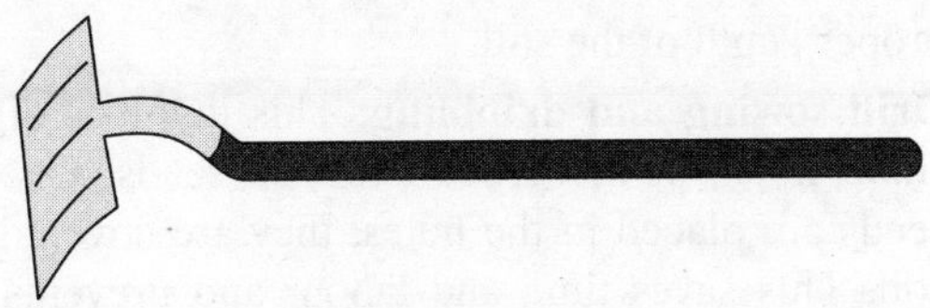

Fig. 9.5 Hoe

- **Cultivator:** This tool is used for loosening and turning the soil in the fields quickly. The cultivator is a machine designed to stir the soil as it matures around the crop to destroy weeds and promote growth; it saves labour and time.

Selection and Sowing of Seeds

Farmers today are aware of the fact that good quality of seeds is a must for a good crop. The farmers must, therefore, select seeds that are

- Healthy
- Of good quality
- Dried properly and which have no dampness that may encourage the growth of fungi
- Not mixed with the seeds of weeds
- Free from any infection
- Stored in sturdy bags
- Free from insects and pests
- Not contaminated with waste, rotting food materials, excreta of animals, etc.

The process of spreading seeds in the soil to germinate is known as **sowing**. For sowing seeds, the following points must be kept in mind:

- Proper spacing of the seeds must be ensured, because overcrowding may result in shortage of nutrients, air, sunlight and water and can kill the seedlings.
- The seeds must be sown in regular rows so that watering or transplanting them is easy. Cotton, potato, maize and soybean are all row crops.
- The seeds must be sown at reasonable depth. If they are sown too deep, they may suffocate without air and if they are planted too close to the surface, they may be eaten by birds.
- Seeds must be sown in the appropriate season.

- The soil must be prepared well so that it has adequate moisture content for the germination of seeds.

Methods of Sowing Seeds

Seeds can be sown using the following methods:

- **Broadcasting:** In this method, seeds are sown manually. They are scattered in the field with no uniformity of rows, no proper spacing and not at the proper depth of the soil.
- **Drill sowing and dribbling:** This involves digging small holes in the ground for the seeds. Once the seeds are placed in the holes, they are covered with soil. This saves time and labour and prevents seed damage by birds.
- **Seed drill:** Another method of sowing seeds is with the help of a simple device consisting of a bamboo tube with a funnel on it attached to a plough. As the plough moves over the field, the tube attached to it plants the seeds kept in the funnel at equal distance and depth. The plough keeps making furrows in the soil in which the seeds are dropped by the seed drill. This ensures that the seeds are also covered by soil after sowing. Traditionally, bullock-driven seeds drills were used; they had only one long tube with a funnel. The tractor-driven seed drill is a modern form of this method (Fig. 9.6). This has 5–6 iron tubes joined together with a common funnel at the top.

Fig. 9.6 Seed drill

(John Deere tractor with Kuhn seed drill…; werktuigendagen; 2005; https://commons.wikimedia.org/wiki/File:Kuhn_seed_drill_at_Werktuigendagen_2005.jpg. Licensed under: CC-BY-SA-2.0)

- **Transplanting:** In some crops like rice, tomato, onion, chilli and brinjal, the seeds are not sown directly in the fields. These seeds are first sown in seed beds called nurseries. After the seeds have grown into small plantlets called seedlings, only healthy and well developed seedlings are transferred and transplanted to the field. This process of transferring seedlings from the nursery to the main field by hand is called transplanting.

Manuring and Adding Fertilisers

Manuring is the process of adding manure to the soil for better yield of crops. As the cultivation of crops is continuous, the nutrients naturally available in the soil get used up. This results in low yield of crops. Therefore, it is important that the nutrients of the soil be replenished. This is done by natural methods or by adding organic manure as fertiliser to the soil. Manure refers to organic substances obtained from dead plant and animal waste. A chemical fertiliser is a salt containing the necessary plant nutrients in concentrated form to provide quick replenishment of nutrients and restore soil fertility. Some examples of fertilisers are urea, ammonium sulphate, super phosphate, potash and NPK (nitrogen, phosphorus, potassium). The excessive use of fertilisers can result in soil and water pollution and affect the fertility of the soil. Therefore, the use of fertilisers should be minimised by adopting natural methods to increase the fertility of the soil.

The differences between manure and fertiliser are listed in Table 9.1.

Table 9.1 Differences between manure and fertiliser

Manure	Fertiliser
Manure is made up of natural substances.	Fertiliser is a synthetic inorganic salt.
Manure is obtained from plant, animal or human waste by the action of microorganisms.	Fertiliser is prepared in factories from chemical salts.
Manure is not very rich in plant nutrients. It is not nutrient-specific.	Fertiliser is rich in plant nutrients such as nitrogen, potassium and phosphorus. It is nutrient specific.
Manure is cheap and easily available.	Fertiliser is costly.
Manure adds humus to the soil.	Fertiliser does not add humus to the soil.

Natural Methods to Maintain Soil Fertility

- **Field left fallow:** In this method, the field is left free or fallow for one or more seasons. The dead plants, animals and other organic matter that are collected in the field are decomposed by microorganisms. Therefore, nutrients are returned to the soil.

  Mixed cropping: The practice of growing two or more crops simultaneously in the same field is known as mixed cropping (Fig. 9.7). It is an age-old practice in India. Crops are chosen in such a way that the products and waste materials from one crop help in the growth of the other crop. For example, if a cereal crop such as wheat is grown along with a leguminous crop such as pulse (example, gram), the uptake of nitrogen from the soil by the wheat is compensated for by the addition of nitrogen to the soil by nitrogen-fixing bacteria like *Rhizobium* present in the pulse.

Nitrogen fixation: This is the initial stage of the nitrogen cycle. It is an essential biological process where bacterial species like *Azotobacter* and *Rhizobium* along with natural phenomena like lightning convert atmospheric nitrogen into ammonia. Plants acquire nitrogen from the soil and use it in protein synthesis. However, the nitrogen gas present in the atmosphere cannot be directly utilised by plants in the way that they use carbon dioxide or oxygen. This makes the nitrogen cycle very important.

Certain prokaryotes such as *Azotobacter*, *Rhodospirillum*, *Cyanobacteria* and *Rhizobium* that live in the root nodules of legumes have a symbiotic relationship with plants, where the bacteria fix nitrogen and the plants provide them with food. These bacteria can convert atmospheric nitrogen to ammonia; this is known as biological nitrogen fixation and the enzyme nitrogenase is used to convert dinitrogen to ammonia. This form of nitrogen dissolves in the soil and plants readily absorb it from the soil; this is why crop rotation or mixed cropping is practised by farmers. Lightning is a natural phenomenon that breaks nitrogen and converts the non-absorbable form to one that can be used by plants.

Fig. 9.7 Mixed cropping with barley and wheat
(Mixed cropping of barley (*Hordeum vulgare*) and wheat (*Triticum aestivum*)…; David Wright; 2009; https://commons.wikimedia.org/wiki/File:Mixed_cropping_of_barley_(Hordeum_vulgare)_and_wheat_(Triticum_aestivum_-_geograph.org.uk_-_1386935.jpg. Licensed under: CC-BY-SA-2.0)

- **Crop rotation:** Wheat crop is planted in November and harvested in March and April. Rice crop is planted in June/July and harvested in October/November. Between these two seasons, the land is left empty. Instead, farmers can plant a pulse crop in this period. So, by the time the farmer needs to plant the cereal crop, the pulse is ready to be harvested. The process of growing a different crop between two similar crops is called crop rotation (Fig. 9.8). This method maintains soil fertility, prevents crop diseases and pests, and ultimately gives better crop yield.

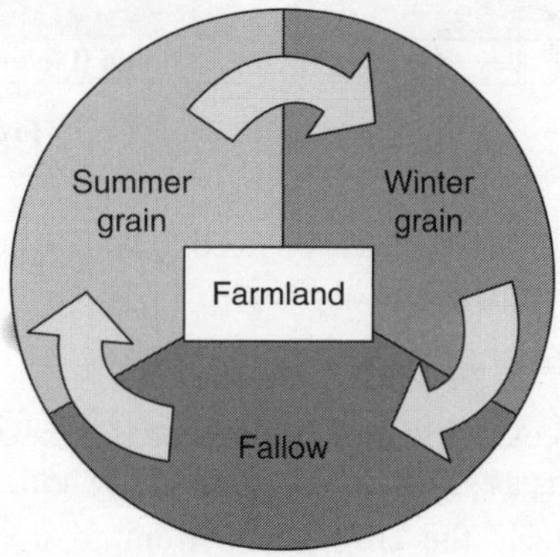

Fig. 9.8 Crop rotation

However, these natural methods alone may not be enough to maintain the fertility of the soil and farmers may sometimes need to add manure and fertilisers.

Advantages of Manure

Organic manure is considered better than fertilisers because:

- It enhances the water holding capacity of the soil.
- It makes the soil porous, due to which exchange of gases becomes easy.
- It increases the number of useful microbes.
- It improves the texture of the soil.

Irrigation

Irrigation is the supply of water to agricultural fields at different intervals. Sufficient supply of rainwater is not always available; therefore, farmers need to supply water to the fields using irrigation. Some crops like paddy require more water, and some like wheat require less water. Sometimes, the water requirement depends on the type of soil used for growing crops. Sandy soil requires more water than clay soil. Loamy soil requires only moderate water supply due to its texture. The crops grown in summer need more water due to excess evaporation in comparison to the crops grown in winter. So, the type of

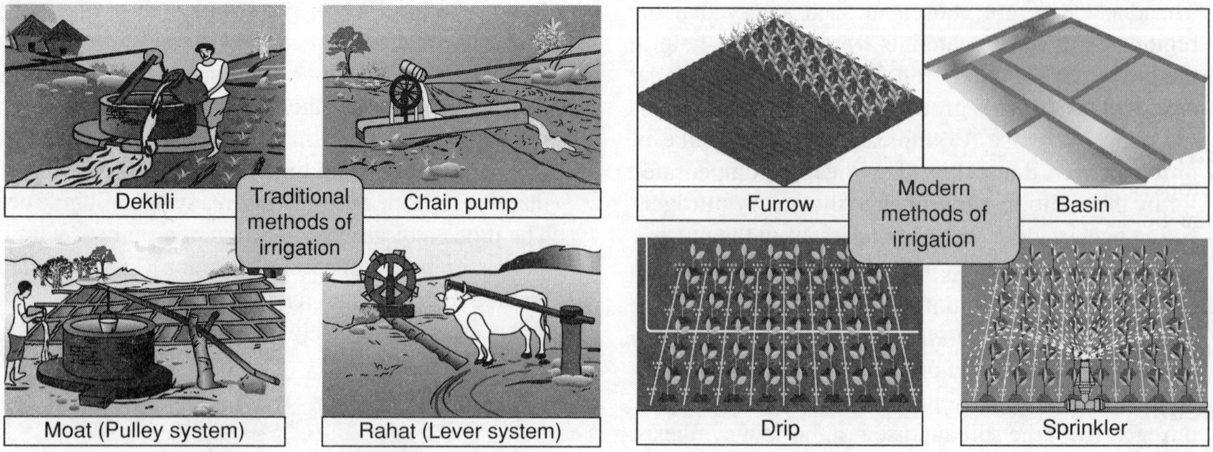

Fig. 9.9 Methods of irrigation

crop, type of soil and season need to be considered while planning for irrigation of fields.

Importance of Irrigation

- It is necessary to provide moisture for the germination of seeds as seeds do not grow in dry soil.
- It protects the crop plants from frost and hot air currents.
- It helps in the absorption of nutrients by the plants from the soil.

Sources of Irrigation

Each crop needs a specific quantity of water at various stages of growth and fruiting. The source of the irrigation can be wells, tube wells, ponds, lakes, rivers, dams and canals.

Methods of Irrigation

This can be classified as traditional and modern (Fig. 9.9).

- **Traditional:** The various traditional methods of drawing water out from wells, lakes and canals were moat (pulley system), chain pump, dhekli and rahat (lever system). Cattle or human labour is used in these methods. These methods are cheaper but less efficient. Pumps are commonly used for lifting water. These pumps are run by diesel, biogas, electricity and solar energy.
- **Modern:** It is important to use water economically. Some modern methods of irrigation have been developed for this purpose:
- **Sprinkler system:** This method is useful in sandy soil and uneven land, where the soil cannot retain water for long or where sufficient water is not available. Rotating nozzles are attached to perpendicular pipes, which are attached to the main pipeline at regular intervals (Fig. 9.10). This system is used for lawns, coffee plantations and several crops.

Fig. 9.10 Irrigation through sprinklers
(Farm field view at ICAR-DGR; Abhay iari; 2004; https://commons.wikimedia.org/wiki/File:Irrigation_through_sprinkler.jpg. Licensed under: CC-BY-SA-4.0)

- **Drip system:** This system involves providing water drop by drop at the roots of the plants (Fig. 9.11).

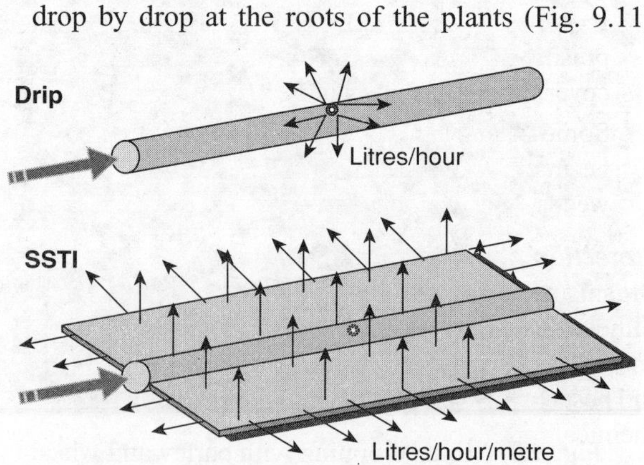

Fig. 9.11 Drip and SSTI (subsurface textile irrigation) system of irrigation

Thus, water is not wasted. This system is practised in regions where water availability is low. However, this method is very expensive. It is the best technique for watering fruit plants, gardens and trees.

Waterlogging

In crops such as wheat and maize, excess water supply to the fields results in a condition called waterlogging. Waterlogging reduces the air in the soil, thereby damaging the roots. Even soil organisms are killed if waterlogging persists for a long time. Waterlogging increases the amount of salts in the soil and damages soil fertility.

Weeding and Crop Protection

The unwanted plants that grow on their own along with the cultivated plants are known as weeds. They are a threat to commercial crops as they compete for sunlight, air, nutrients and water. Some weeds interfere even in harvesting and may be poisonous to animals and humans. In comparison to cultivated crops, the seeds of weeds germinate easily and flower early. They produce a large number of seeds. They also spread diseases to cultivated plants as they become hosts to many insects and microorganisms. The process of removing weeds from the fields is known as weeding. The best time for weeding is before the weeds produce flowers and seeds. The following methods are used for getting rid of weeds:

- Weeding is carried out by pulling out the weeds manually or by using a hoe, harrow or seed drill.
- The chemicals used to kill the weeds are known as weedicides. A common weedicide is 2,4-D (2,4-dichlorophenoxyacetic acid). This chemical kills only the weeds and not the cultivated crops. It prevents seed formation. Since weedicides are poisonous, while spraying them, farmers should cover their nose and mouth with a piece of cloth.
- Some insects that do not cause harm to the crops can be made to breed and grow. They then destroy the weeds, leaving the crops safe.

Protection from Pests and Diseases

Organisms that cause diseases in plants are called pathogens while those that damage or destroy crops are known as pests. Crops can be protected from pathogens and pests by spraying chemicals at regular intervals. Such chemicals are known as pesticides. Some commonly used pesticides are DDT, BHC and malathion. They are either sprayed by hand-operated machines or by low flying aircraft, if the area to be covered is large.

Though pesticides are very effective, care must be taken while using them. Excessive use of pesticides is very harmful to human beings as they irritate the skin and cause gastric and respiratory disorders.

Harvesting, Threshing, Winnowing

The process of cutting and collecting the mature crops from the fields is known as harvesting. Almost all crops complete their cycle of maturity in 3–4 months. The mature crops are cut manually using a sickle. Harvesting is carried out using a machine known as the harvester.

Once the crop is harvested, the grains are separated from the chaff by a process known as threshing. Threshing can be carried out manually by beating the crops against a solid surface or by using animals to stamp on the crops. The harvested crop is spread on the ground and bullocks or camels are made to walk over them again and again till the grains separate from the chaff. Nowadays, tractors are used for this job. The leaves and stems of crops are crushed into smaller pieces. The grains are then separated from the chaff, dried leaves and stems of the crops in a machine or by winnowing. Harvesting and threshing can be done simultaneously with the help of a machine called the combine.

After threshing, some grain may still be mixed with the chaff and hay and needs to be cleaned before it can be used. The process of separating grain from chaff and hay with the help of wind is called winnowing.

Storage

The harvested grains have high moisture content, so the seed grains are left in the field to dry in the sun before they are stored. Otherwise, the moisture may spoil the grains by promoting the growth of fungi and moulds. The grains must be stored properly in jute bags or metallic bins. Large-scale storage of grains is done in silos and granaries to protect them from pests like rats and insects. Care must be taken to ensure that the grains are kept:

- Away from moisture
- Safe from rain
- Protected from pests and birds
- In air-tight godowns
- In proper bags for the purpose of transportation

Dried neem leaves are used for storing food grains at home. For example, when wheat is stored in iron drums, some dried neem leaves are placed in it. For storing large quantities in big godowns, specific chemical treatments are required to protect the grains from pests and microorganisms.

CROP IMPROVEMENT

The improvements in various agricultural practices will be of no use if the quality of the grains is poor. Scientists are now working on producing new varieties of crops with desirable features such as

- Crops with more yield
- Crops with better quality of grains
- Crops which are disease resistant

This has been carried out using **cross-breeding**. Two plants with different desired features are taken as parent plants. Cross-breeding between them results in a new generation of plants with the desired features. With increasing population, the land available for agriculture is decreasing; therefore, there is a need for high yielding crops. Scientists are now trying to develop new varieties of crops that are more resistant to climatic changes such as drought, heat, cold and winds and also to various soil conditions. Some of the **high yielding varieties** of crops are:

- Wheat: Sonalika, Sonora-64, Kalyan sona
- Rice: Jaya, Padma, IR8, Pusa-205
- Maize: Ganga, Ranjit, Deccan

Various approaches have also been used to improve crop variety and these methods are known as **plant breeding methods**. These techniques include:

- **Introduction:** The transfer of plants from their place of cultivation to a place they have never been grown is known as plant introduction. The plants adapt to their new environment through a process called acclimatization.
- **Hybridisation:** This is one of the ways of incorporating desirable characters into crop varieties. The process of crossing between two genetically dissimilar plants is known as hybridisation.
- **Genetic engineering:** The introduction of genes carrying the desired characteristics in the crop is known as genetic engineering. These are called genetically modified (GM) crops.
- **Mutation breeding:** Changes may be induced by chemical agents, known as mutagens, or physical agents such gamma radiation and selective hybridisation.
- **Selection:** The offspring of crops with the most desirable traits are selected for controlled propagation.
- **Polyploidy:** Increasing the number of chromosomes in an individual is known as polyploidy.

FOOD FROM ANIMALS

As with plants, we obtain many food items from some animals too. They provide us with edible products like milk, cheese, curd, meat, honey and eggs. So, in order to obtain quality products, we need to provide animals with proper food, care and shelter. The science that deals with breeding, feeding and caring for domestic animals is called **animal husbandry**.

The rearing of fish is called **pisciculture**. **Fishery** is the farming of fish (Fig. 9.12). The fish found in marine waters are called sea fish or marine fish (examples are salmon and hilsa) while the fish found in freshwater are called freshwater fish (examples, rohu, catla).

Fig. 9.12 Fish farming
(Fish farming in Iran; M Hossein Movahedinejad; 2019; https://commons.wikimedia.org/wiki/File:Fish_farming_Iran6.jpg. Licensed under: CC-BY-4.0)

The rearing of honeybees is called **apiculture**. This is carried out on a large scale in farms. Honeybees produce honey by sucking the nectar of flowers, which is rich in minerals and sugar.

SUMMARY

- Food is the most vital source of energy for all organisms.
- We know that among living beings, plants prepare their own food while all other animals, including human beings, depend on plants for food.
- The word agriculture comes from the Latin word '*ager*' meaning 'field' and '*cultura*' meaning 'care'.
- Agriculture is the growing of plants or raising of animals for human use. Quality in agriculture comprises those activities that are essential for improving the environment for the growth of plants and animals, thereby improving their usefulness to humans.
- The practice or business of agriculture is called farming and the place where the work is carried out is called a farm.
- There are two main crop varieties depending on the seasons. Kharif crops are sown in the rainy season from the month of June and harvested around October. For example, paddy, pigeon pea, maize, cotton, soybean and groundnut. Rabi crops are sown in winter from October and harvested around April. For example, wheat, gram, pea, mustard, linseed, oilseed and barley.
- The steps involved in agriculture which are followed by farmers for good crop yield are known as agriculture practices. The various agriculture practices are soil preparation, selection and sowing of seeds, manuring or adding fertiliser, irrigation, weeding, harvesting, threshing, winnowing and storage.
- Cross-breeding between two plants with desired characteristics results in a new plant with the desired features.
- Plant breeding methods are used for crop improvement like introduction, hybridisation, genetic engineering, mutation breeding, selection and polyploidy.
- The science that deals with breeding, feeding and caring for domestic animals is called animal husbandry.
- The rearing of fish is called pisciculture.
- The rearing of honeybees is called apiculture.

PRACTICE QUESTIONS

Analyse Your Concepts (School Exam Based)

Multiple Choice Questions

1. The crops that are sown in the rainy season are called:
(a) Kharif crops (b) Rabi crops
(c) Cash crops (d) None of these

2. Which of the following is not a rabi crop?
(a) Wheat (b) Gram
(c) Cotton (d) Mustard

3. The process of loosening and turning of the soil is called:
(a) Ploughing (b) Levelling
(c) Manuring (d) None of these

4. Animals kept on the farm are called:
(a) Livestock (b) Pets
(c) Farm animals (d) None of these

5. Cultivating crops and rearing animals in individual farms is called:
(a) Farming (b) Agriculture
(c) Mixed farming (d) None of these

6. A farmer performs some tasks for the cultivation of a crop, called:
(a) Sowing (b) Preparation of soil
(c) Agricultural practices (d) None of these

7. ________ is used for removing weeds and for loosening the soil.
(a) Hoe (b) Plough
(c) Cultivator (d) Tiller

8. ________ are organic substances obtained from dead plant and animal waste.
(a) Fertilisers (b) Pesticides
(c) Manures (d) Insecticides

9. _______ is the practice of growing two or more crops simultaneously in the same field.
(a) Skip cropping (b) Mixed cropping
(c) Endemic cropping (d) Field left fallow

10. ______ provides water drop by drop at the roots of the plants:
(a) Sprinkler system (b) Traditional
(c) Waterlogging (d) Drip system

11. Increasing the number of chromosomes in an individual is known as:
(a) Haploploidy (b) Diploploidy
(c) Polyploidy (d) Monoploidy

Match the Following

Each question contains statements given in two columns which have to be matched. Statements in column I have to be matched with statements in column II.

1. Choose the correct option:

Column I	Column II
i. Crop rotation	a. Seed sowing
ii. Broadcasting	b. Soil fertility
iii. Cultivator	c. Irrigation
iv. Drip	d. Agriculture implements

2. Choose the correct option:

Column I	Column II
i. Weedicide	a. Rainy season
ii. Apiculture	b. *Rhizobium*
iii. Nitrogen fixation	c. 2,4-D
iv. Kharif	d. Bees

3. Choose the best match:

Column I	Column II
i. Wheat	a. DDT
ii. Kharif	b. Rabi
iii. Pesticide	c. Soybean

True or False

Read the following statements and write your answer as true or false.

1. Earthworm is a friend of the farmer.
2. Weedicides make weeds grow well.
3. Canals, wells and rains are the sources of water for irrigation.
4. Crop rotation helps to increase the fertility of soil.
5. Crop can be damaged by excess supply of water.
6. Cross-breeding results in a new generation of plants with the non-desired features.
7. Sprinkler system is useful in sandy soil and uneven land, where the soil can retain water for a long time.
8. The enzyme nitrogenase is used to convert dinitrogen to ammonia.
9. The loosening and turning of a few inches of topsoil is known as levelling.
10. The practice or business of agriculture is called farming.

Fill in the Blanks

Complete the following statements with an appropriate word/term to be filled in the blank spaces.

1. The same kind of plants grown and cultivated on a large scale at a place is called __________.
2. The first step before growing crops is __________ of the soil.
3. Damaged seeds would __________ on top of water.
4. For growing a crop, sufficient sunlight and __________ and __________ from the soil are essential.
5. The steps involved in agriculture, which are followed by farmers for good crop yield, are known as __________.
6. Big lumps of soil are crushed using wooden or iron planks called __________.
7. __________ is used for loosening and turning the soil in fields quickly.
8. In __________ method, seeds are sown manually.
9. Bacteria can convert atmospheric nitrogen to ammonia; this is known as __________.
10. The science that deals with breeding, feeding and caring for domestic animals is called __________.

Assertion–Reason Questions

The following questions consist of two statements: Assertion (A) and Reason (R). Answer these questions selecting the appropriate option given below:

(a) Both A and R are true and R is the correct explanation of A
(b) Both A and R are true but R is not the correct explanation of A
(c) A is true but R is false
(d) A is false but R is true

1. Assertion (A): First, the soil needs to be prepared and made ready for cultivation.
Reason (R): This step involves basic tasks such as ploughing or tilling and levelling.
2. Assertion (A): Proper spacing of the seeds must be ensured.
Reason (R): Overcrowding may result in a shortage of nutrients, air, sunlight and water and can kill the seedlings.
3. Assertion (A): Bacteria can convert atmospheric nitrogen to ammonia using enzymes.
Reason (R): The enzyme nitrocellulase is used to convert dinitrogen to ammonia.
4. Assertion (A): Sandy soil requires more water than clay soil.
Reason (R): Loamy soil requires excess water supply due to its texture.
5. Assertion (A): Waterlogging reduces the air in the soil, thereby damaging the roots.
Reason (R): Waterlogging decreases the amount of salts in the soil and damages soil fertility.

One-word Answers

1. Which crops are sown in the rainy season?
2. The ploughshare is attached to a long log of wood, known as?
3. What is the method of sowing seeds manually?
4. In transplanting, seeds are first sown in seed beds known as?
5. Which enzyme is used to convert dinitrogen to ammonia?
6. What protects the crop plants from frost and hot air currents?
7. In mutation breeding, changes may be induced by which agent?

Define the Term

1. Rabi crop
2. Levelling
3. Cultivator
4. Seed drill
5. Transplanting

Groups of Terms

Given below are groups of terms. In each group the first pair indicates the relationship between the two terms. Rewrite and complete the second pair on a similar basis.

1. Bees: apiculture :: Fish : _____
2. Wheat: rabi :: Soybean:_____
3. Manual: plough :: Machine :_____
4. Insect: insecticide :: Weed : _____
5. Crop: farming :: Livestock : ____

Identify the Odd One Out

1. Plough, hammer, cultivator, hoe
2. Levelling, broadcasting, seed drill, transplanting
3. Chain pump, furrow, rahat, moat
4. DDT, BHC, malathion, auxin
5. Temperature, sunlight, moisture, humidity

Image-Based Questions

1. What is this agricultural implement?

(a) Irrigator (b) Cultivator
(c) Hoe (d) Plough

2. What is this method to maintain soil fertility?

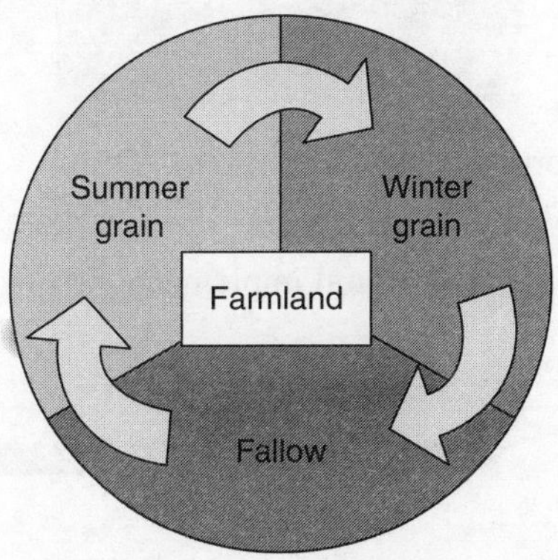

(a) Field left fallow (b) Mixed cropping
(c) Crop rotation (d) None of the above

3. What is this type of animal husbandry?

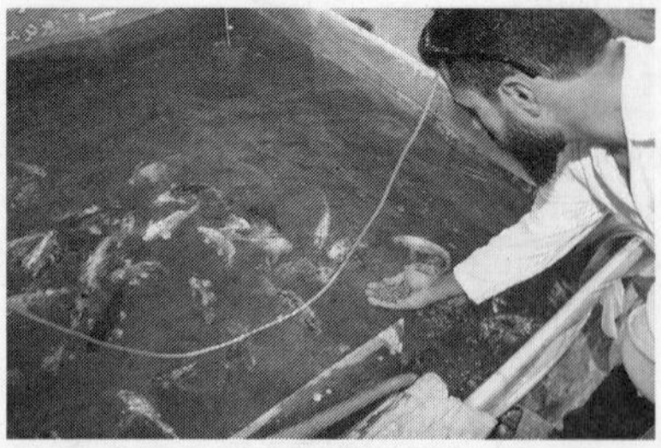

(a) Pisciculture (b) Apiculture
(c) Aquaculture (d) Cattle farming

4. Which type of threshing does this picture show?

(Farmers threshing paddy manually in Kolli Hills, 2017; P Jeganathan; https://commons.wikimedia.org/wiki/File:Paddy_Manual_threshing_Kolli_Hills_JEG3107_.jpg. Licensed under: CC-BY-SA-3.0)

(a) Crop rotation (b) Machine
(c) Manual (d) Sowing

5. Choose the type of irrigation:

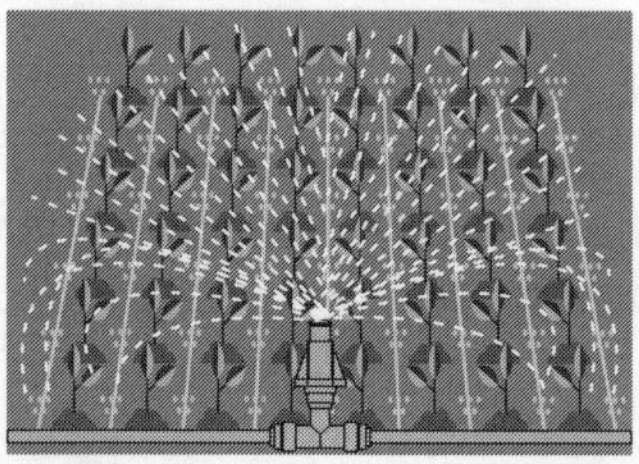

(a) Sprinkler (b) Drip
(c) Traditional (d) Canal

6. Select the agricultural implement:

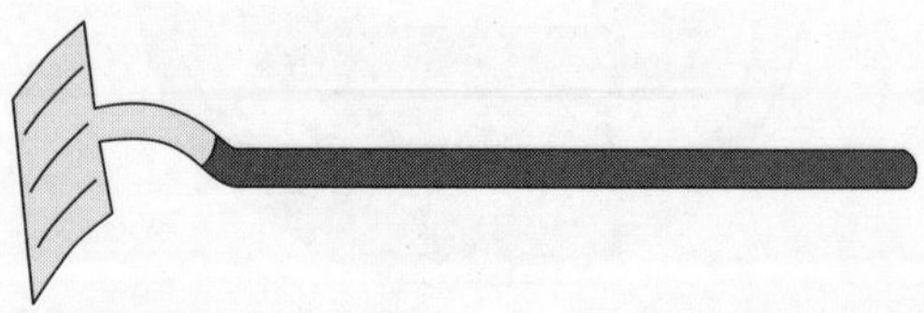

(a) Plough (b) Hoe
(c) Cultivator (d) None of the above

Very Short Answer Questions

1. What do you understand by agriculture?
2. What is the traditional method of sowing seeds?
3. What is mixed cropping?
4. Define harvesting.
5. Explain how fertilisers are different from manure.
6. What is irrigation? Name two sources of irrigation.
7. If wheat is sown in the kharif season, what would happen?
8. What is a cultivator?
9. What is hybridisation?
10. What are weeds?

Short Answer Questions

1. What are the two varieties of crops divided on the basis of seasons? Give examples of each.
2. What are the steps of agriculture practices?
3. What is the importance of levelling?
4. Explain transplanting.
5. What is a field left fallow?
6. What are the advantages of manure?
7. What is drip system of irrigation?
8. Define pisciculture and apiculture.
9. What are the parameters that should be kept in mind before storage of crops?
10. What are the methods used to get rid of weeds?

Long Answer Questions

1. Explain agriculture and the steps in the practice of agriculture.
2. Write a note on soil preparation.
3. What is seed sowing? What are the methods of seed sowing?
4. What are fertilisers and manure? Write their importance and the differences between them.
5. What are the natural methods to maintain soil fertility?
6. What is irrigation? What are its advantages? Explain the methods of irrigation.
7. Write a note on crop improvement.
8. Write a note on food from animals.

9. What are weeding and crop protection?
10. What are the parameters for the selection and sowing of seeds?

Case Study-Based Questions

1. Anchal was visiting her village in the rainy season and saw farmers sowing soybean. This made her wonder why specific crops are grown in specific seasons. Can you help her?
   (a) What are rabi and kharif crops?
   (b) What is an example of fruits of rabi and kharif?
2. Jay while visiting his farm with his grandfather saw farmers ploughing. Can you explain this to him?
   (a) What is the plough's purpose?
   (b) What are the two functions of a plough?
3. Kesar saw two farms on her way home. She saw one small field being prepared manually while the big one was being prepared by a machine. Can you explain this to her?
   (a) What is the manual way of soil preparation?
   (b) What is the machine used for soil preparation?
4. Romi while visiting a farm saw sacks full of materials labelled manure and fertilisers. She did not understand these terms. Can you help her?
   (a) What is manure?
   (b) What is fertiliser?
5. In summers, while visiting the coastal area farms, Palak saw farmers cultivating sea animals. Can you help her understand the terminologies associated with it?
   (a) What is animal husbandry?
   (b) What is pisciculture?

NCERT Corner

1. Select the correct word from the following list and fill in the blanks: float, water, crop, nutrients, preparation
   (a) The same kind of plants grown and cultivated on a large scale at a place is called _____.
   (b) The first step before growing crops is _______ of the soil.
   (c) Damaged seeds would ______ on top of the water.
   (d) For growing a crop, sufficient sunlight and ______ and ______ from the soil are essential.
2. Match items in column A with those in column B:

Column A	Column B
(i) Kharif crops	(a) Food for cattle
(ii) Rabi crops	(b) Urea and superphosphate
(iii) Chemical fertilisers	(c) Animal excreta, cow dung, urine and plant waste
(iv) Organic manure	(d) Wheat, gram, pea
	(e) Paddy and maize

3. Give two examples of each.
   (a) Kharif crop (b) Rabi crop
4. Write a paragraph in your own words on each of the following:
   (a) Preparation of soil (b) Sowing
   (c) Weeding (d) Threshing
5. Explain how fertilisers are different from manure.
6. What is irrigation? Describe two methods of irrigation which conserve water.
7. If wheat is sown in the kharif season, what would happen? Discuss.
8. Explain how soil is affected by the continuous plantation of crops in a field.
9. What are weeds? How can we control them?
10. Arrange the following boxes in the proper order to make a flow chart of sugarcane crop production.

Sending crop to sugar factory	Irrigation	Harvesting
1	2	3
Sowing	Preparation of soil	Ploughing the field
4	5	6
Manuring		
7		

11. Complete the following word puzzle with the help of clues given below.

Down

1. Providing water to crops.
2. Keeping crop grains for a long time under proper conditions.
5. Certain plants of the same kind grown on a large scale.

Across

3. A machine used for cutting mature crop.
4. A rabi crop that is also one of the pulses.
6. A process of separating grain from chaff.

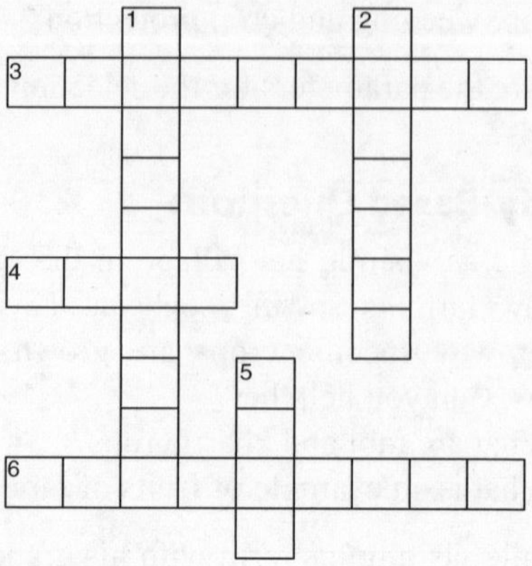

COMPETITION WINDOW (OBJECTIVE TYPE)

1. Mutations in plant cells can be induced by:
   (a) Gamma rays (b) Infrared rays
   (c) Zeatin (d) Kinetin
2. Breeding crops with higher levels of vitamins and minerals is called:
   (a) Biomagnification (b) Bioaccumulation
   (c) Biofortification (d) Bioremediation
3. Which crops are cultivated in a CO_2-rich environment?
   (a) Sugar beet and cabbage
   (b) Carrots and tomatoes
   (c) Wheat and sugar beet
   (d) Tomatoes and bell pepper
4. Which of the following is related to the Blue Revolution?
   (a) Development of water reservoirs
   (b) Crops and by-products
   (c) Fishery industry
   (d) Honey and its by-products
5. Which of the following is not the objective of biofortification?
   (a) Improving vitamin content
   (b) Disease resistance
   (c) Improving protein content
   (d) Improving micronutrient content
6. A system of rotating crops with legume or grass pasture to improve soil structure is called:
   (a) Contour farming (b) Strip farming
   (c) Shifting farming (d) Ley farming
7. Jaya and Ratna developed for green revolution are varieties of:
   (a) Rice (b) Wheat
   (c) Bajra (d) Maize

Answer Keys

Multiple Choice Questions

1. (a) 2. (c) 3. (a) 4. (a) 5. (a)
6. (c) 7. (a) 8. (c) 9. (b) 10. (d)
11. (c)

Match the Following

1. i (b), ii (a), iii (d), iv (c)
2. i (c), ii (d), iii (b), iv (a)
3. i (b), ii (c), iii (a)

True or False

1. T 2. F 3. T 4. T 5. T
6. F 7. F 8. T 9. F 10. T

Fill in the Blanks

1. Crop
2. Preparation
3. Float
4. Water, nutrients

5. Agriculture practices
6. Leveller
7. Cultivator
8. Broadcasting
9. Biological nitrogen fixation
10. Animal husbandry

Assertion–Reason Questions

1. (b) 2. (a) 3. (c) 4. (d) 5. (c)

One-word Answers

1. Kharif crops
2. Ploughshaft
3. Broadcasting
4. Nurseries
5. Nitrogenase
6. Irrigation
7. Mutagen

Define the Term

1. These are sown in the winter season, starting from October and harvested around April.
   Example: Barley, mustard, wheat, linseed, gram and pea.
2. Even after ploughing, big lumps of soil (called crumbs) may remain in the field. These are crushed using wooden or iron planks called levellers and this process is called levelling. A leveller is driven by a machine or an animal.
3. This tool is used for loosening and turning the soil in the fields quickly. The use of cultivator saves labour and time.
4. Another method for sowing seeds is with the help of a simple device consisting of a bamboo tube with a funnel on it attached to a plough. As the plough moves over the field, the tube attached to it drops the seeds kept in the funnel at an equal distance and depth. The plough keeps making furrows in the soil in which the seeds are dropped by the seed drill and it ensures after sowing, the seeds to get covered by soil.
5. In crops like rice, tomato, onion, chilli and brinjal, the seeds are not directly sown in the fields. Nurseries are where the seeds are first sown in seed beds. After the seeds have grown into small plantlets called seedlings, only healthy and well-developed seedlings are transferred and transplanted into the field. This process of transferring the seedlings from the nursery to the main field by hand is called transplanting.

Groups of Terms

1. Pisciculture
2. Kharif
3. Cultivator
4. Weedicide
5. Animal husbandry

Identify the Odd One Out

1. Hammer
2. Levelling
3. Furrow
4. Auxin
5. Sunlight

Image-Based Questions

1. (d) 2. (c) 3. (a) 4. (c) 5. (a)
6. (b)

Very Short Answer Questions

1. Agriculture refers to the growing of plants or raising of animals for human use and is mostly dependent on those activities that are essential for improving the environment for the growth or rearing of plants and animals, thereby improving their usefulness to humans.
2. Broadcasting is the traditional method of sowing seeds manually.
3. The practice of growing two or more crops simultaneously in the same field is known as mixed cropping.
4. The process of cutting and collecting mature crops from the fields is known as harvesting.
5. Manure are organic substances obtained from dead plant and animal waste. A chemical fertiliser is a salt containing the necessary plant nutrients in concentrated form to provide quick replenishment of nutrients and restore soil fertility.
6. Irrigation is the supply of water to agricultural fields at different intervals. Sufficient supply of rainwater is not always available; therefore, farmers need to supply water to the fields using irrigation. The two sources of irrigation are traditional and modern.
7. Kharif crops are those that are sown in the rainy season. Crops like wheat require less water which will lead to damage to crops.
8. This tool is used for loosening and turning the soil in the fields quickly. The cultivator is a machine designed to stir the soil as it matures around the crop to destroy weeds and promote growth that saves labour and time.

9. This is one of the ways of incorporating desirable characters into crop varieties. The process of crossing between two genetically dissimilar plants is known as hybridisation.
10. The unwanted plants that grow on their own along with the cultivated plants are known as weeds.

Short Answer Questions

1. Kharif crops are sown in the rainy season, starting in the month of June, and harvested around October. For example, paddy, pigeon pea, maize, cotton, soybean and groundnut.

   Rabi crops are sown in winter, starting in the month of October, and harvested around April. For example, wheat, gram, pea, mustard, linseed, oilseed and barley.
2. The steps involved in agriculture, which are followed by farmers for good crop yield, are known as agriculture practices. These steps are as follows: Soil preparation, Selection and sowing of seeds, Manuring or adding fertilisers, Irrigation, Weeding, Harvesting, Threshing and winnowing, and Storage.
3. It helps in preventing loss of moisture from the ploughed soil.

   It enables the soil to settle so that it is not eroded by wind or water.

   It brings uniformity to the field so that water and manure can be distributed evenly to all the parts of the field.
4. In some crops like rice, tomato, onion, chili and brinjal, the seeds are not sown directly in the fields. These seeds are first sown in seed beds called nurseries. After the seeds have grown into small plantlets called seedlings, only healthy and well-developed seedlings are transferred and transplanted to the field. This process of transferring seedlings from the nursery to the main field by hand is called transplanting.
5. In this method, the field is left free or fallow for one or more seasons. The dead plants, animals and other organic matter that are collected in the field are decomposed by microorganisms. Therefore, nutrients are returned to the soil.
6. It enhances the water-holding capacity of the soil.

   It makes the soil porous, due to which the exchange of gases becomes easy.

   It increases the number of useful microbes.

   It improves the texture of the soil.
7. This system involves providing water drop by drop at the roots of the plants. Thus, water is not wasted. This system is practised in regions where water availability is low. However, this method is very expensive. It is the best technique for watering fruit plants, gardens and trees.
8. The rearing of fish is called pisciculture. Fishery is the farming of fish. The fish found in marine waters are called sea fish or marine fish (examples are salmon and hilsa) while the fish found in freshwater are called freshwater fish (examples, rohu, catla).

   The rearing of honeybees is called apiculture. This is carried out on a large scale on farms. Honeybees produce honey by sucking the nectar of flowers, which is rich in minerals and sugar.
9. Away from moisture, safe from rain, protected from pests and birds, in air-tight godowns, in proper bags for the purpose of transportation.
10. The following methods are used for getting rid of weeds:
    - Weeding is carried out by pulling out the weeds manually or by using a hoe, harrow or seed drill.
    - The chemicals used to kill the weeds are known as weedicides. A common weedicide is 2,4-D (2,4-dichlorophenoxyacetic acid). This chemical kills only the weeds and not the cultivated crops. It prevents seed formation. Since weedicides are poisonous, while spraying them, farmers should cover their noses and mouth with a piece of cloth.
    - Some insects that do not cause harm to the crops can be made to breed and grow. They then destroy the weeds, leaving the crops safe.

Long Answer Questions

1. Hint: Write the definition and explain all the steps involved in agriculture and its practices.
2. Hint: Write all the methods involved in soil preparation.
3. Hint: Explain seed sowing and the methods involved in it.
4. Hint: Explain the composition of fertilisers and manure, and their importance with their differences, as mentioned in this chapter.
5. Hint: Explain all the natural methods involved in maintaining soil fertility.
6. Hint: Explain irrigation, its advantages and the methods of traditional and modern irrigation.
7. Hint: Explain crop improvement, its importance and the various methods involved.
8. Hint: Explain the types of agriculture involved in obtaining food from animals.

9. Hint: Explain weeds, weeding and methods for crop protection.
10. Hint: Explain all the parameters involved in the selection and sowing of seeds.

Case Study-Based Questions

1. (a) Kharif crops: These are sown in the rainy season starting from the month of June and harvested around October. Rabi crops: These are sown in the winter season starting from the month of October and harvested around April.
   (b) Muskmelon is a kharif crop, pomegranate is a rabi crop.
2. (a) To make the soil ready for cultivation, the soil needs to be prepared. The preparation of soil involves some basic steps like levelling, ploughing or tilling.
   (b) It is also easy to mix the fertilisers and manure in the loose soil, and it helps in removing the weeds by uprooting them.
3. (a) This tool is used for removing weeds, and for loosening the soil. It consists of a long rod of wood or iron. There is a handle at one end of the hoe. A broad, strong and bent plate of iron is fixed and works like a blade below the handle while the other end that is attached to the beam is kept on the neck of bullocks, and the hoe is pulled by animals.
   (b) This tool is used for loosening and turning the soil in the fields quickly. The use of cultivators saves labour and time.
4. (a) Manure are organic substances obtained from dead plant and animal wastes. This is an external source of a small quantity of nutrients and a large quantity of organic matter to the soil.
   (b) A chemical fertiliser is a salt containing necessary plant nutrients in concentrated form, to provide quick replenishment of nutrients and restore its fertility. Some examples of fertilisers are urea, ammonium sulphate, superphosphate, potash, NPK.
5. (a) Animal husbandry is the science that deals with feeding, breeding and taking care of domestic animals.
   (b) The rearing of fish is called pisciculture. Fishery is the farming of fish. The fish found in marine water are called sea fish or marine fish, such as salmon and hilsa, and the fish found in freshwater are called freshwater fish, such as rohu and catla.

NCERT Corner

1. (a) Crop
   (b) Preparation
   (c) Float
   (d) Water, nutrients
2. (i) (e)
   (ii) (d)
   (iii) (b)
   (iv) (c)
3. (a) Kharif crops: Paddy and maize
   (b) Rabi crops: Wheat and gram
4. (a) Preparation of soil: Soil preparation is necessary before growing a crop. It involves tilling and loosening the soil. This allows the roots to penetrate deep in the soil and to breath easily even when they are deep.
   (b) Sowing: The process of putting seeds into the soil is called sowing. The tool used traditionally for sowing seeds is funnel-shaped. Nowadays a seed drill is used for sowing with the help of tractors. This tool sows the seed uniformly at a proper distance and depth.
   (c) Weeding: Some undesirable plants grow along with crop and these unwanted plants are called weeds. The process of removing these unwanted plants is called weeding.
   (d) Threshing: The process of separating the grain seeds from the chaff is called threshing.
5.

Fertiliser	Manure
(i) A fertiliser is an inorganic salt.	(i) Manure is a natural substance obtained by the decomposition of cattle dung, human waste and plant residues.
(ii) A fertiliser is prepared in factories.	(ii) Manure can be prepared in the fields.
(iii) A fertiliser does not provide any humus to the soil.	(iii) Manure provides a lot of humus to the soil.
(iv) Fertilisers are very rich in plant nutrients like nitrogen, phosphorus and potassium.	(iv) Manure is relatively less rich in plant nutrients.

6. The artificial method of watering the plants for assisting in their growth is called irrigation. The main sources of irrigation are wells, tube-wells, ponds, lakes, rivers.

   Two methods which help us to conserve water are:

   (i) Sprinkler irrigation system: This irrigation system has an arrangement of vertical pipes with rotating nozzles on the top. It is more useful in uneven and sandy land where sufficient water is not available.

   (ii) Drip irrigation system: This irrigation system has an arrangement of pipes or tubes with very small holes in them to water plants drop by drop just at the base of the root. It is very efficient as water is not wasted at all.

7. Wheat crop is sown from November/December to March/April. It is grown in winter and requires less water. If wheat is sown in kharif season, its production will be decreased considerably.

8. Continuous plantation of crops makes the soil poorer in certain nutrients as the crops take up nutrients from the soil. The soil becomes infertile. It does not get enough time to replenish the nutrients.

9. The undesirable and unwanted plants which grow naturally along with the crop are called weeds. The growth of weeds can be controlled by adopting many ways. Tilling before sowing of crops helps in the uprooting and killing of weeds, which may then dry up and get mixed with the soil. Weeds are also controlled by using certain chemicals, called weedicides. Weedicides are sprayed in the fields to kill the weeds.

10.

Preparation of soil 5 | Ploughing the field 6 | Sowing 4

Manuring 7 | Irrigation 2 | Harvesting 6

Sending crop to sugar factory 1

11.

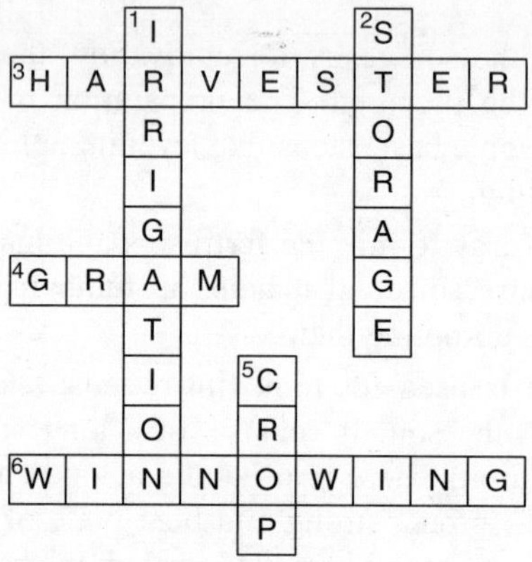

Competition Window

1. (a) 2. (c) 3. (d) 4. (c) 5. (b)
6. (d) 7. (a)

CHAPTER 10

Ecosystem and Biodiversity

PREREQUISITES

You must have knowledge of:

- Biotic and abiotic components
- Basic terminology about the ecosystem
- How the environment maintains balance

LEARNING OBJECTIVES

At the end of this chapter, you will be able to:

- Define ecosystem
- Explain the link between an ecosystem's biotic and abiotic components
- Describe the role of the food chain, food web, trophic levels and energy flow in an ecosystem
- Compare and contrast the advantages of collaboration in other forms of associations, such as mutualism, commensalism and symbiosis

INTRODUCTION

Everything that surrounds us – our house, garden, town, shops, hills, rivers, ocean, air, soil, sunlight, and other plants and animals – is referred to as the environment. The place, people, things and nature that surround any living organism is called its environment. It is a combination of natural and human-made phenomena. The natural environment refers to both biotic (living) and abiotic (non-living) conditions.

Land, water, air, plants and animals comprise the natural environment. You must be familiar with the meaning of lithosphere, hydrosphere, atmosphere and biosphere. Let us learn more about these domains in this chapter.

STRUCTURE OF THE EARTH

Among the eight planets of the Solar System, the planet that we live on is the third closest to the Sun. The Earth has all the conditions required for the survival of life. The outer layer of the Earth is solid and can support life; it is known as the **lithosphere.** Minerals and other essential nutrients for plants are present in this layer. About 70% of the Earth's surface is covered by another natural element called water in the form of lakes, rivers and oceans. This aquatic envelope is known as the **hydrosphere.** The envelope of gas or air that surrounds the lithosphere is known as the **atmosphere.**

The lithosphere is the solid crust or hard top layer of the Earth. It is made up of rocks and minerals and covered by a thin layer of soil (Fig. 10.1). It is an irregular surface with various landforms such as mountains, plateaus, plains and valleys. Land forms are found over the continents and also on the ocean floor. The lithosphere is the domain that provides forests, grasslands for grazing, land for agriculture and human settlements. It is also a source of mineral wealth.

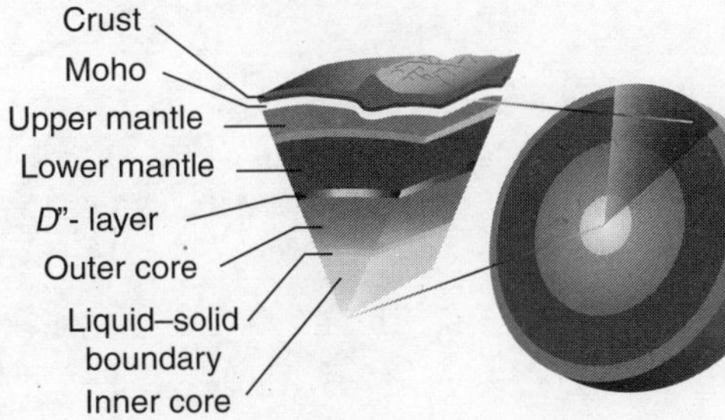

Fig. 10.1 Layers of the Earth

All the water on the planet makes up the hydrosphere (Fig. 10.2). It includes the water that is present in the air or underground. It could be in any form: liquid, ice or vapour. On Earth, liquid water exists in the form of lakes, rivers and oceans. The basic components of the hydrosphere are:

- Surface water
- Freshwater
- Oceans
- Groundwater
- Atmospheric water vapour
- Glacial water

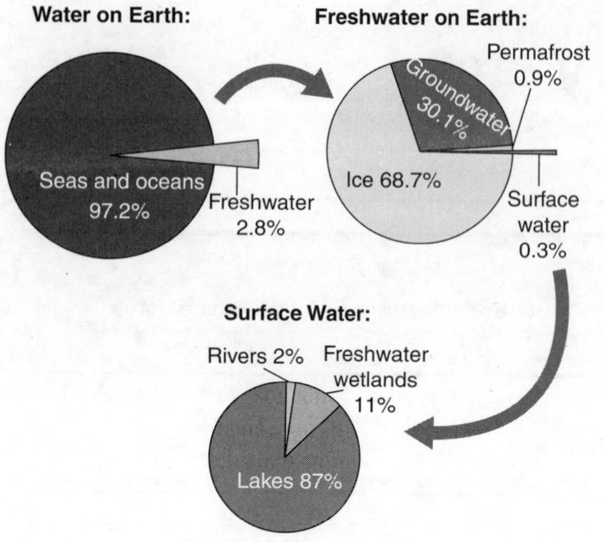

Fig. 10.2 Hydrosphere

The atmosphere protects us from the harmful rays and scorching heat of the sun (Fig. 10.3). It consists of a number of gases, dust and water vapour. Changes in the atmosphere produce changes in the weather and climate.

The plant and animal kingdoms together make up the biosphere or the living world. It is a narrow zone of the Earth where land, water and air interact to support life. All plants, animals and human beings

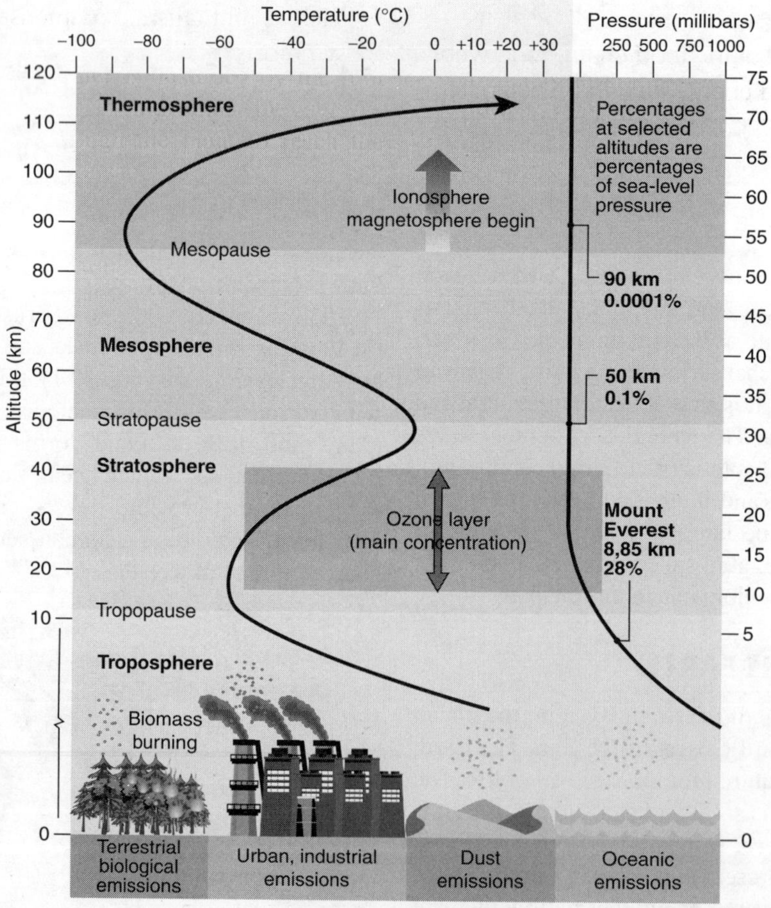

Fig. 10.3 Atmosphere

depend on their immediate surroundings. They are also interdependent on each other. This relation between living organisms, as well as the relation between organisms and their surroundings, form an ecosystem. Thus, large rainforests, grasslands, deserts, mountains, lakes, rivers, oceans and even small ponds can be called ecosystems.

THE HUMAN ENVIRONMENT

Human beings interact with the environment and modify it according to their needs. Early humans adapted themselves to their natural surroundings. They led a simple life and fulfilled their requirements from the nature around them. With time, our needs grew and became more varied. Humans learned new ways to use and change the environment. They learn to grow crops, domesticate animals and lead a settled life. The wheel was invented, surplus food was produced, a barter system emerged, trade began and commerce developed. The Industrial Revolution enabled large-scale production. Transportation became faster. The Information Revolution made communication easier and more speedy across the world.

Have you ever thought about why you love eating a juicy watermelon in summer and hot roasted peanuts in winter? A perfect balance is necessary between the natural and human environment. Humans must learn to live and use their environment in a harmonious way.

THE ECOSYSTEM AND ITS COMPONENTS

The environment is where living organisms obtain their sustenance and oxygen. Plants and animals interact with one other as well as with their physical environment. As a result, an ecosystem can be defined as "a biological environment that includes all the species residing in a certain area, as well as the non-living physical components of the environment with which the organisms interact." AG Tansley proposed the notion of ecosystem in 1935. The word ecosystem comes from the Greek '*oikos*' meaning 'house' and '*systema*' meaning 'system'.

An ecosystem is a self-contained unit of living things (plants, animals, decomposers), and the non-living environment (soil, air and water). For example, a forest, pond, lake or grassland. In an ecosystem, energy and matter are exchanged between living and non-living components (Fig. 10.4).

An ecosystem can be natural or man-made. Some examples of natural ecosystems are grasslands, forests, rivers, deserts, mountains, ponds and lakes. Deserts, grasslands and mountains represent the terrestrial ecosystem (land-based ecosystem), while ponds, rivers, lakes and seas form the aquatic ecosystem (water-based ecosystem). Examples of artificial ecosystems are gardens, crop fields, parks and aquariums.

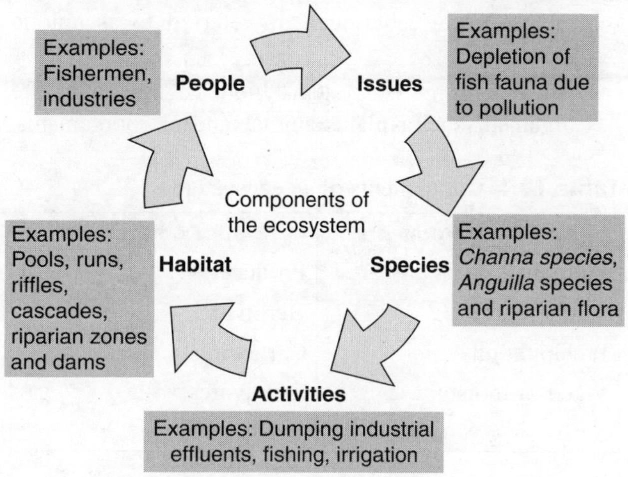

Fig. 10.4 Components of the ecosystem

Types of Ecosystems

- **Aquatic ecosystem:** This is of two basic types: freshwater, which includes streams, rivers and lakes, and marine, which is present in oceans, seas and marshes. Individual ecosystems could be very limited in area.
- **Terrestrial ecosystem:** This land-based ecosystem is characterised by limited water supply for flora and fauna. They include:
  - **Forest:** From tropical to high-altitude, forests vary by physical location and include the different plants found in an area.
  - **Grassland:** Native grasses and small shrubs make up the savannas.
  - **Desert:** This type of ecosystem experiences cold nights and hot days.
  - **Tundra:** Minimal rainfall and cold temperature makes life in the tundra tough. This is seen in the Arctic regions and snow-covered mountain ranges with no trees.

Fun Fact!

The term ecosystem was coined by AG Tansley.

Relationship between the Components of an Ecosystem

An ecosystem is made up of both non-living and living things (Table 10.1). They are referred to as abiotic and biotic components, respectively.

- **Abiotic:** The non-living physical and chemical factors in an ecosystem are referred to as abiotic components.
- **Biotic:** In an ecosystem, biotic components are organisms such as plants, animals and microorganisms.

Table 10.1 Components of an ecosystem

Abiotic Components	Biotic Components
Sunlight	Producers
Temperature	Herbivores
Precipitation	Carnivores
Water or moisture	Omnivores
Soil	Detritivores

All of these elements change across time and space. You have probably noticed that the vegetation of coastal regions, hilly areas and deserts are all rather different. Do you have any idea why? It is because each has its own set of abiotic factors, such as temperature, soil and moisture. As a result, we can examine how abiotic elements affect various creatures in the ecosystem.

Biotic Components

The biotic components of an ecosystem can be classified as follows:

- **Producers:** All green plants and blue-green algae can produce their own food (sugar and starch) from inorganic substances using light energy (photosynthesis). Green plants take water from the soil and carbon dioxide from the air to prepare food with the help of sunlight. Thus, plants are called **autotrophs.**

  Autotrophs are organisms that can produce their own food. We know that all plants (with the exception of a few parasitic plants) can make their own food through photosynthesis, but did you know that some organisms can make their own food through chemosynthesis? This category includes a number of bacteria. Plants are also known as producers because they supply food to all creatures, directly or indirectly. Autotrophs are the foundation of any biotic system because they capture solar energy and use it to produce food for everyone. Plankton are microscopic organisms that float freely on the surface of the water in ponds, lakes, rivers and oceans. They are of two types: phytoplankton and zooplankton. Phytoplankton are microscopic aquatic plants while zooplankton are microscopic aquatic animals. Protozoa are an example of zooplankton. Autotrophs in terrestrial habitats are primarily rooted plants, but in aquatic ecosystems, they include phytoplankton and shallow-water rooted plants known as macrophytes.

> **Fun Fact!**
>
> Did you know that half the world's species live in the tropical rainforests?

- **Consumers:** These are organisms that consume other organisms or their products as their food. All animals belong to this category. Consumers depend on producers for food, directly or indirectly. They are known as **heterotrophs**. Animals that depend on plants for food are known as **herbivores** (examples are cow, goat and deer). Animals that eat other living animals are called **carnivores** (examples are leopard, tiger and lion). Some animals feed on dead animals and are known as **scavengers** (examples are eagle and vulture). There are also animals that eat both other animals and plants; they are known as **omnivores** (examples are crow and humans).
- **Decomposers:** Microbes that break down (decompose) dead plants and animals from complex compounds into simpler ones are known as decomposers. They help in regeneration of natural resources. They return many nutrients from dead plants and animals to the soil, water and air. There is a cyclic exchange of materials between an ecosystem's environment and its biotic community in which nutrients are constantly recycled and used. Decomposers are also known as saprotrophs or reducers of the natural environment.

> **Fun Fact!**
>
> Did you know that it has been estimated that more than 100,000 different decomposers exist?

Importance of decomposers:

- Decomposers help in degrading waste and the dead bodies of plants and animals. Therefore, they also clean the environment and create space for newer generations of organisms.
- Decomposers release minerals trapped in organic matter. These are taken up by plants and help to maintain the fertility of the soil.
- They produce organic acids which are useful in solubilisation of minerals.

- They help in replenishing the materials in the biosphere, which helps in the processes of life.

Abiotic Components

Abiotic components are also known as non-living components. They include the air, light and soil. Climatic factors like rainfall, wind and temperature are also abiotic components.

- **Soil:** The uppermost layer of the Earth's crust is called soil and it is formed by the weathering of rocks. Soil contains water, minerals and humus, which are essential for the growth of plants. Plants are the main source of food for most animals and the soil supports their growth. Topsoil is dark and just below it is a layer of subsoil; at the bottom is a layer of rocks known as **bedrock.** A variety of organisms live in soil such as fungi, bacteria, insects and worms.
- **Air:** Oxygen and carbon dioxide are the most important components of air as all living organisms need oxygen for respiration and use carbon dioxide for photosynthesis. This cyclic movement of gases to maintain balance in nature occurs in the atmosphere.
- **Water:** Rainwater is the primary source of water as all organisms need water for survival. Water is known as the universal solvent and even minerals are absorbed by plants in the form of solutions. Plants take water from the soil for photosynthesis. The amount of water that is available in a geographical area determines the habitat and the flora and fauna.
- **Temperature:** The degree of coldness or heat is known as temperature; it keeps varying from place to place and determines the distribution of flora and fauna. Animals like penguins or polar bears are adapted to cold places while camels are known to survive in high temperatures. The leaves of plants in a hot area become spiny to reduce water loss and animals like snakes and frogs go underground when they **hibernate**.

Fun Fact!

Did you know that 57.8°C is the highest temperature and −89.2°C is the lowest temperature recorded on Earth?

- **Light:** Plants manufacture food in the presence of solar energy and this is stored as chemical energy. Light influences the closing and opening of stomata. Light also affects flowering, seed germination and movement. Similarly, the behaviour of animals is also affected by light; most animals are active during the day but some like millipeded and earthworm do not like light and are found in dark places.

Relation between Organisms

The interaction of organisms depends on evolution and the environmental conditions. They can be utilised as a framework to dissect the network of environments. These interactions can be of two types:

- **Intraspecific:** Interaction between the same species
- **Interspecific:** Interactions with different species

The different types of interactions are as follows:

- **Competition:** Two species use the same resources which are limited and have a negative impact on each other
- **Mutualism:** A symbiotic relationship benefitting both organisms
- **Parasitism:** A symbiotic relationship in which one benefits and the other is harmed
- **Commensalism:** A symbiotic relationship in which one benefits and the other remains unaffected

FOOD CHAIN AND FOOD WEB

It is clear that organisms establish an 'eating' and 'being eaten' chain. Small fish feed on phytoplankton and are then devoured by larger fish. The food chain works in this way.

The sequence of living organisms in a community in which one organism consumes another is a **food chain**. It is unidirectional as transfer of energy takes place in only one direction. In a food chain, flow of energy takes place.

Producers, herbivores and carnivores are the essential components of a simple food chain. In a terrestrial ecosystem, a simple food chain connects the trees and shrubs (producers), giraffes (herbivores that consume trees and shrubs) and lions, much like in a pond (carnivores that eat the herbivores). Each link in the chain provides food for the next level and is classified as belonging to a specific trophic level ('*trophos*' means 'feeding'). The producers in this case are trees and shrubs, which are at trophic level I, giraffe at trophic level II and lion at trophic level III.

A food chain can be defined as "the succession of creatures in an ecological community that forms a passing on of food energy from one organism to another as each consumes a lower member and is in turn preyed upon by a higher member of the food chain" (Fig. 10.5).

Types of Food Chains

In an ecosystem, different types of food chains can exist:

- In a terrestrial ecosystem food chain, grass is eaten by deer. Deer are in turn eaten by lions.
- In an aquatic ecosystem food chain (Fig. 10.6), weeds or phytoplankton are eaten by zooplankton, which in

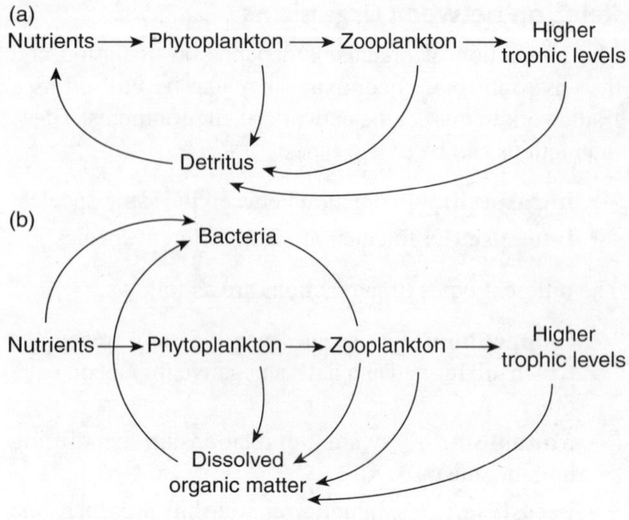

Fig. 10.5 Simplified scheme for (a) food chain and (b) food web

turn are eaten by small fish. The small fish are eaten by large fish and they in turn are eaten by whales.

Fun Fact!

Did you know that the predators at the top of a food chain are known as top predators?

- A detritus food chain begins with biological material that has died. Microorganisms such as fungi and bacteria break down the decaying organic waste into basic nutrients. Smaller carnivores devour these simple nutrients and decomposers, and eventually provide food for larger carnivores.

Significance of Food Chains

- The study of food chains helps in understanding food relationships and interactions among the organisms in an ecosystem. In a food chain, energy is transferred between the various living components of an ecosystem.
- Food chains show the dynamic nature of an ecosystem or biosphere.
- The movement of toxic substances like pesticides and weedicides through food chains can be harmful.

Most animals are members of many food chains and consume a variety of foods in order to meet their nutritional and energetic needs. A food web is formed by these interwoven food networks. The inter-connected food chains operating in an ecosystem that establish a network of relationships between various species is called a **food web** (Fig. 10.7). In a food web, one organism may occupy a position in more than one type of food chain. An

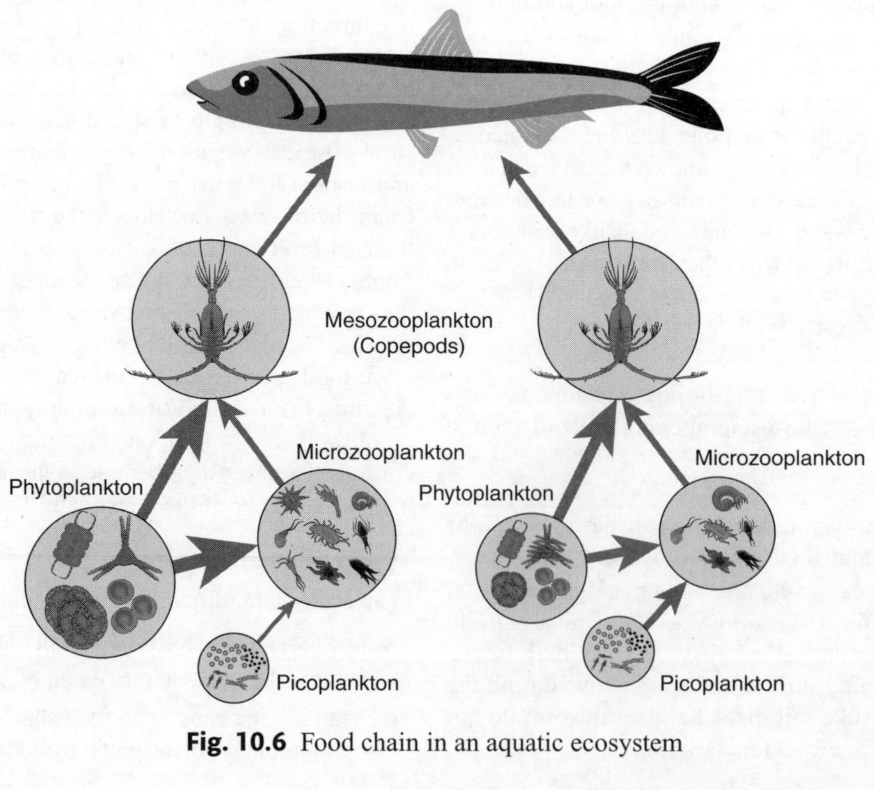

Fig. 10.6 Food chain in an aquatic ecosystem

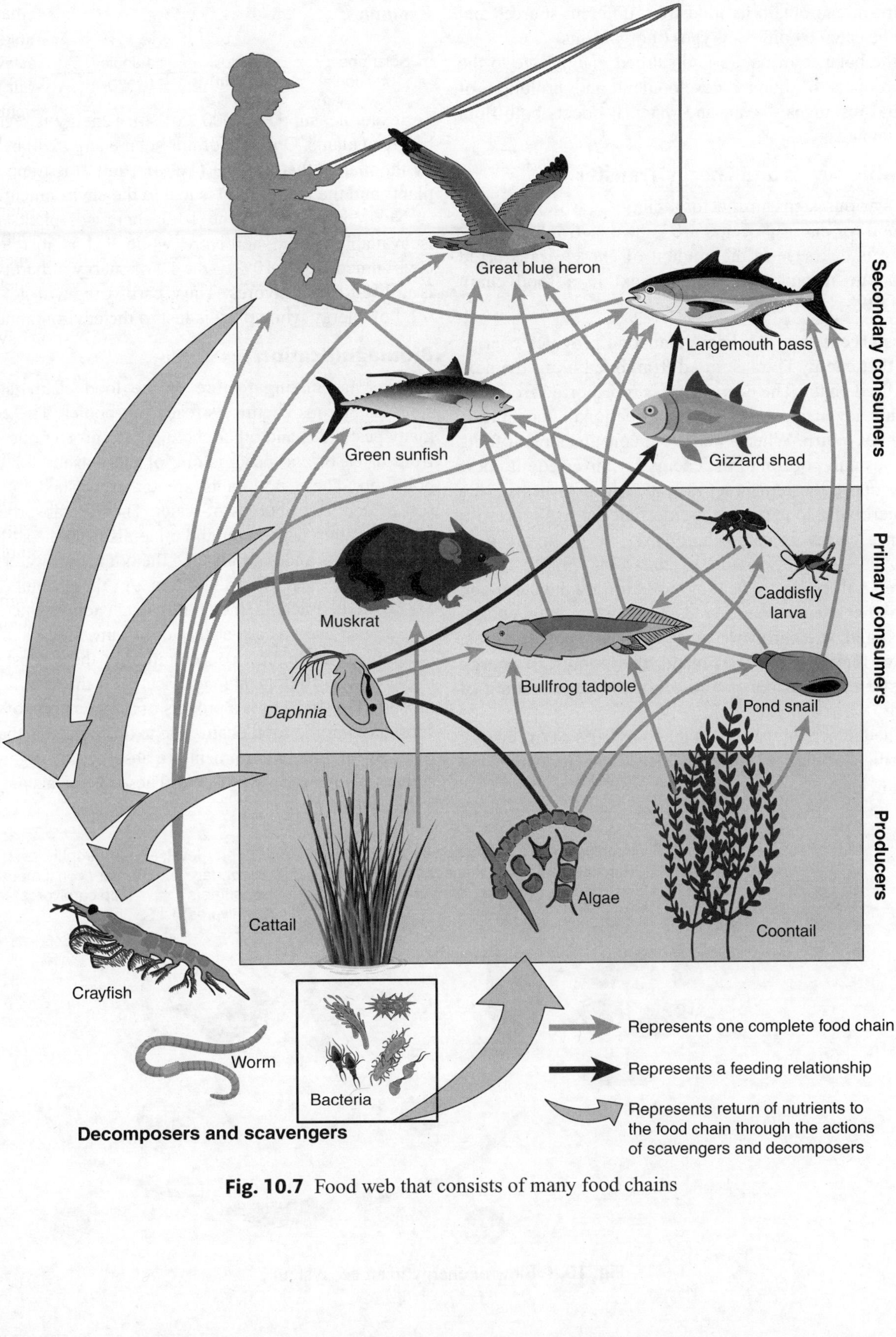

Fig. 10.7 Food web that consists of many food chains

organism can obtain its food from different sources and may be eaten by different types of organisms.

The balance of nature is disturbed mainly due to the by-products of human development and pollution of natural resources like air and water. It affects both flora and fauna.

Trophic Levels and Energy Transfer

The various steps in the food chain at which transfer of food (or energy) occurs are known as trophic levels. There is decrease in the amount of energy transferred from one trophic level to the next in a food chain (Fig. 10.8).

Ten percent law: We know that food provides energy, and that energy is transmitted from one link to the next in a food chain. The energy that is transported from one trophic level to the next is unidirectional, meaning it does not return. When a herbivore consumes plant food, only a small portion of the energy is converted into new body mass; the remainder is lost as heat or utilised by the herbivore to carry out its life functions. As a result, when a carnivore eats a herbivore, the carnivore only receives a small portion of the total energy. Some energy will be lost as heat or "used up" by the carnivore from the energy transferred. To gain enough energy to expand, the carnivore must eat a lot of herbivores. As we move up the food chain, the amount of energy transferred decreases as a result of the huge amount of energy wasted at each link.

The ten percent law states that only 10% of the energy entering a trophic level is available for transfer to the next trophic level.

Examples:

Solar energy (1000 Joule) $\xrightarrow{1\%}$ 1000 J (Plant) $\xrightarrow{10\%}$ 1 Joule (Deer) $\xrightarrow{10\%}$ 0.1 Joule (Lion)

For example, suppose 1000 J of solar energy is received by green plants. Only 1% of this solar energy will be used by the plants, that is, 10 J (1% of 1000 J) is trapped by plants and the rest (990 J) is lost to the environment.

Next, only 10% of the 10 J energy of plants (1 J) is available to the herbivore while 9 J is lost to the environment. Only 10% of the 1 J of energy of herbivores is utilised by carnivores. Thus, carnivores will get only 0.1 J of energy while 0.9 J is lost to the environment.

Biomagnification

Another fascinating feature of the food chain is how some dangerous chemicals reach our bodies. The use of many pesticides and other chemicals to protect our crops from diseases and pests is one of the reasons for water pollution. These pollutants are either washed into the soil or end up in bodies of water. They are absorbed by plants together with water and minerals from the soil, and aquatic plants and animals take them up from water; this is how they enter the food chain. As these compounds are not biodegradable, they build up at each trophic level.

Because humans are at the top of any food chain, we accumulate the highest concentration of these substances in our bodies. Biological magnification is the term for this process (Fig. 10.9). It is a process of accumulation of toxic substances in the food chain. The toxic chemicals that are released into the environment are absorbed by organisms such as plants, earthworms, etc. These chemicals are then

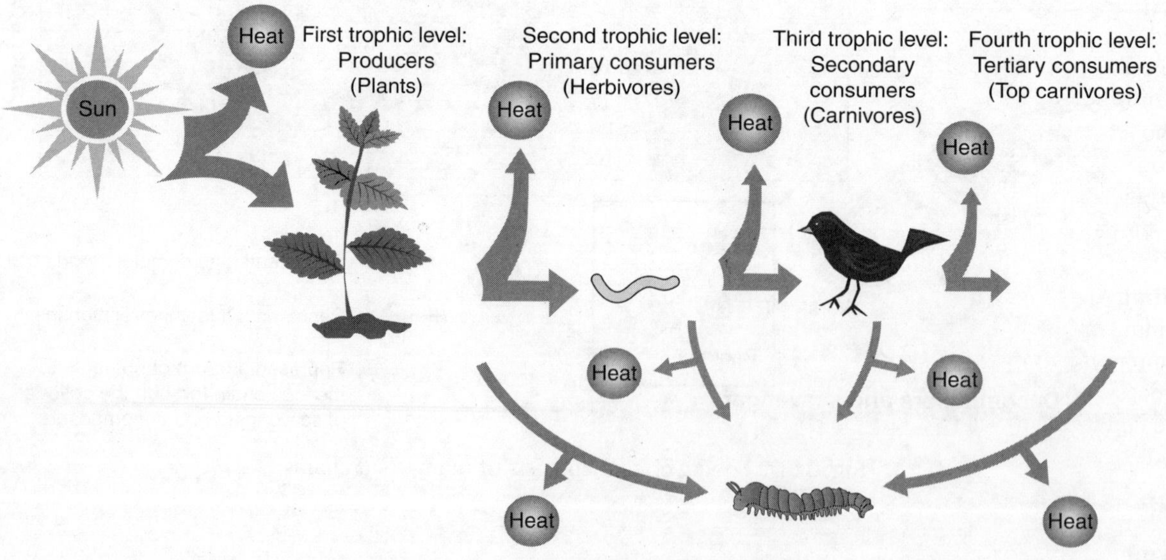

Fig. 10.8 Flow of energy in an ecosystem

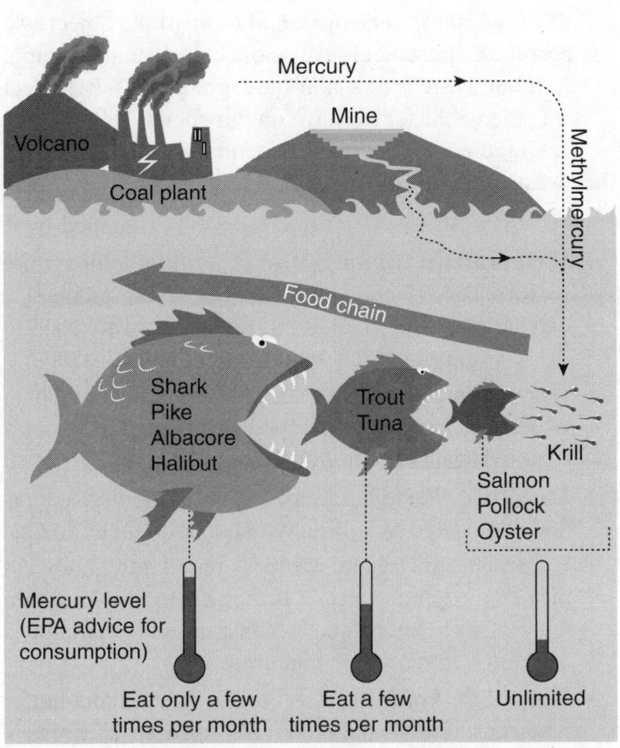

Fig. 10.9 Biomagnification

transferred to different trophic levels when these organisms are eaten by other organisms. This is why pesticide residues can be found in food grains like wheat and rice, as well as vegetables and fruits, and even meat. They cannot be got rid of easily by washing or other methods.

Following are a few substances involved biomagnification:

- **Pesticides:** Pesticides and chemicals such as DDT and mercury released into lakes and rivers are ingested by aquatic life. These are accumulated in their body tissues and transferred to other organisms who feed on them. Since the pesticides are industrially processed, they contain heavy metals such as lead, arsenic, cadmium, etc. These metals have been found in the bodies of animals and humans and have an adverse effect on them.
- **Mining activities in the ocean:** In the deep sea, mining takes place to extract metals like zinc, aluminium, cobalt, silver and gold. The mining process produces some selenium and sulphide, which are deposited in water and which destroys the oceans and coastal regions.

Interdependence between Organisms

- **Symbiosis** is a long-term and close biological interaction between two organisms of different species which could be obligatory (where they have to depend on each other for survival) or facultative (where they can generally live independently as well). It is also defined as a physical attachment where two or more symbionts form a single body; this is known as conjunctive symbiosis.
- **Mutualism** is a type of symbiosis. For example, the relationship between the Ritteri sea anemones on whose tentacles the Ocellaris clownfish dwell. This territorial fish protects the sea anemone from predators and in turn the tentacles protect the clownfish from its predators.
- **Parasitism** is a situation where one organism lives either on or inside the host organism. Examples are the malaria-causing protozoans that feed on humans and are transmitted from mosquitoes, vampire bats, amoebic dysentery, ringworm causing fungus. An example of plant parasitism is the honey fungi that grow in the roots of plants and big trees and eventually kill them; later, the fungi become saprophytic, living on the dead wood. Parasitism is seen in humans, insects and other organisms as well.
- **Predation** is an interaction between organisms where one kills (predator) and eats the other (prey). This type of interdependence is very common.

WHAT EFFECT DO OUR ACTIONS HAVE ON THE ENVIRONMENT?

We are a vital element of the natural world. Changes in the environment have an impact on humans, and our actions have an impact on the environment. We will now look at two environmental issues in depth: ozone depletion and trash disposal.

The Ozone Layer and How it is being Depleted

Ozone (O_3) is a molecule made up of three oxygen atoms. While O_2, commonly known as oxygen, is required for all aerobic forms of life, ozone is a poisonous gas. It protects the Earth's surface from the Sun's ultraviolet (UV) radiation (Fig. 10.10). This radiation is extremely harmful to creatures; for example, it has been linked to skin cancer in humans.

UV radiation with greater energy splits some molecular oxygen (O_2) into free oxygen (O) atoms. These atoms interact with molecular oxygen to produce ozone.

$$O_2 \xrightarrow{UV} O + O$$

$$O + O_2 \longrightarrow \underset{\text{(Ozone)}}{O_3}$$

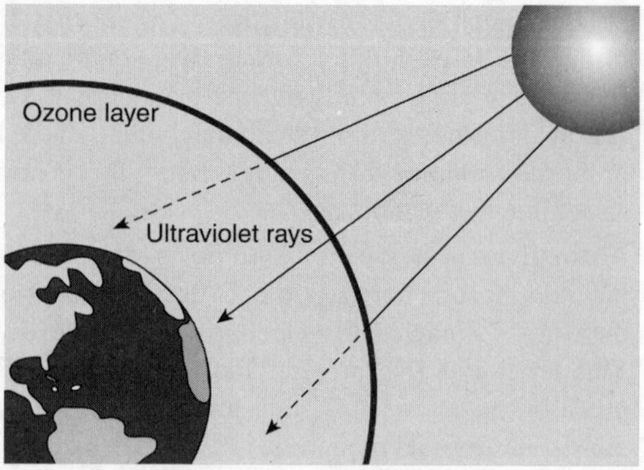

Fig. 10.10 Ozone layer

In the 1980s, the amount of ozone in the atmosphere began to fall. Synthetic chemicals such as chlorofluorocarbons (CFCs), which are used as refrigerants and in fire extinguishers, have been linked to this decline. The United Nations Environment Programme (UNEP) was successful in negotiating a deal in 1987 to limit CFC production to 1986 levels.

Managing the Waste we Create on Earth

Any town or city we visit will undoubtedly have mountains of waste strewn about. Many tourist attractions are often cluttered with empty food wrappers.

Increases in trash output have resulted from improvements in our way of life. Changes in mentality are also important, since more and more of the objects we use become disposable. Because of changes in packaging, much of our waste is now non-biodegradable. What effect do you believe these will have on our society?

The method to be used depends on the nature of the waste. Some of the important modes of waste disposal are:

- **Incineration:** Burning of waste at high temperature to form ash is called incineration. This process is carried out in an incinerator. Incinerators are used to destroy household, chemical and biological waste.
- **Open dumping:** This is a conventional method in which solid waste is dumped in selected areas of a region. It causes pollution.
- **Landfills:** Waste is dumped in less populated areas and compacted by rolling with the help of bulldozers.
- **Composting:** Organic waste is poured into a compost pit (2 m × 1 m × 1 m). It is then covered with a thin layer of soil. After three months, the same garbage is converted into organic manure.
- **Recycling:** Solid waste is broken down into simpler materials. These materials are used to make new items. Non-biodegradable solid waste like plastic and metal can be recycled.
- **Reuse:** A simple conventional technique of using an item again and again is known as reuse. For example, paper can be reused to make envelopes.

SUMMARY

- The various components of an ecosystem are interdependent on each other.
- Producers obtain energy from sunlight and this energy is made available to the rest of the ecosystem.
- There is loss of energy as we move from one trophic level to the next.
- Human activities have an impact on the environment.
- The use of chemicals like CFCs has endangered the ozone layer. Since the ozone layer protects against ultraviolet radiation from the Sun, this can cause serious threats to the environment.
- The waste we generate can be biodegradable or non-biodegradable. Disposal of the waste we generate is causing serious environmental problems.

PRACTICE QUESTIONS

Analyse Your Concepts (School Exam Based)

Multiple Choice Questions

1. _______ is the domain that provides forests, grasslands for grazing and land for agriculture:
(a) Hydrosphere (b) Ionosphere
(c) Lithosphere (d) Atmosphere

2. Animals that feed on dead animals are known as:
(a) Herbivores (b) Scavengers
(c) Omnivores (d) Carnivores

3. _______ produce organic acids which are useful in the solubilisation of minerals.
(a) Decomposers
(b) Producers
(c) Consumers
(d) None of the above

4. _______ is a symbiotic relationship benefitting both organisms.
(a) Mutualism (b) Commensalism
(c) Parasitism (d) Predation

5. Physical attachment where the symbionts form a single body are known as:
(a) Disjunctive symbiosis
(b) Constructive symbiosis
(c) Destructive symbiosis
(d) Conjunctive symbiosis

6. Synthetic chemicals such as _____ have been linked to the decline _____.
(a) CFCs, Oxygen
(b) CFCs, Ozone
(c) Ozone, CFCs
(d) Carbon dioxide, Ozone

7. _______is the process of accumulation of toxic substances in the food chain.
(a) Commensalism (b) Mutualism
(c) Biomagnification (d) Biotoxification

Match the Following

Each question contains statements given in two columns which have to be matched. Statements in column I have to be matched with statements in column II.

1. Choose the correct option:

Column I	Column II
i. Mutualism	a. Abiotic
ii. Light	b. Interaction
iii. Consumer	c. Food chain law
iv. Energy flow	d. Biotic

2. Choose the correct option:

Column I	Column II
i. Lion	a. Decomposer
ii. Deer	b. Autotroph
iii. Grass	c. Predator
iv. Earthworm	d. Herbivore

3. Choose the best match:

Column I	Column II
i. CFC	a. Organisation for environment
ii. DDT	b. Ozone depletion
iii. UNEP	c. Pesticide

True or False

Read the following statements and write your answer as true or false.

1. Landforms are found over continents and also on ocean floors.
2. An ecosystem may be natural or man-made.
3. The aquatic envelope of the earth is known as the atmosphere.
4. Decomposers release minerals trapped in organic matter.
5. The microscopic aquatic plants that float freely on the surface of water are called zooplanktons.

6. Animals that feed on dead animals are known as omnivores.

7. Rainwater is the primary source of water.

8. Multipedes or earthworms like light and are found in bright places.

9. The grazing food chain can be found in both aquatic and grassland ecosystems.

10. The ten percent law states that only 1 percent of the energy entering a trophic level is available for transfer to the next trophic level.

11. Ozone is a poisonous gas.

12. Eutrophication is the process of accumulation of toxic substances in the food chain.

13. Non-biodegradable solid waste like plastic and metal can be recycled.

Fill in the Blanks

Complete the following statements with an appropriate word/term to be filled in the blank spaces.

1. _______ is the biological environment that includes all species residing in a certain area, as well as the non-living physical components of the environment with which the organisms interact.

2. __________ act as a cleansing agent for the environment.

3. The top soil is __________ and just below it is a layer of __________, then at bottom is a layer of rocks known as __________.

4. ________ are known to survive at high temperatures.

5. __________ and __________ are what savannas include between desert and forest.

6. The various steps in a food chain at which the transfer of food (or energy) occurs are known as __________.

7. __________ protects the earth's surface from the sun's ultraviolet (UV) radiation.

8. The __________ was successful in negotiating a deal in 1987 to limit CFC production at 1986 levels.

9. __________ is the process of accumulation of toxic substances in the food chain.

10. Burning of waste at high temperature to form ash is called _______.

Assertion–Reason Questions

The following questions consist of two statements: Assertion (A) and Reason (R). Answer these questions selecting the appropriate option given below:

(a) Both A and R are true and R is the correct explanation of A
(b) Both A and R are true but R is not the correct explanation of A
(c) A is true but R is false
(d) A is false but R is true

1. Assertion (A): Burning of waste at high temperatures to form ash is called incineration.
Reason (R): This process is carried out in an incinerator.

2. Assertion (A): The interaction between organisms where one kills and eats the other is known as predation.
Reason (R): This type of interdependence is the most common of all but the same as scavenging or parasitism.

3. Assertion (A): There are various steps in the food chain at which the transfer of water occurs.
Reason (R): These levels are known as trophic levels.

4. Assertion (A): Green plants take water from the soil and carbon dioxide from the air to prepare food with the help of sunlight.
Reason (R): These plants are called autotrophs.

5. Assertion (A): Microbes break down (decompose) dead plants and animals into simple compounds.
Reason (R): The organisms that decompose organic matter are known as decomposers.

One-word Answers

1. Which layer of the earth has gases present in it?
2. What relationship does a malarial parasite have with humans?
3. Which law states that only 10 per cent of the energy entering a trophic level of organisms is available for transfer to the next trophic level?
4. Which food chain includes goat and lion?
5. In which method is the solid waste broken down into its simpler materials?

Define the Term

1. Symbiosis
2. Hydrosphere
3. Incineration

4. Producer
5. Consumer

Groups of Terms

Given below are groups of terms. In each group, the first pair indicates the relationship between the two terms. Rewrite and complete the second pair on a similar basis.

1. Goat: herbivore :: Eagles: _____
2. DDT: biomagnification :: CFC: ______
3. Zooplankton: aquatic :: Earthworms: _____
4. Oxygen: atmosphere :: Rocks: _____
5. Compost: composting :: Ashes: ______

Identify the Odd One Out

1. Atmosphere, lithosphere, hydrosphere, rocksphere
2. Incineration, reuse, recycle, trash
3. CFC, oxygen, nitrogen, carbon dioxide
4. Zooplankton, whale, earthworm, phytoplankton
5. Parasitism, predation, decomposer, mutualism

Image-Based Questions

1. What is this effect?

(Courtesy: Ralf Vetterle; https://pixabay.com/photos/industry-environmental-pollution-1761801/)

(a) Cold (b) Global warming
(c) Rain (d) Soil erosion

2. What type of food chain is this?

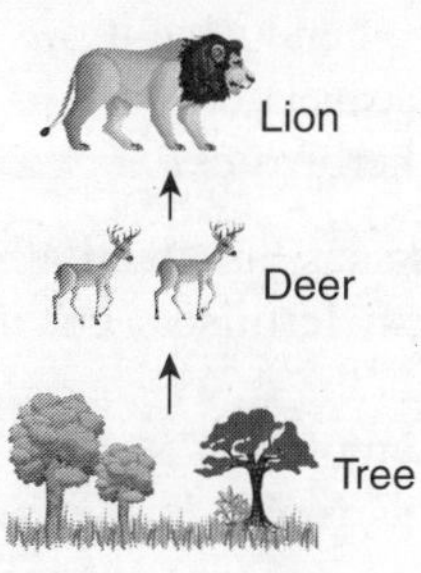

(a) Marine (b) Terrestrial
(c) Aquatic (d) Freshwater

3. What type of food chain is this?

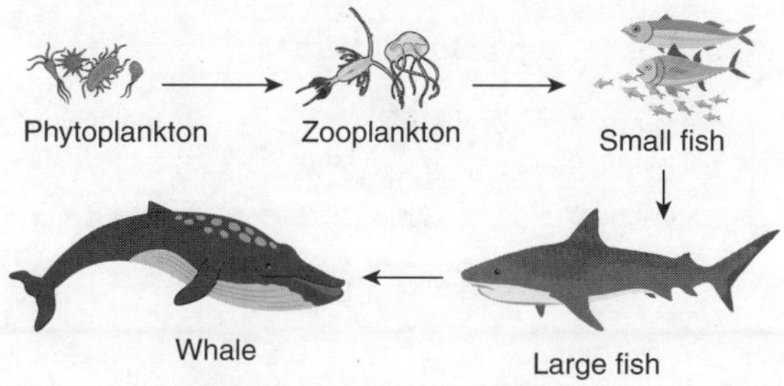

(a) Grassland (b) Inland
(c) Freshwater (d) Marine

4. Which component of the ecosystem is this?

(Courtesy: Stefan Schweihofer; https://pixabay.com/photos/wheat-crops-barley-crops-cereals-143373/)

(a) Producer (b) Consumer
(c) Heterotroph (d) Decomposer

5. Choose the organisms with the most energy:

(a) Tree (b) Lion
(c) Deer (d) Earthworm

6. According to biomagnification, which one will have highest chemical accumulation?

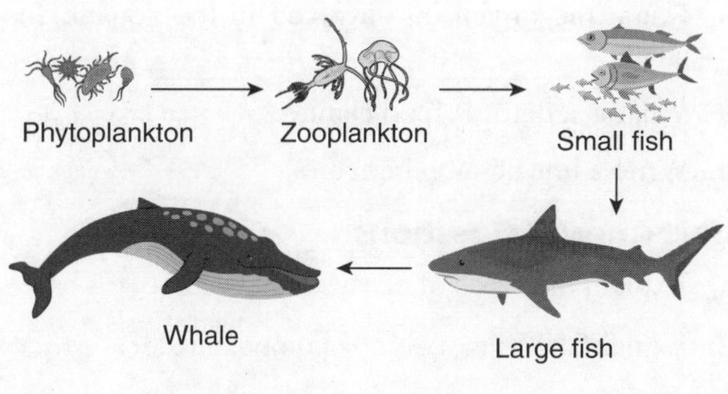

(a) Phytoplankton (b) Small fish
(c) Large fish (d) Whale

7. Which is the top predator here?

(a) Tree (b) Deer
(c) Lion (d) Sun

8. Choose the mode of nutrition:

(a) Parasitic (b) Saprophytic
(c) Heterotroph (d) Autotroph

Very Short Answer Questions

1. Define ecosystem.
2. What are components of ecosystem?
3. Give the types of ecosystems.
4. What is interaction between organisms?
5. What function does ozone play?
6. What are the methods of waste disposal?
7. What is a food chain?
8. Name the organisms involved in the aquatic food chain.
9. What is a detritus food chain?
10. Write a line about hibernation.

Short Answer Questions

1. Explain the types of ecosystems.
2. Explain the role of abiotic components.
3. What is the ten per cent law?
4. Write about
   (a) Pollution
   (b) Ozone
5. Write about symbiosis and mutualism in the interdependent relationship between organisms.
6. Write about pesticide's role in biomagnification.
7. Write about any two important modes of waste disposal.
8. Write a short note on energy flow.
9. What is biomagnification?
10. What is the significance of the food chain?

Long Answer Questions

1. Explain the structure of the earth.
2. Explain the ecosystem in detail.
3. Write an essay on the components of the ecosystem.
4. Explain food chain and its types.
5. Define the types of interactions between organisms.
6. Write a brief account on global warming.
7. Write an essay on ozone.
8. Explain biomagnification.
9. Write a note on waste disposal.
10. Write an essay on the human environment.
11. Write a short note on
    (a) Light
    (b) Temperature
    (c) Decomposers

Case Study-Based Questions

1. While roaming around the countryside, Suman observed goats eating plants and an eagle eating a dead mouse. She wondered why the preferences of food are different and what it was called.
   (a) What are scavengers?
   (b) What are herbivores?
2. Raman always sees his grandfather watering plants and giving them fertilisers but no solid food. Can you help Raman?
   (a) What are autotrophs?
   (b) What is the trophic level of plants?
3. While walking on the road, Ritesh sees cars producing smoke that goes up in the sky. This makes

him think about which layer it goes to. Can you help him with that?
(a) What is the atmosphere?
(b) How many layers does the earth have?

4. Rohan and his friends were playing when they saw air conditioners and refrigerators of his house were generating heat more than usual. After reading about it, they encountered some confusing terms. Can you help them with that?
(a) What are CFCs?
(b) What is the ozone layer?

5. Sandeep sees his Mom separating wet and dry waste to compost and for recycling, respectively. He wonders why she does that. Can you help him with that?
(a) What is recycling?
(b) What is composting?

NCERT Corner

1. What is irrigation?
2. What is crop rotation?
3. How can we enable land to regain nutrients naturally?
4. Explain mixed cropping.
5. Write a note on irrigation.

COMPETITION WINDOW (OBJECTIVE TYPE)

1. The amount of nutrients such as carbon, phosphorous and calcium present in the soil is referred to as:
(a) Nutrient of soil (b) Standing state
(c) Standing crop (d) Mineral state

2. Which of the following statement is not correct?
(a) Pyramid of energy is always upright
(b) Pyramid of numbers in grassland is upright
(c) Pyramid of biomass in sea is generally inverted
(d) Pyramid of biomass in sea is generally upright

3. Which of the following statements is correct?
(a) Biomass decreasees from first to fourth trophic level
(b) Energy content gradually increases from first to fourth trophic level
(c) Number of individuals decreases from first trophic level to fourth trophic level
(d) Energy content gradually decreases from first to fourth trophic level

4. The rate of decomposition is faster in an ecosystem due to following factors, except:
(a) Detritus rich in sugar
(b) Warm and moist environment
(c) Presence of aerobic soil microbes
(d) Detritus rich in lignin and chitin

5. Which of the following ecological pyramids is generally inverted?
(a) Pyramid of biomass in sea
(b) Pyramid of numbers in grassland
(c) Pyramid of energy
(d) Pyramid of biomass in forest

6. World Ozone Day is celebrated on:
(a) 5 June (b) 21 April
(c) 16 September (d) 22 April

7. Presence of plants arranged in well-defined layers depending on their height can be best seen in:
(a) Tropical rain forest (b) Grassland
(c) Temperate forest (d) Tropical savannah

8. The term ecosystem was coined by:
(a) AG Tansley (b) E Haeckel
(c) E Warming (d) EP Odum

9. Vertical distribution of different species occupying different levels in abiotic community is:
(a) Divergence (b) Stratification
(c) Zonation (d) Pyramid

10. Most animals are tree dwellers in:
(a) Coniferous forest
(b) Thorn woodland
(c) Temperate deciduous forest
(d) Tropical rainforest

Answer Keys

Multiple Choice Questions

1. (c) 2. (b) 3. (a) 4. (a) 5. (d)
6. (b) 7. (c)

Match the Following

1. i (b), ii (a), iii (d), iv (c)
2. i (c), ii (d), iii (b), iv (a)
3. i (b), ii (c), iii (a)

True or False

1. T 2. T 3. F 4. T 5. F
6. F 7. T 8. F 9. T 10. F
11. T 12. F 13. T

Fill in the Blanks

1. Ecosystem
2. Decomposers
3. Dark, subsoil, bed rock
4. Camels
5. Native grasses, small shrubs
6. Trophic levels
7. Ozone
8. United Nations Environment Programme (UNEP)
9. Biomagnification
10. Incineration

Assertion-Reason Questions

1. (b) 2. (c) 3. (d) 4. (a) 5. (a)

One-word Answers

1. Atmosphere
2. Parasitism
3. Ten per cent law
4. Terrestrial food chain
5. Recycle

Define the Term

1. Symbiosis is a long-term and close biological interaction between two organisms of different species which could be obligatory (where they have to depend on each other for survival) or facultative (where they can generally live independently as well). It is also defined as a physical attachment where two or more symbionts form a single body; this is known as conjunctive symbiosis
2. Minerals and other essential nutrients for plants are present in this layer. About 70% of the Earth's surface is covered with another natural element called water in the form of lakes, rivers and oceans. This aquatic envelope is known as the hydrosphere. The amount of total water on a planet is known as the hydrosphere. It includes water that is present in the air or underground. The water on the plant could be in any form, liquid, ice or vapour. On Earth, liquid water exists in the form of lakes, rivers, and oceans. The five basic components of a hydrosphere are surface water, freshwater, oceans, groundwater, atmospheric water vapour and glacial water.
3. Burning of waste at high temperature to form ash is called incineration. This process is carried out in an incinerator. Incinerators are used to destroy household, chemical and biological waste.
4. All green plants and blue-green algae can produce their own food (sugar and starch) from inorganic substance using light energy (photosynthesis). Green plants take water from the soil and carbon dioxide from the air to prepare food with the help of sunlight. Thus, plants are called autotrophs.
5. These are organisms that consume other organisms or their products as their food. All animals belong to this category. Consumers depend on producers for food, directly or indirectly. They are known as heterotrophs. Animals that depend on plants for food are known as herbivores (examples are cow, goat and deer). Animals that eat other living animals are called carnivores (examples are leopard, tiger and lion). Some animals feed on dead animals and are known as scavengers (examples are eagle and vulture). There are also animals that eat both other animals and plants; they are known as omnivores (examples are crow and humans).

Groups of Terms

1. Scavengers
2. Ozone depletion
3. Detritus
4. Lithosphere
5. Incineration

Identify the Odd One Out

1. Rocksphere
2. Trash
3. CFC
4. Earthworm
5. Decomposer

Image-Based Questions

1. (b) 2. (b) 3. (d) 4. (a) 5. (a)
6. (d) 7. (c) 8. (d)

Very Short Answer Questions

1. Ecosystem can be defined as a biological environment that includes all the species residing in a certain area, as well as the non-living physical components of the environment with which the organisms interact.
2. There are two components of ecosystem: abiotic and biotic components.
3. Two types of ecosystem are present: aquatic and terrestrial ecosystem.
4. Five types of interactions are present between organisms: competition, predation, mutualism, parasitism and commensalism.
5. Ozone is a poisonous gas but it protects the earth's surface from the sun's ultraviolet (UV) radiation.
6. Incineration, open dumping, landfills, composting and recycling are the different methods of waste disposal.
7. A food chain is unidirectional where transfer of energy takes place in only one direction. In a food chain, flow of energy takes place.
8. Phytoplankton → Zooplankton → Small fish → Large fish → Whale
9. This type of food chain begins with biological materials that have died. Microorganisms such as fungi and bacteria break down decaying organic waste into basic nutrients. Smaller carnivores devour these simple nutrients and decomposers, and eventually provide food for larger carnivores.
10. The leaves of plants in a hot area become spiny to reduce water loss and animals like snakes and frogs go into hibernation where they go underground.

Short Answer Questions

1. Aquatic ecosystem: This is of two basic types: freshwater system which includes streams, rivers and lakes; and marine ecosystem, which is present in ocean, seas and marshes worldwide. The individual ecosystems could be very limited in the territory.

   Terrestrial ecosystem: This is land-based and characterised by limited water support of flora and fauna life. They include:

   (a) Forest: From tropical to high-altitude forests, it varies by physical location and includes the different plants found in an area

   (b) Grassland: Native grasses and small shrubs are what savannas include between desert and forests

   (c) Desert: Cold night and hot days of deserts are found throughout the world

   (d) Tundra: Minimal rainfall and cold temperature makes life tough in the arctic regions, with snow-covered mountain ranges and no trees

2. Abiotic components are also known as non-living components like air, light and soil. The climatic factors are also abiotic components like rainfall, win and temperature.

   Soil: The uppermost layer of the earth's crust is soil and is formed by the weathering of rocks. Soil contains water, minerals and humus that are essential for plants growth. The main source of food is plants for most animals on earth and soil supports their growth. Topsoil is dark and below it is a layer of subsoil followed by a layer of rocks known as bedrock.

   Air: Oxygen and carbon dioxide are the most important components of air as all living organisms need oxygen for respiration and use carbon dioxide for photosynthesis. This cyclic movement of gases to maintain balance in nature occurs in the atmosphere.

   Water: Rainwater is the primary source of water. All organisms need water for survival; hence it is known as the universal solvent and even minerals are absorbed in form of solution. The amount of water that is available in a geographical area determines the habitat.

   Temperature: The degree of coldness or heat is known as temperature and it keeps varying from place to place and controls the distribution of flora and fauna on earth. Animals like penguins or polar bears are adapted to cold places while camels are known to survive at high temperatures.

3. The ten percent law states that only 10 percent of the energy entering a trophic level of organisms is available for transfer to the next trophic level.

   For example, if 1000 J of solar energy is received by green plants, only 1% of solar energy available on earth will be used by plants. So only 10 J (1% of 1000 J) is trapped by plants and the rest (990 J of energy) is lost to the environment. Next, only 10% of the 10 J energy of plant (1 J) is available to herbivores. Only 10% of the 1 J of energy is utilised by carnivores, while 0.9 J is lost to the environment.

4. (a) The balance of nature is disturbed mainly by the by-products of human development and the pollution of natural resources like air and water. It affects both flora and fauna.

   (b) Ozone (O_3) is a molecule made up of three oxygen atoms. While O_2, commonly known as oxygen, is required for all aerobic forms of life. Ozone is a poisonous gas. Ozone, on the other hand, plays an important role at higher altitudes in the atmosphere. It protects the earth's surface from the sun's ultraviolet (UV) radiation.

5. Symbiosis: This is the long term and close biological interaction between two organisms of different species which could be obligatory where they have to depend on each other for survival or facultative where they can generally live independently as well. It is also defined as physical attachment where the symbionts from a single body is known as conjuctive symbiosis.

   Mutualism is a part of symbiosis. For example, the relationship between the Ritteri sea anemones on whose tentacles the Ocellaris clownfish dwells. This territorial fish protects the sea anemone from predators and in turn the tentacles protect the clownfish from predators.

6. Pesticides and chemicals such as DDT and mercury released into lakes and rivers are ingested by the aquatic life. These are accumulated in their body tissues and transferred to other organisms who feed on them. Since the pesticides are industrially processed, they contain heavy metals such as lead, arsenic, cadmium, etc. These metals have been found in the bodies of animals and humans and have an adverse effect on them.

7. Incineration: Burning of waste at high temperature to form ash is called incineration. This process is carried out in an incinerator. Incinerators are used to destroy household, chemical and biological wastes.

   Open dumping: A conventional method in which solid waste are dumped in selected areas of a region. It causes pollution.

8. The energy that is transported from one trophic level to the next is unidirectional, meaning it does not return. When a herbivore consumes plant food, only a small portion of the energy is converted into new body mass; the remainder is lost as heat or utilised by the herbivore to carry out its life functions (movement, digestion, respiration, reproduction). As a result, when a carnivore eats a herbivore, the carnivore only receives a small portion of the total energy. Some energy will be lost as heat or 'used up' by the carnivore from the energy transferred from the herbivore to the carnivore. To gain enough energy to expand, the carnivore must eat a lot of herbivores. As we move up the food chain, the amount of energy transferred decreases as a result of the huge amount of energy wasted at each link.

9. Biomagnification is the process of accumulation of toxic substances in the food chain. The toxic chemicals that are released into the environment are absorbed by the organisms such as plants, earthworms, etc. These chemicals are then transferred to different trophic levels when these organisms are eaten by other organisms. The following are the few substances responsible for biomagnification:

   Pesticides: The pesticides and chemicals such as DDT and mercury released into lakes and rivers are ingested by the aquatic life. These get accumulated in their body tissues and are transferred to other organisms that feed on them. Since the pesticides are industrially processed, they contain heavy metals such as lead, arsenic, cadmium, etc. These metals have been found in the bodies of animals and humans and have an adverse effect on them.

   Mining Activities in the Ocean: In the deep sea to extract metals like zinc, aluminium, cobalt, silver and gold. The mining process produces some amount of selenium and sulphide, which are deposited in water and destroys the oceans and coastal regions.

10. Significance of food chains:

    (a) The study of food chains helps in understanding the food relationships and interactions in the organisms in an ecosystem. The food chains, transfer the energy between various living components of an ecosystem.

    (b) The food chains show the dynamic nature of an ecosystem or biosphere.

    (c) The movement of toxic substances like pesticides, and weedicides through food chains, can be harmful.

Long Answer Questions

1. Hint: Explain about the layers of the earth with their significance.
2. Hint: Define ecosystem, and describe its components in detail along with the food chain.
3. Hint: Write about the essential components of the ecosystem and their roles.
4. Hint: Describe the types of food chains present in an ecosystem with details of organisms involved and energy flow with the ten per cent law.
5. Hint: Describe the interdependent relationships between organisms mentioned in the chapter with an example.
6. Hint: Explain global warming with the reason behind the increase in temperature and pollution.
7. Hint: Define ozone, its characteristics, its depletion and the reasons for it.
8. Hint: Describe the modes of waste disposal in detail.
9. Hint: Write about the environment, ecosystem, pollution and ozone depletion.
10. Hints: (a) Describe light in detail, (b) Describe temperature in detail, (c) Describe decomposers in detail with their involvement in the detritus food chain.

Case Study-Based Questions

1. (a) There are some animals that feed on dead animals and are known as scavengers like eagle, vulture.

   (b) Animals that depend on plants for food are known as herbivores like cow, goat and deer.
2. (a) Plants are also known as producers since they supply food to all creatures, either directly or indirectly. Autotrophs are the foundation of any biotic system because they capture solar energy and use it to produce food for everyone.

   (b) The highest trophic level is of plants.
3. (a) The envelope of gas or air that surrounds the lithosphere is known as the atmosphere.

   (b) This planet consists of three layers: lithosphere, hydrosphere and atmosphere.
4. (a) Synthetic chemicals such as chlorofluorocarbons (CFCs) are used as refrigerants and in fire extinguishers.

   (b) Ozone, on the other hand, plays an important role at higher altitudes in the atmosphere. It protects the earth's surface from the sun's ultraviolet (UV) radiation. This radiation is extremely harmful to creatures; for example, it has been linked to skin cancer in humans.
5. (a) Solid wastes are broken down into simpler materials. These materials are used to make new items. Non-biodegradable solid wastes like plastic and metal can be recycled.

   (b) Organic wastes are filled into compost pits (2 m × 1 m × 1 m). It is then covered with a thin layer of soil. After three months, the same garbage is converted to organic manure.

NCERT Corner

1. The word agriculture comes from the Latin 'ager' meaning 'field' and 'cultura' meaning 'care'. Agriculture means the growing of plants or raising animals for human use. Quality in agriculture comprises mainly of activities that are essential in improving the environment for the growth of plants and animals and thereby improving the usefulness of plants and animals. Farming is the practice of agriculture and farm is known as the place where the work is done.
2. Wheat crop is planted in November and harvested in March and April. In June and July, rice crop is planted and in October-November it is harvested. The land lies empty in between two seasons. Instead of leaving it as it is, farmers plant a pulse crop during this time. So, by the time the farmer has to plant the cereal crops, the pulse is ready to be harvested. This is called crop rotation. This method maintains soil fertility, prevents crop diseases and pests, and ultimately gives a better crop yield.
3. To allow the land to regain the nutrients naturally, the practice is to either leave it free or fallow for one or more seasons. The dead plants, animals and other organic matter that get collected on the field are decomposed by microorganisms. Thus, the nutrients return to the soil.
4. Mixed cropping is the practice of growing two or more crops simultaneously on the same field. It is an age-old practice in our country. Farmers mix crops in such a way that the products and waste materials from one crop help in the growth of the other crop. For example, if a cereal crop such as wheat is grown along with a leguminous crop such as pulse (like gram), the uptake of nitrogen from the soil by the cereal is compensated for by the addition of nitrogen to the soil by the Rhizobium present in the legume.

5. Irrigation is the supply of water to agricultural fields at different intervals. A sufficient supply of rainwater is not always present; therefore, farmers need to supply water to the fields. Some crops require a lot more water than others. Sometimes the requirement for water depends upon the type of soil used for growing crops. Sandy soil requires more water than clayey soil. In summer, crops need more water due to excess of evaporation in comparison to crops grown in winter. Thus, other factors should be considered too while planning for irrigating fields.

Competition Window

1. (b)	**2.** (d)	**3.** (b)	**4.** (d)	**5.** (a)
6. (c)	**7.** (a)	**8.** (a)	**9.** (b)	**10.** (d)

CHAPTER **11**

Conservation of Plants and Animals

PREREQUISITES

You must have knowledge of:

- Ecosystem
- General structure of plants and animals
- Conservation
- Species and genera

LEARNING OBJECTIVES

At the end of this chapter, you will be able to:

- Explain deforestation, its causes and consequences
- Describe flora and fauna
- Describe methods of conservation of forests and wildlife
- Explain recycling
- Describe the Red Data Book and list the endemic species

INTRODUCTION

Wildlife conservation helps ensure that future generations enjoy the Earth's resources. Conservation can include harvesting natural resources, activities such as hunting, fishing, trapping and harvesting timber as well as non-consumptive activities such as bird watching, photography and hiking. Conservation must strike a balance between the wildlife and human populations. Conservation of wildlife implies ensuring that threatened and endangered species receive special management to protect their existence in the future.

Conservation may include the preservation or protection of natural resources that emphasizes non-consumptive activities. A habitat or ecosystem can be preserved by manipulation and an area may also be managed by doing nothing at all. For example, a forest can be conserved by allowing it to mature without any human manipulation such as timber harvesting, grazing or tree planting. The phenomenon of efficient use or preservation of various resources under physical laws is known as conservation. Conservation that is more focused on the protection of species with enhancement of ecosystem services to restore and maintain the habitats of the species is known as nature conservation. Carl F Jordan defined the term biological conservation as a philosophy of environment management that does not spoil, exhaust or extinguish our resources.

The natural forest ecosystem, comprising semi-natural and natural forests, can be protected and managed in many ways. Earlier, forest reserves and national parks were the only ways of conserving biological diversity. However, lately, multiple forest management methods, including harvesting of forest products, are more inclined towards a sustainable approach that includes both conservation and supplying benefits to the people, thereby maintaining the economy.

DEFORESTATION: ITS CAUSES AND CONSEQUENCES

For the survival and well-being of humans, it is essential that different types of plants and animals co-exist on this planet. Deforestation is the main threat against organisms and their existence in the biosphere (Fig. 11.1). It is also known as forest clearance or the removal of trees or forests

Fig. 11.1 Deforestation
(Irreparable environmental threats occur from deforestation...; USAID Indonesia; 2015; https://commons.wikimedia.org/wiki/File:Deforestation_site_(18418058486).jpg. Licensed under: PD USAID)

for the conversion of forest land to non-forest use. The non-forest use could be to create ranches, farms or for urban activities. Most of the concentrated deforestation occurs in the tropical rainforests. Today, only 31% of the Earth's land surface is forest. Every year, around 15–18 million hectares of forest are cleared.

Fun Fact!

Forest area net change (FANC) is often confused with the term deforestation, but they are not the same. FANC is the sum of all the forests gained along with all the forests lost in a given period of time. This change could be positive or negative, indicating gain or loss.

Deforestation causes imbalance in nature by disturbing many natural cycles and phenomena. Increase in temperature along with pollution and water scarcity are some of the most well-known consequences of deforestation. Cutting of trees appears to result in a significant increase in the level of carbon dioxide, which traps the heat rays from the Sun and increases the temperature of Earth, leading to global warming. The groundwater level also tends to decrease due to loss of vast areas of forests and trees due to low frequency of rainfall. This leads to loss in fertility of soil and increases the chances of calamities like droughts and floods. Soil erosion results in desertification as the land is devoid of humus and becomes less fertile due to exposure of the top layer to wind and conversion of the land into a hard and rocky one.

Deforestation takes place when forest land is cleared for other purposes like construction or mining. The natural causes of deforestation are:

- Severe draught
- Forest fires

The man-made causes of deforestation are:

- Cultivation: Removal of wood for fuel contributes to up to 5% of deforestation; 14% is contributed by logging; while commercial agriculture and subsistence farming contribute approximately 32% and 48%, respectively.
- Construction due to increased population density
- Exploitation of wood for purposes like furniture or fuel
- Overpopulation

The **consequences** of deforestation based on present conditions or from fossil records are:

- Loss of biodiversity
- Habitat loss
- Increase in aridity
- Increase in extinction
- Climate change
- Population displacement
- Desertification
- Global warming
- Reduction in biosequestration of atmospheric carbon dioxide
- Increase in negative feedback cycles

Let us examine the main consequences in detail.

Drought

When there is prolonged shortage of water supply in an area due to atmospheric conditions such as low precipitation and shortage of ground or surface water, it is known as drought (Fig. 11.2). It can last for days, months or years with a significant impact on the agriculture and the ecosystem of the affected region. It affects not only the ecosystem but also the local economy.

Fig. 11.2 Drought
(Courtesy: Sven Lachmann; https://pixabay.com/photos/dryness-drought-cracks-dry-3618653)

In the tropics, the annual dry seasons contribute significantly towards increasing the chances of drought and bush fires. Conditions can worsen due to increased evaporation of water caused by longer periods of heat. Drought has been a recurring phenomenon, but it has become unpredictable and extreme due to the climate change caused by deforestation. The productivity of most arid ecosystems is very low.

Droughts can be of three types (Fig. 11.3):

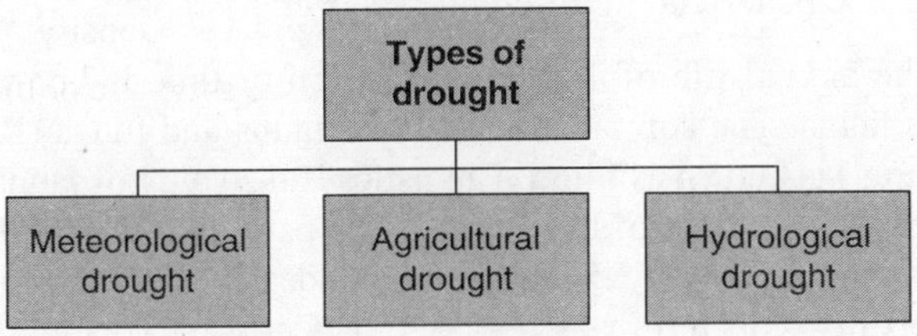

Fig. 11.3 Types of drought

- **Meteorological:** When the precipitation of the atmosphere is lower than average for a long period of time, it is known as meteorological drought. This is the most common type of drought.
- **Agricultural:** When crop production along with the ecology of the range is affected, it is known as agricultural drought. This can arise due to soil erosion caused by deforestation and poorly managed agriculture or due to increased irrigation leading to change in precipitation.
- **Hydrological:** When the level of water in reserves of different sources like lakes, aquifers and reservoirs falls below a significant value, it is known as hydrological drought. This can be triggered for reasons such as overuse of water or diversion of water to other areas, leading to depletion or drought.

Soil Erosion or Aridity

The displacement of the upper layer of the soil that leads to soil degradation is known as soil erosion. Soil erosion is a slow process of serious loss of the top-most layer of soil that may occur at an alarming rate and go unnoticed. There are various natural erosive agents like air, snow, water, animals and plants. Soil erosion can be of different **types**, depending on the causative agents:

- Glacial erosion
- Water erosion
- Wind erosion
- Snow erosion
- Zoogenic erosion
- Anthropogenic erosion

Soil erosion leads to aridity (Fig. 11.4) and this results in lower water levels, reduced crop production and damage to drainage networks. Worldwide, the rate of soil erosion has increased by 10–50 times because of excessive offsite and onsite problems. Desertification is the end result of soil erosion. The two most common causative agents of erosion are wind and water. They are responsible for land degradation, which is about 84% of all degradation in the world. The activities that stimulate soil erosion are climate change, deforestation, intensive agriculture and excessive urbanisation.

Fig. 11.4 Aridity
(Courtesy: Photocurry; https://pixabay.com/photos/dry-arid-yellow-nature-landscape-856544)

Desertification

In the drylands, the type of degradation of land in which loss of biological productivity is observed to be due to human-induced activities or natural processes that increase the aridity in fertile areas is known as desertification. A variety of factors lead to desertification of land, such as overexploitation of soil and climate change. Deserts form and develop naturally. The factors that influence desertification are:

- Deforestation
- Climate change
- Improper land management
- Continuous use of fertilisers and pesticides

Deforestation or loss of vegetation is an immediate cause of desertification and is connected to other factors such as climatic shifts, drought, overgrazing and excessive loss of vegetation and trees for purposes like construction materials or fuel. Deforestation leads to excessive erosion and run-off that makes the soil infertile, with the lower layers of soil baking in the sun, thereby becoming hard and unproductive.

Fun Fact!

Scientists claim that the world's largest desert, Sahara, was formed due to natural climate change that caused the area to lack water in different eras, thereby converting it from savanna to desert.

Climate Change

The impact of global warming on the Earth's weather patterns is a part of climate change. Though change in climate is natural, the changes taking place currently are rapid and more frequent and caused by the emission of greenhouse gases. These greenhouse gases, primarily carbon dioxide and methane, are produced by burning of fossil fuels, deforestation, cement manufacture and agriculture. Greenhouse gases are generally transparent to sunlight and are known to trap the heat near the surface of the Earth. Infrared rays emitted by the Earth as heat are absorbed by greenhouse gases, leading to heating of the planet and changes such as increased global warming.

On an average, the temperature on Earth has increased two-fold over the last few years. As a result, heat waves, intense storms and forest waves are more frequent and deserts are expanding at the cost of vegetation. The increase in temperature due to global warming, especially in the Arctic, has contributed to the melting of glacial, permafrost and sea ice. Climate change not only threatens us with scarcity of water and food with loss in agriculture and economy, but also with more frequent floods, extreme heat and diseases. The extremes of global warming will be warmer weather with a huge rise in sea levels and more acidic oceans. These changes in the environment affect all species and lead to extinction.

Fun Fact!

Negative climate feedback is a phenomenon where the severity of the initial change in the climate is decreased by it. These changes lead to a secondary change to reduce the effect of the former. Increase in temperature, leading to accelerated formation of clouds, which in term could reduce the warming is an example of negative feedback.

FLORA AND FAUNA

The native plant life present in a particular area or time is known as flora. Flora includes some bacteria and fungi like skin or gut flora. In Roman mythology, Flora is the goddess of flowers, plants and fertility. According to the region, climate, environment and period, plants are categorised into different types of flora:

- **Agricultural and horticultural flora:** Grown by humans.
- **Native flora:** Indigenous and native flora of an area.
- **Weed flora:** These comprise three types of undesirable plants: weedy species, native species and invasive species. Some weeds have been observed to be beneficial to different ecosystems.

The animal life of a particular region or time is known as fauna. The combination of flora, fauna and funga (for fungi) is known as **biota**. Faunistics is the study of fauna, that is, the study of the animals of a particular region. In Roman mythology, Fauna is the goddess of animals and fertility. This term was first used by Carl Linnaeus in his book *Fauna Svecica*.

Indian Flora and Fauna

In India, there are around 2000 species of birds, 500 species of mammals, 10,000 species of insects and a wide variety of reptiles, amphibians and fish. The most popular animals in India are the Royal Bengal tiger, elephant, lion (especially white lion), rhino and wild bison (Fig. 11.5). There are also a great variety of monkeys, cats and dogs. The animals are found in different areas based on their habitat; for example, lions are found in the rocky hills of Gujarat while elephants are found in the forests of Kerala, Odisha and Karnataka. Among reptiles, crocodiles, lizards and snakes are common, especially the king cobra and krait.

Fig. 11.5 Indian fauna: Royal Bengal tiger, elephant, wild bison and rhino
(Courtesy: Karsten Paulick, https://pixabay.com/photos/tiger-bengal-tiger-royal-tiger-330148/; Michael Siebert, https://pixabay.com/photos/elephants-herd-family-calf-2923917; Lubos Houska, https://pixabay.com/photos/bison-european-bison-animal-big-2118538/; Pexels,https://pixabay.com/photos/animal-africa-wilderness-wildlife-1281695)

Apart from land animals, multiple varieties of birds are found in India with different colours and patterns. The peacock is the National Bird of India and is very popular. India has 441 wildlife sanctuaries, 80 national parks and 35 zoological parks to conserve and protect these fauna.

Among all the countries in the world, India has a wide range of climatic conditions to support vegetation of different types. The floristic regions of India are Assam, the Andamans, the Ganga plains, the Eastern Himalayas, the Western Himalayas, the Malabar, the Indus plains and the Deccan. Due to its tropical location, India is rich in deciduous forests in which sal and teak are the most important. Other important trees are coconut, sandalwood, pine and cedar.

CONSERVATION OF FORESTS AND WILDLIFE

The flora and fauna along with their habitats can be protected in three ways, based on area (Table 11.1):

- Wildlife sanctuaries
- National parks
- Biosphere reserves

In these areas, different activities that harm flora and fauna like cultivation, deforestation, grazing, plantation, poaching and hunting are prohibited.

Table 11.1 Types of protection areas

Wildlife sanctuary	Animals and their habitats are protected here without any disturbance
National park	Animals are allowed to use natural resources and an area is reserved for their habitat
Biosphere reserve	Both flora and fauna along with animal resources and tribal people are protected in this large area

SCIENCE IN DAILY LIFE

Note down the names of the biosphere reserves, wildlife sanctuaries, national and zoological parks in your area and visit them to know about the flora and fauna protected there.

Biosphere Reserves

An area where all the living organisms coexist to support life is known as a biosphere. The variety of organisms with their inter-relationships and their relationship with the environment is known as biological diversity or biodiversity.

Fun Fact!

Did you know that Pachmarhi Biosphere Reserve (Madhya Pradesh) has the same type of flora and fauna as in the upper Himalayan peaks and lower Western Ghats?

The areas that are meant to conserve and protect animals and plants along with tribal people are called biosphere reserves. In India, there are around 18 biosphere reserves established by the Government. To promote sustainable development, the Man and the Biosphere programme by UNESCO established these reserves in 1971.

In this programme, the areas that are recognised are:

- Those where all living organisms coexist and work in harmony with each other and the environment
- The most typical coastal and terrestrial ecosystems
- Those that conserve the biodiversity and natural ecosystem to promote sustainable development

Biosphere reserves are divided into three zones:

- **Core Zone:** This is an area with no human intervention and legally protected by the prohibition of any human activity. This is the innermost area of the reserve with an undisturbed ecosystem. It provides information that helps to maintain the quality of the environment in the surrounding areas with access to sustainable activities.
- **Buffer Zone:** This is the area that surrounds the core zone; only research and education activities by humans can take place here. The conservation objectives of the core area are not obstructed by this zone. The activities allowed in this area help to manage and enhance the production of the agricultural land, natural vegetation, forests and fisheries. Sometimes, the zone may include tourism and recreation facilities, but human activities are less intensive in this zone.
- **Manipulation Zone:** This is the outermost or peripheral area of the biosphere reserve where human activities like recreation, cropping, forestry and settlements are allowed with the cooperation of the local people and reserve management. Degraded areas are restored to their natural state by these activities. In this zone, various professionals like

conservation agencies, scientists, local communities and other stakeholders work in a sustainable way for the benefit of the people living in that area.

Biosphere reserves (Table 11.2) are importance for the following reasons:

- **Conservation:** The main purpose of these reserves is the conservation of flora and fauna of the ecosystem along with protection of genetic diversity and the landscape of the area without affecting the inhabitants.
- **Restoration:** To restore damaged ecosystems and habitats.
- **Development:** To ensure development in different areas like social, cultural and economic.
- **Land usage:** Farmers, landowners, industries, public institutions, scientists and conservation groups can work together for better land management.
- **Research and education:** The information gathered from these areas regarding conservation, development and restoration of the ecosystem makes it possible to recreate the landscape that was once destroyed by human exploitation.

Wildlife Sanctuaries

The area in which animals and their surroundings are protected from any damage and disturbance is known as a wildlife sanctuary. The actions of humans that are forbidden in these sanctuaries include poaching, capturing and hunting of wildlife. The animals are provided with comfortable living conditions with rivers, mountains and dense forests. It is one of the finest ways of protecting endangered species. Wildlife sanctuaries do not allow tourism as their main aim is to educate people about how the ecosystem should be untouched by human exploitation. In India, the wildlife sanctuaries are responsible for protecting threatened animals like golden cat, gharial, white-eyed buck, marsh crocodile, blackbuck, python, elephant and rhinoceros.

Wildlife sanctuaries (Table 11.3) are important for a number of reasons:

- To protect animals in their natural habitat without relocating them
- To protect endangered species
- To carry out research and other biological activities to learn about the fauna
- To monitor endangered species so they grow and reproduce while protecting and enabling the breeding of some other species
- To rehabilitate injured and abandoned animals before releasing them back into the forest
- To protect and conserve animals from humans and predators

Fun Fact!

The Fleurieu Peninsula Sanctuary protects the she-oak habitat of the glossy black cockatoo that is on the verge of extinction.

The human activities that can take place in the wildlife sanctuary are:

- Wine production
- Electricity production and distribution
- Organic horticulture
- Grazing and farming enterprises

Table 11.2 Main biosphere reserves in India

Biosphere reserve	Year of establishment	Place (India)
Sundarbans Biosphere Reserve	2001	West Bengal
Nanda Devi Biosphere Reserve	2004	Uttarakhand
Nilgiri Biosphere Reserve	2000	Tamil Nadu, Karnataka, Kerala
Great Nicobar Biosphere Reserve	2013	Nicobar
Gulf of Mannar Biosphere Reserve	2001	Tamil Nadu
Simlipal Biosphere Reserve	2009	Odisha
Nokrek Biosphere Reserve	2009	Meghalaya

Table 11.3 Main wildlife sanctuaries in India

Wildlife sanctuary	Place in India
Gir National Park and Wildlife Sanctuary	Gujarat
Chinnar Wildlife Sanctuary	Kerala
Bharatpur Bird Sanctuary	Rajasthan
Periyar Wildlife Sanctuary	Kerala
Chilika Lake Bird Sanctuary	Odisha

National Parks

A natural park created and protected by the Government has the objective of conserving the flora and fauna of the area. National parks are developed on semi-natural and natural land and are declared as such by the state.

The first national park, Yellowstone, was established in the US in 1872. According to the 2006 report of the IUCN, there are around 6,555 national parks globally. IUCN stands for International Union for the Conservation of Nature and Natural Resources. It manages protection of the ecosystem. The biggest challenge for the management of national parks is to maintain a balance between conservation and development of the ecosystem. They protect and preserve the whole landscape, fauna, flora and historic objects of that particular area.

The first national forest in India was Satpura National Park with the finest teak trees. Today, there are more than 100 national parks in India (Table 11.4), but even the protected parks are not safe from human exploitation and damage.

Table 11.4 Smallest and largest national parks and wildlife sanctuaries in India

Category	Name of the national park or wildlife sanctuary	Place in India
Smallest wildlife sanctuary	Bor Tiger Reserve	Maharashtra
Largest wildlife sanctuary	Rann of Kutch	Gujarat
Smallest national park	South Button Island National Park	Andaman and Nicobar
Largest national park	Hemis National Park	Jammu and Kashmir

In Satpura National Park, evidence of prehistoric human life in the form of a rock shelter has been found. Paintings in these rock shelters often give us a glimpse into the life of the primitive people. They often show figures of people fighting, animals fighting, people dancing, playing a musical instrument and hunting. Around 55 of these rock shelters have been found in Pachmarhi Biosphere Reserve. A number of tribal people also still reside in these shelters.

Project Tiger

Tigers are slowly disappearing from our forests. In April 1973, the Government of India launched a tiger conservation programme known as Project Tiger. This project was headed by Indira Gandhi and aimed at protecting the Bengal tiger from extinction along with declaring and preserving its natural habitat as a natural heritage of biological importance. These areas represent the diversity in ecosystem across the tiger's range in India. The main focus was to increase the tiger population and ensure their migration to the adjacent forests. Based on the survey results, the population of tigers in 2018 increased from 2,603 to 3,346.

Satpura Tiger Reserve has a played a primary role in increasing the population of tigers. It contains a significant number of animals like wild buffalo, lion, barasingha and elephant but their population is diminishing and many are on the verge of extinction. Disturbance in the habitat of animals is the main cause of the decrease in survival rate. The smaller animals in the ecosystem like frog, snake and lizard are at greater danger of extinction. They may be small but their role in the food chain of the ecosystem is significant.

RED DATA BOOK

The Red Data Book is a public document created to record the rare and endangered species of animals, plants and fungi along with local subspecies of a particular region. It helps to provide complete information regarding different programmes on rare and endangered species for research, study and monitoring purposes. This book helps to protect and identify the species that are on the verge of extinction. The book originated in Russia and was known as the Red Data Book of the Russian Federation (RDBRF) or Russian Red Data Book. It was based on research conducted by Russian biologists between 1961 and 1964. Currently, IUCN maintains the Red Data Book as a complete record of every species. The pages of the book are colour-coded and arranged based on the risk of extinction. The colours used are listed in Table 11.5.

Table 11.5 Colours of the Red Data Book

Colour	Indication
Black	Species confirmed to be extinct
Red	Endangered species
Amber	Vulnerable species
White	Rare species
Green	Earlier endangered but now recovering
Grey	Rare, endangered or vulnerable species with insufficient information to be properly classified

The Red Data Book tags the species as non-threatened, threatened and unknown. It also provides information on why the species became extinct along with data about its population trends and distribution. The advantages of the Red Data Book are:

- Evaluation of the population of a particular species
- Estimation of risk of taxa becoming extinct
- Identification of all birds, animals and other species and information regarding their conservation status
- Framework of guidelines to implement measures for the protection of endangered species

Red Data Book of India

The conservation status of plants and animals endemic to the Indian subcontinent can be found in the Red Data Book of India. The Zoological Survey of India (ZSI) and Botanical Survey of India (BSI) provide the data under the guidance of the Ministry of Environment, Forest and Climate Change. In the Red Data Book of India, the species that are included in different categories are as listed in Table 11.6.

Table 11.6 Conservation status of some organisms in India

Category	Species
Critically endangered	Kashmiri stag, Malabar civet, river dolphin (Fig. 11.6)
Critically endangered arthropods	Peacock, tarantula (Rameshwaram Parachute spider)
Critically endangered amphibians and reptiles	Toad-skinned frog, white-spotted bush frog, gharial
Critically endangered fish	Pondicherry shark, Ganges river shark, Pookode lake barb

Fig. 11.6 Gharial and river dolphin
(Courtesy: Bishnu Sarangi, https://pixabay.com/photos/gharial-crocodile-chambal-river-4883131; Paul Brennan,https://pixabay.com/photos/dolphin-wildlife-wild-river-ocean-2074048

> **Fun Fact!**
>
> Did you know there is a Green Data Book with data about the environment of more than 200 economies including indicators such as forestry, agriculture, pollution, biodiversity and sanitation?

Types of Species

Endangered species: These are species that are on the verge of extinction either worldwide or in a particular region. The factors that lead to endangerment are poaching, habitat loss and invasive species. The IUCN Red List contains the global conservation status of various species. Laws have been enacted by many nations to protect the conservation-reliant species by measures such as restriction of land development, forbidding hunting or creating protected areas. Captive breeding and habitat restoration are the target of extensive conservation efforts. The Siberian tiger (Fig. 11.7) and Saimaa ringed seal are examples of endangered species.

Fig. 11.7 Siberian tiger
(Courtesy: Marcel Langthim, https://pixabay.com/photos/tiger-siberian-tiger-snow-predator-1975790/)

Rare species: These are organisms that are very scarce, uncommon or infrequently encountered. They are different from endangered or threatened species. The species that fall under this category are classified by the official bodies of national or state governments. The IUCN does not make this designation. If there are fewer than 10,000 of a species in the world, it will come under this category. Endemic species also influence the concept of rare species as almost 75% of known species can be classified as rare. IUCN uses the term 'rare' for isolated geographical location as they are not endangered but 'at risk'. If the species is distributed in large areas, it may be endangered or vulnerable, but not rare. However, a small population size would not be able to recover from ecological disasters. Examples of rare species are Fennec fox, hornbill, Himalayan brown bear (Fig. 11.8) and wild Asiatic buffalo.

Fig. 11.8 Himalayan brown bear
(Courtesy: Gerhard, https://pixabay.com/photos/bear-predator-brown-bear-dangerous-1376361)

Fig. 11.9 Arctic tern
(Courtesy: Matthias Kost, https://pixabay.com/photos/arctic-tern-bird-flying-tern-6285345/)

Vulnerable species: These species are categorised by the IUCN as threatened with extinction in the future if the circumstances leading to decline in reproduction and survival remain the same. Habitat loss or destruction of the home of the species leads to vulnerability. An example is the military macaw. Around 6789 plants and 596 animals have been classified as vulnerable. The conservation of vulnerable breeds has improved with the use of techniques such as cryoconservation of animal genetic resources.

Endemic species: These species are found in a particular and defined location such as a state, island or country. The state of these species is known as endemism. The organisms that are also found somewhere else come under the category of indigenous species and are not endemic. For example, the Cape sugarbird is only found in southwestern South Africa and is endemic to that part of the world. The species restricted to only one geographical area are also known as precinctive. Endemic species are more prone to becoming extinct or endangered as they are already confined to a small area of distribution. The areas with residing endemic species are considered to be top priorities in term of conservation as endemism is a factor used to measure the biodiversity of a given region.

The opposite of endemic species is **cosmopolitan species** that are found in a widespread range.

Migration

The large-scale movement of species to a different environment is known as migration. It is a natural behaviour and a necessity for the life cycle of many species. There are two types of migration: animal and bird. Migration is often based on seasons or sometimes takes place on a daily basis. The members of a species migrate to take advantage of better and favourable conditions for their survival based on availability of food, mating, safety from predators and other environmental conditions. The route of migration is fixed on their first journey by the older members of the group that pass the information to other members. Considerable similarity is seen in behaviour and migratory cues of the different phases of migration. Photoperiod and weather conditions play a role as environmental cues along with hormone levels to determine the time of migration. Olfaction and magnetoreception are used by migratory species to navigate the route or orientation, respectively.

Seasonal changes and events are factors that determine the migration methods. The energy used and rate of migration of animals is directly related to the safety of the members while moving from one location to another. The rate of migration is higher when the species migrates to a region with high habitat or food competition. Birds, fish, insects and plants (seed dispersal) practice migration under different environmental constraints.

In a particular time of the year when the climate changes, migratory birds fly from faraway areas to lay eggs in warm areas when their natural habitat becomes cold and inhospitable. The migration route is often north and south along a flyway. The rate of predation and mortality is high in migration and includes hunting as well. The Arctic tern (Fig. 11.9) is known for its long-distance migration record. It travels between the Arctic breeding grounds and the Antarctic every year. The change in the length of day is one of the primary control factors for migration as the birds use celestial cues, mental maps and the Earth's magnetic field.

Recycling

The process of converting waste materials into new objects and materials is known as recycling. The ability of a material to reacquire its original state is the key factor in ensuring its recyclability. This not only reduces waste and the consumption of raw materials, but it also helps to reduce greenhouse gas emissions along with energy use and pollution. The famous 3Rs of modern waste reduction are Reduce, Reuse and Recycle (Fig. 11.10).

Recyclable materials include many types of metal, glass, cardboard, tyres, paper, electronics and batteries. The reuse of biodegradable waste and compost like food and garden waste is also a form of recycling. The recyclable materials are sorted, then cleaned and reprocessed into new materials. For example, used office paper could be converted to new office paper by recycling them. Metals can be remanufactured repeatedly without any loss in purity. Recycling can be difficult sometimes or too expensive and so reuse is an alternative.

Fig. 11.10 3 Rs
(Cover of the game "reduce reuse recycle"; Nadine3013; 2017; https://commons.wikimedia.org/wiki/File:Reduce_Reuse_Recycle.jpg. Licensed under: CC-BY-SA-4.0)

Plastic recycling: The process of scrap or waste plastic recovery along with its reprocessing into useful materials is known as plastic recycling. The products could be completely different from the original state such as melting a soft drink bottle and casting it as a plastic table. The same piece of plastic can be used about 2–3 times before its quality degrades. There are two types of plastic recycling:

- **Physical recycling:** The plastics are melted to form new objects like polyester for clothing made by recycling PET water bottles. The molecular weight of the plastic can change and unwanted side products could be obtained in this process.
- **Chemical recycling:** The conversion of plastic to monomers such as using dialkyl terephthalate as a catalyst with alcohol on PET. The product obtained could be used with ethylene glycol to make new polyester.

Paper recycling: The process by which new paper products are obtained from waste paper is known as paper recycling. There are a number of advantages in paper recycling such as preventing methane production from waste paper as it breaks down (paper fibre contains carbon and recycling keeps the carbon in it locked for a longer period). Virgin fibres are added in the pulp recipe of paper production as the repeated processing of fibres makes it too short for the production of new paper. Three types of paper are used as feedstock to make recycled paper:

- **Mill broke:** The trimmings of paper and scrap from paper manufacturers which are recycled in a paper mill.
- **Pre-consumer:** The material left in the mill but discarded before consumption.
- **Post-consumer:** Discarded material such as old magazines and newspapers (Fig. 11.11) after de-inking.

Fig. 11.11 Post-consumer discarded newspaper bundles
(Courtesy: Pexels, https://pixabay.com/photos/bundle-jute-rope-newspaper-1853667)

Reforestation

The intentional or natural restocking of existing woodlands and forests that are currently under the effect of deforestation is known as reforestation. Rectification of the effects of deforestation and improving the quality of life by decreasing effects such as global warming and pollution, as well as harvesting of resources are the main objectives of reforestation. Significant attention has been paid to reforestation to reduce the effect of climate change. The net area of forest loss has been decreased substantially but the target set by the UN is to increase the area by 3% by the end of 2030. According to the data, the global forest area lost between 1990 and 2020 was around 178 million hectares, making reforestation a necessity and not a choice. The planting of new trees using seedlings has a failure rate of 90%. In deforested areas, the existing root system is often present. The growth of these plants can be accelerated by coppicing or pruning. The new seeds of these plants are not planted, thereby making this process more economical. The survival rate will be higher and as the root system already exists, it can trap groundwater easily. This method is known as **farmer-managed natural regeneration.**

The Indian Forest Acts that were implemented under the British served as a base for the Indian Forest Act, 1927. This act reserves and consolidates the forest-covered areas and the significant wildlife with regulation in movement and transition on the products of the forests such as timber and other products. It also states the procedures that should be followed to declare an area as a Reserved Forest or Protected or Village Forest. It also defines the prohibited activities with penalties and forest offences on violation of the provisions of this act. The area of land notified under the Indian Forest Act and State Forest Act with complete protection is known as Reserved Forest. All activities are forbidden in these areas unless permitted by the officials.

SUMMARY

- Deforestation is the main cause of many environmental problems such as global warming, soil erosion, drought, climate change and pollution.
- Increase in deforestation has severe consequences like loss of biodiversity.
- The loss of flora and fauna worldwide lead to the need for protection and conservation and therefore wildlife sanctuaries, national parks and biosphere reserves were set up.
- The Red Data Book has different coloured sheets indicating the type of species such as endangered, rare, vulnerable and extinct.
- Endemic species are more prone to becoming extinct.
- Many species tend to migrate in order to increase their survival rate for better breeding; deforestation leading to climate change affects these species.
- Reforestation and recycling are methods used to rectify the damage inflicted by deforestation and pollution, in order to conserve and protect the flora and fauna along with the whole ecosystem.

PRACTICE QUESTIONS

Analyse Your Concepts (School Exam Based)

Multiple Choice Questions

1. Agricultural flora is:
   (a) Native flora (b) Indigenous flora
   (c) Grown by humans (d) Undesirable plants

2. An area where all living organisms coexist to support life is known as _____:
   (a) Biosphere (b) Wildlife sanctuary
   (c) Zoological Park (d) Wildlife

3. ________ does not allow tourism.
   (a) Biosphere (b) Wildlife sanctuary
   (c) Zoological Park (d) Wildlife

4. The first national forest in India was:
   (a) South Button Island National Park
   (b) Hemis National Park
   (c) Rann of Kutch
   (d) Satpura National Park

5. The ________is a public document created to record the rare and endangered species of animals, plants and fungi along with local subspecies of a particular region.
   (a) Yellow data book (b) Green data book
   (c) Red Data Book (d) Black data book

6. Kashmiri stag and river dolphin belong to:
   (a) Critically endangered species
   (b) Critically rare species
   (c) Vulnerable species
   (d) Extinct species

7. ________ species are found in a particular and defined location.
   (a) Cosmopolitan (b) Rare
   (c) Endemic (d) Migratory

8. Chemical recycling involves the conversion of plastic to monomer using ____ as a catalyst.
   (a) Hydrochloric acid
   (b) Dialkyl terephthalate
   (c) Dialkyl ketonce
   (d) Carboxylic acid

9. The area of land notified under the Indian Forest Act and State Forest Act with complete protection is known as _______.
   (a) Endemic forest (b) Deforest
   (c) Reserved Forest (d) Unreserved forest

10. The Arctic tern is known for its ________migration.
   (a) Short distance (b) Average distance
   (c) Long-distance (d) None of the above

Match the Following

Each question contains statements given in two columns which have to be matched. Statements in column I have to be matched with statements in column II.

1. Choose the correct option:

Column I	Column II
i. Black	a. Rare
ii. Red	b. Extinct
iii. Amber	c. Vulnerable
iv. White	d. Endangered

2. Choose the correct option:

Column I	Column II
i. Smallest wildlife sanctuary	a. Rann of Kutch
ii. Smallest national park	b. Hemis National Park
iii. Largest wildlife sanctuary	c. Bor Tiger Reserve
iv. Largest national park	d. South Button Island National Park

3. Choose the best match:

Column I	Column II
i. Core Zone	a. The outermost or the peripheral area of the biosphere reserves
ii. Buffer Zone	b. With no human intervention and legally protected
iii. Manipulation Zone	c. Only permits the activities of research and education by humans

4. Choose the correct option:

Column I	Column II
i. Sunderbans Biosphere Reserve	a. Tamil Nadu
ii. Gulf of Mannar Biosphere Reserve	b. West Bengal
iii. Simlipal Biosphere Reserve	c. Meghalaya
iv. Nokrek Biosphere Reserve	d. Odisha

True or False

Read the following statements and write your answer as true or false.

1. Deforestation causes imbalance in many natural cycles.
2. Agricultural drought is the most common type of drought.
3. Desertification is the end result of soil erosion.
4. Carbon dioxide and methane are produced by burning of fossil fuels.
5. Native flora is the type of flora that includes undesirable plants.
6. India is rich in tropical forests.
7. Core Zone is the area in the biosphere reserve with no human intervention.
8. Sunderban Biosphere Reserve is located in West Bengal.
9. Wildlife sanctuaries protect animals and plants along with tribals.
10. The organisms that are either very scarce or uncommon are known as rare species.
11. Yellowstone National Park was the first national park to be established.
12. The species that are found in a particular and single defined location are known as endangered species.
13. Magnetoreception is used by migratory species to navigate the route for migration.
14. Indian Forest Act states the procedures that should be followed to declare an area as a Reserved Forest.
15. Flamingo is known for its long-distance migration record for travelling between Arctic breeding grounds and the Antarctic every year.

Fill in the Blanks

Complete the following statements with an appropriate word/term to be filled in the blank spaces.

1. The part of the Earth where all the living organisms coexist to support life is known as ___________.
2. _________is the largest wildlife sanctuary of India located in Gujarat.
3. In April 1973, the government of India launched a tiger conservation programme known as ___________.
4. The combination of flora, fauna and funga is known as ___________.
5. ________ are the species that are found in widespread range
6. The first national forest of India was __________ with the finest teak trees.
7. Rock shelters often give us the brief idea of the primitive people's life by ________.
8. ________ is a book that helps to protect and identify the species that are on the verge of becoming extinct.
9. The conservation of vulnerable breeds are increased by techniques like ___________ of animal genetic resources.
10. ___________ is the end result of the soil erosion.

Assertion-Reason Questions

The following questions consist of two statements: Assertion (A) and Reason (R). Answer these questions selecting the appropriate option given below:

(a) Both A and R are true and R is the correct explanation of A
(b) Both A and R are true but R is not the correct explanation of A
(c) A is true but R is false
(d) A is false but R is true

1. Assertion (A): Deforestation causes imbalance in nature.
   Reason (R): It disturbs many natural cycles and phenomena.
2. Assertion (A): When the level of water in reserves of different sources falls below a significant value, it is known as hydrological drought.
   Reason (R): This can be triggered by overuse of water.
3. Assertion (A): Desertification is the end result of soil erosion.
   Reason (R): Soil erosion leads to floods.
4. Assertion (A): The fauna in an environment is known as biota.
   Reason (R): Faunistics is the study of fauna.
5. Assertion (A): Rectification of the effects of deforestation is known as reforestation.
   Reason (R): Significant attention has been paid to reforestation to reduce the effect of climate change.

One-word Answers

1. Severe drought and forest fires are natural causes of which effect?
2. Which is the most common type of drought?
3. Which type of plants are categorised under weed flora?
4. What is the study of the animals of a particular region called?
5. What colour denotes extinct species in Red Data Book?
6. In which protection area animals are allowed to use natural resources?
7. What is the innermost area of the reserve with an undisturbed ecosystem?
8. What does IUCN stands for?
9. What is the natural restocking of existing woodlands that are currently under deforestation called?
10. What is the forest under Indian Forest Act known as?

Define the Term

1. Drought
2. Desertification
3. Flora
4. Manipulation zone
5. Vulnerable species

Groups of Terms

Given below are groups of terms. In each group the first pair indicates the relationship between the two terms. Rewrite and complete the second pair on a similar basis.

1. Black: extinct :: Red:_____
2. River dolphin: critically endangered :: Himalayan brown bear: _____
3. Smallest: Bor Tiger Reserve :: Largest: _____

4. Gir national park: Gujarat :: Chilika national park : ______

5. Local communities: manipulation zone :: Research : ______

6. Lion: Gujarat :: Elephant: ______

Identify the Odd One Out

1. Birds, fishes, insects, mammals
2. Red, amber, violet, green
3. Bor Reserve, Rann of Kutch, South Button Island, Gir
4. Reuse, Reduce, Recycle, Recall
5. Aridity, drought, soil erosion, rain

Image-Based Questions

1. What is the endangered fauna shown in the image?

(a) Siberian tiger (b) Sunderban tiger
(c) Nicobar tiger (d) Gir tiger

2. By what name is this migratory bird known?

(a) Flamingo (b) Arctic tern
(c) Antarctic tern (d) Iceland tern

3. What is the term used to describe this consequence of deforestation?

(a) Aridity (b) Flood
(c) Drought (d) Infertility

4. What is the name of this Indian fauna?

(a) Rhino (b) Water buffalo
(c) Wild hare (d) Wild bison

5. Choose the category to which this fauna belongs:

(a) Critically endangered
(b) Critically vulnerable
(c) Rare
(d) Endangered

6. What is the name of this biosphere reserve in West Bengal?

(Images from Bali Island, Sundarbans, Manoj Karingamadathil, 2019, https://commons.wikimedia.org/wiki/File:Images_from_Bali_Island_Sunderbans_IMG_20171112_062731.jpg. Licensed under: CC-BY-SA-4.0)

(a) Gorumara
(b) Sunderbans
(c) Gir
(d) Jim Corbett

7. What does this sign symbolise?

(a) 3Rs: Recycle. Reuse, Reduce
(b) 3Rs: Reduce, Reuse, Recycle
(c) 3Rs: Recycle, Reduce, Reforestation
(d) 3Rs: Reuse, Refuse, Recycle

8. To which category does this fauna belong?

(a) Vulnerable (b) Endangered
(c) Rare (d) Extinct

Very Short Answer Questions

1. What is the after-effect of aridity?
2. Name all the major biosphere reserves.
3. Which of the protected areas for animals is divided into three zones?
4. Which book has a list of all the species of animals and plants?
5. What are the two ways of recycling paper?
6. What are the different ways of protecting flora and fauna?
7. What are different categories of species based on the survival of species?
8. Name two animals known for migration.
9. What is the project launched in India for protection of the tiger?
10. Name the three Rs important for the environment.

Short Answer Questions

1. Name the consequences of deforestation.
2. What is migration?
3. What is the Red Data Book?
4. What is the difference between rare, vulnerable and endangered species?
5. Explain the importance of biosphere reserves.
6. What are rock paintings? Where are they found?
7. What occupations are allowed in protected areas for fauna and flora?
8. Discuss the importance of recycling.
9. Write the different colours and explain their significance in the Red Data Book.
10. Describe
    (a) Deforestation
    (b) Reforestation

Long Answer Questions

1. Describe deforestation and its effects on the environment.
2. Describe the consequences of global warming.
3. What is the Red Data Book? Describe all the colours and their symbolism with an example of each.
4. What are the various ways to protect fauna and flora?
5. Explain Project Tiger.
6. Explain the Indian Forest Act.
7. What is migration? What is the purpose of migration? What are the techniques used by different animals for migration?
8. What is recycling? What are the different ways of recycling paper and plastic?
9. Write a short note explaining the three zones of biosphere reserves in detail.
10. What is a wildlife sanctuary? What are the different reasons that makes wildlife sanctuaries important?

Case Study-Based Questions

1. Vinita visited Kerala and Alka visited Gujarat. They visited a national park and biosphere reserve, respectively. They had different observations while sightseeing. Later, they had a conversation about their experiences. Can you help them answer the following questions?
   (a) What is the difference between national park and biosphere reserve?
   (b) What fauna must they have encountered while visiting two different states?
   (c) Name one famous biosphere reserve of Kerala and one National Park of Gujarat.
2. Rinky visited Rajasthan to meet her old friend Raju. There they visited the desert area and Raju told Rinky that the desert shifts and also enlarges and has

flora and fauna of its own. Can you help Rinky get more information about it?
(a) What is desertification?
(b) How does soil erosion affects desertification?
(c) Name the famous flora and fauna of deserts.

3. Ramesh has been watching rag pickers picking plastic bottles from his locality and is curious about why they collect the items and how they are treated. Can you help him?
(a) What are the 3Rs?
(b) What is recycling?
(c) What are two types of plastic recycling?

4. Anju has been reading the newspaper daily and the editorial column has news about deforestation and its effects on flora and fauna. Can you help her understand different things related to it?
(a) What is deforestation?
(b) What are the causes of deforestation?
(c) How does deforestation extinction?

5. Pranjal always sees his parents selling old newspapers to local vendors. Can you help him understand the concept behind it?
(a) Why is paper recycling important?
(b) What are three types of paper used in recycling?

NCERT Corner

1. Fill in the blanks:
(a) A place where animals are protected in their natural habitat is called ______.
(b) Species that are found only in a particular area is known as ______.
(c) Migratory birds fly from one place to another because of ______changes.

2. Differentiate between the following:

A	B
i. Flora	Fauna
ii. Wildlife sanctuary	Biosphere Reserve
iii. Zoological park	Wildlife Sanctuary
iv. Endangered species	Extinct species

3. What are the effects of deforestation on:
(i) Wild animals (ii) Villages
(iii) Earth (iv) Environment
(v) Cities (vi) Next generation

4. What will happen?
(i) If the habitat of an animal is disturbed
(ii) If we keep destroying trees and forests
(iii) If the top layer of soil is exposed

5. Answer briefly:
(a) Why should we conserve biodiversity?
(b) How are some tribals dependent on forests?
(c) What is the Red Data Book?
(d) Why are the protected forests not completely safe for wild animals?
(e) What is migration?

6. To meet the ever-increasing demand of factories as well as for shelter, trees are being continually cut. Is it justified to cut trees for such projects? Discuss briefly.

7. How can you contribute to the maintenance of green wealth in your locality?

8. How does deforestation leads to less rainfall?

9. Why should paper be saved? What are the ways to save paper?

COMPETITION WINDOW (OBJECTIVE TYPE)

1. The National Aquatic Animal of India is:
(a) Gangetic shark (b) Sea horse
(c) Blue Whale (d) River dolphin

2. Red List contains information on:
(a) All economically important plants
(b) Threatened species
(c) Plants with products for international trade
(d) Marine vertebrates only

3. In a biosphere reserve, the region which is legally protected with no human activity involved is known as:
(a) Buffer zone (b) Restoration zone
(c) Core zone (d) Transition zone

4. Which pair represents endangered organisms of India?
(a) Blackbuck and banyan
(b) Leopard and Cinchona

(c) Red panda and *Bentinckia nicobarica*
(d) Rhesus monkey and tamarind

5. Which is not in situ conservation?
(a) Botanical garden (b) National park
(c) Sanctuary (d) Biosphere reserve

6. Endemic plants are those:
(a) Found in the Arctic regions
(b) Cosmopolitan in distribution
(c) Gregarious in habit
(d) Restricted to an area

7. Red Data Book gives information on:
(a) Red coloured fish
(b) Red flower plants
(c) Red eyed birds
(d) Endangered flora and fauna

8. Which of the following is most dangerous to wildlife?
(a) Man-made forest
(b) Over exploitation
(c) Habitat destruction
(d) Introduction to foreign species

9. What is preserved in national parks?
(a) Fauna (b) Flora
(c) Both (a) and (b) (d) None of the above

10. Red Data Book is maintained by:
(a) IUCN
(b) WPSI
(c) IUCNNR
(d) The Bombay Natural History Society

11. Which of the following species is restricted to an area?
(a) Sibling species (b) Allopatric species
(c) Endemic species (d) Sympatric species

12. The number of ecological hotspots in India is:
(a) 1 (b) 2
(c) 4 (d) 3

13. An example of ex situ conservation is:
(a) National park (b) Seed bank
(c) Wildlife sanctuary (d) Sacred grove

14. The factor that helps to maintain biodiversity:
(a) Biography (b) Geography
(c) Ecology (d) Biosphere reserves

15. Which of the following is an effect of timber logging:
(a) Soil erosion (b) Flood
(c) Afforestation (d) (a) and (b)

16. Which of the following is a living resource in a forest?
(a) Soil (b) Flora
(c) Fauna (d) (b) and (c)

17. What is the greatest threat to biodiversity?
(a) Man's interference with nature
(b) Habitat destruction
(c) Introduction of alien species
(d) All of these

18. Which of the following is a consequence of human interference with nature?
(a) Afforestation (b) Poaching
(c) Recycling (d) Rain harvesting

19. Which of the following is a biodiversity hotspot?
(a) River (b) Ocean
(c) Forest (d) Desert

20. Which of the following is not on the verge of extinction?
(a) Fox (b) Nilgiri langur
(c) Leopard (d) Sloth

Answer Keys

Multiple Choice Questions

1. (c) 2. (a) 3. (b) 4. (d) 5. (c)
6. (a) 7. (c) 8. (b) 9. (c) 10. (c)

Match the Following

1. i (b), ii (d), iii (c), iv (a)
2. i (c), ii (d), iii (a), iv (b)
3. i (b), ii (c), iii (a)
4. i (b), ii (a), iii (d),iv (c)

True or False

1. T 2. F 3. T 4. T 5. F
6. F 7. T 8. T 9. F 10. T
11. T 12. F 13. F 14. T 15. F

Fill in the Blanks

1. Biosphere 2. Rann of Kutch
3. Project Tiger 4. Biota
5. Cosmopolitan species

6. Satpura National park
7. Rock paintings
8. Red Data Book
9. Cryoconservation
10. Desertification

Assertion-Reason Questions

1. (a) 2. (a) 3. (c) 4. (d) 5. (b)

One-word Answers

1. Deforestation
2. Meteorological draught
3. Weedy, native, invasive
4. Faunistics
5. Black
6. National park
7. Core zone
8. International Union for the Conservation of Nature and Natural Resources
9. Reforestation
10. Reserved forest

Define the Term

1. When the shortage of water supply is prolonged in an area due to atmospheric conditions like low precipitation and shortage of ground or surface water, it is known as drought. It can last for days, months or years with a significant impact on the agriculture and ecosystem of the particular affected region.
2. In drylands, the type of degradation of land in which loss of biological productivity is observed due to human activities or natural processes increasing the aridity of the fertile areas is known as desertification. A variety of factors result in desertification of the land such as overexploitation of soil and climate change.
3. The native plant life in a particular area or time is known as flora. Flora includes a few bacteria and fungi like skin and gut flora. The term comes from Flora, the Roman goddess of flowers, plants and fertility.
4. This is the outermost or peripheral area of biosphere reserves where human activities like recreation, cropping, forestry and settlements are allowed with the cooperation of local people and reserve management. In the zone, various professionals like conservation agencies, scientists, local communities and other stakeholder work in a sustainable way for the benefit of the people of that area.
5. The IUCN category of those species threatened with extinction in the future if the circumstances leading to decline of reproduction and survival remain the same. Habitat loss or destruction of the home of the species leads to vulnerability. Around 6789 plants and 596 animals have been classified as vulnerable.

Groups of Terms

1. Endangered
2. Rare
3. Rann of Kutch
4. Odisha
5. Buffer zone
6. Kerala

Identify the Odd One Out

1. Mammals
2. Violet
3. Gir
4. Recall
5. Rain

Image-Based Questions

1. (a) 2. (b) 3. (c) 4. (d) 5. (a)
6. (b) 7. (b) 8. (c)

Very Short Answer Questions

1. Desertification is the end result of aridity.
2.

Biosphere Reserves	Places (India)
Sunderbans Biosphere Reserve	West Bengal
Nanda Devi Biosphere Reserve	Uttarakhand
Niligiri Biosphere Reserve	Tamil Nadu, Karnataka, Kerala
Great Nicobar Biosphere Reserve	Nicobar
Gulf of Mannar Biosphere Reserve	Tamil Nadu
Simlipal Biosphere Reserve	Odisha
Nokrek Biosphere Reserve	Meghalaya

3. Biosphere reserve is divided into three zones.
4. Red Data Book has a list of all species of animals and plants.
5. The two ways of recycling paper are physical and chemical recycling.
6. There are different ways of protecting flora and fauna: wildlife sanctuaries, zoological parks, biosphere reserve, national parks.

7. The different categories of species based on the survival of species: endangered, rare, vulnerable, endemic and cosmopolitan.
8. Two animals are known for migration: fish and birds.
9. Project Tiger was launched in India for the protection of tigers.
10. Three important Rs for the environment are Reduce, Reuse and Recycle.

Short Answer Questions

1. There are different consequences of deforestation that are known either due to present conditions or from the past from the fossil records: Loss of biodiversity, Habitat loss, Increase in aridity, Increase in extinction, Climate change, Population displacement, Desertification, Global warming, Reduction in biosequestration of atmospheric carbon dioxide, Increase in negative feedback cycles.
2. The large scale movement of species to a different environment is known as migration. It is a natural behaviour and a necessity for the life cycle of many species. There are two types of migration: animal migration and bird migration. Migration is often based on seasons or sometimes on daily basis. The members of species migrate to take advantage of better and favourable conditions. for their survival based on availability of food, mates, safety from predators and other environmental conditions.
3. A public document created to record the rare and endangered species of animals, plants and fungi along with local subspecies of a particular region is known as Red Data Book. It helps to provide complete information regarding different programs on rare and endangered species for research, studies and monitoring purposes. This book helps to protect and identify the species that are on the verge to get extinct.
4. **Endangered**: The species that is on the verge of extinction in the coming future either worldwide or in a particular region is known as endangered species. The factors that lead to making species endangered are poaching habitat loss and invasive species

   Rare: The organisms that are either very scarce, uncommon or infrequently encountered are known as rare species. They are different from endangered or threatened species. The species that fall into this category are classified by the official bodies of national or state governments.

   Vulnerable: The species categoried by IUCN as threatened with extinction in the future if the circumstances leading to decline of reproduction and survival remains the same. Habitat loss or the destruction of the home of the species leads to vulnerability.
5. **Conservation**: The first importance of these reserves are the conservation of flora and fauna of the given ecosystem along with protection of the genetic diversities and landscapes of the area without affecting the inhabitants.

   Restoration: To restore the damage of ecosystems and habitats

   Development: To ensure a certain development in different areas like social, cultural and economic

   Land usage: The farmers, landowners, industries, public institutions, scientists and conservation groups can work together for better land management

   Research and education: The information gathered from these areas regarding conservation, development and restoration of ecosystem makes it possible to recreate the landscape that was once destroyed by human exploitation.
6. The rock paintings often display figures of men fighting, animals fighting, humans dancing, playing musical instruments and hunting. Around 55 of these rock shelters are found in Pachmarhi Biosphere Reserve. A number of tribals still reside in these shelters.
7. The people who live in the wildlife sanctuary are responsible for wine production, electricity production and distribution, organic horticulture, grazing and farming enterprises
8. The process of conversion of waste materials into new objects and materials is known as recycling. The ability of a material to reacquire its original state is the key factor to ensure its recyclability. It not only reduces the waste and consumption of raw materials but also helps to reduce greenhouse gas emissions along with energy use and pollutions. The famous 3Rs of the modern waste reduction are Reduce, Reuse and Recycle. The environmental sustainability is by redirecting the output of waste and removing the input of raw materials.

Colours	Indication
Black	Species confirmed to be extinct
Red	Endangered species
Amber	Vulnerable species
White	Rare species

Colours	Indication
Green	Earlier endangered species but not recovering
Grey	Rare, endangered or vulnerable species with insufficient information to be properly classified

9. **Reforestation**: The intentional or the natural restocking of the existing woodlands and forests that are under the effect of deforestation is known as reforestation. The rectification of the effects of deforestation and improving the quality of human life by decreasing the effects of it like global warming and pollution, harvesting of resources are the main objectives of reforestation. Significant attention to reforestation to reduce the effect of climate change is required with the motto of promoting sustainable development.

   Deforestation: For the survival and well-being of life, different types of plants and animals must co-exist on this planet. Deforestation is the major threat against the organisms and their existence in the biosphere. Deforestation or forest clearance is the removal of trees or forests for the conversion of forest land to non-forest use. The non-forest use could be to form ranches, farms or for urban activities. Most of the concentrated deforestation occurs in the tropical rainforests.

Long Answer Questions

1. Hint: Explain deforestation, its effects and its consequences.
2. Hint: Define global warming and explain the consequences of global warming mentioned in this chapter.
3. Hint: Describe the Red Data Book, IUCN list, red list and the species included in it.
4. Hint: Explain wildlife sanctuary, zoological parks, biosphere reserve.
5. Hint: Explain the details of Project Tiger mentioned in this chapter.
6. Hint: Explain details of the Indian Forest Act mentioned in this chapter.
7. Hint: Explain migration, its purpose and techniques and animals involved in it.
8. Hint: Explain recycling and the different ways of recycling paper and plastic.
9. Hint: Explain biosphere reserve, its different zones and the different biosphere reserves in India.
10. Hint: Define wildlife sanctuary and write about the different sanctuaries in India, their importance and the activities allowed.

Case Study-Based Questions

1. (a) Refer to the differences mentioned in this chapter.
   (b) Alka would have seen lions while Vinita would have seen elephants.
   (c) Niligiri Biosphere Reserve in Kerala and Gir National Park in Gujarat
2. (a) Soil erosion results in desertification as the land is devoid of humus and becomes less fertile due to exposure of the top layer and and conversion of the land into a hard and rocky one.
   (b) Soil erosion leads to aridity and its effects include lowering of water quality surface, reducing crop production and damaging drainage networks. Desertification is the end result of the soil erosion.
   (c) Camel and cactus
3. (a) 3Rs: Reduce, Recycle and Reuse
   (b) The process of conversion of waste materials into new objects and materials is known as recycling. The ability of a material to reacquire its original state is the key factor to ensure its recyclability.
   (c) Physical and chemical recycling
4. (a) Deforestation or loss of vegetation is one of the immediate causes of desertification which is connected to other factors like climatic shifts, droughts, overgrazing, excessive loss of vegetation and trees for different purposes like construction materials or fuel.
   (b) The natural causes are severe draught and forest fires; the man-made causes are cultivation, construction due to increased population density, exploitation of wood for purposes like furniture or fuel, overpopulation.
   (c) The extremes of global warming due to deforestation will be warmer weather with huge rise in sea levels and more acidic oceans. These changes in the environment affect all the species and lead to extinction.
5. (a) There are various advantages of paper recycling such as preventing methane production. The waste paper breaks down in the form of paper fibres having carbon. This carbon gets locked in for longer period due to continuous recycling.
   (b) There are three types of paper used as feedstock to make recycled paper:

Mill broke: The trimmings of paper and scrap from paper manufacture which is recycled in paper mills.

Pre-consumer: The material left in the mill but nonetheless discarded before consumption.

Post-consumer: Discarded material like old magazines and newspapers after the consumer usage and deinking.

NCERT Corner

1. (a) Wildlife sanctuary
(b) Endemic species
(c) Climatic

2. i. Flora and Fauna

Flora	Fauna
The term comes from the Roman *Flora*, the goddess of flowers, plants and fertility.	The term comes from *Fauna*, the Roman goddess of fertility and Earth.
The native plant life present in a particular area or a time.	The animal life of a particular region or a particular time.
Flora includes few bacteria and fungi like skin flora or gut flora.	Fauna includes all insects, aves and other
Example: Teak trees in Satpura National Park	Example: Royal Bengal tigers of West Bengal (India)

ii. Wildlife sanctuary and Biosphere Reserve

Wildlife Sanctuary	Biosphere Reserve
The area in which the animal habitats along with their surroundings are protected from any damage and distrubance.	The areas that are meant to conserve and protect animals and plants along with tribals in the vicinity.
The animals are provided a comfortable living with large rivers, large mountains and dense forests.	Consists of three zones: Core zone, Buffer zone, Manipulation Zone.
It is one of the finest ways of protecting endangered species.	Protects the genetic diversities and landscapes of the area without affecting the inhabitants.
Example: Bharatpur Bird Sanctuary, Rajasthan, India	Example: Nanda Devi Biosphere Reserve, Uttarakhand, India

(Continued)

iii. Zoological park and wildlife sanctuary

Zoological Park	Wildlife sanctuary
A facility for animals to be housed, cared for within enclosures and displayed to the public.	The area in which the animal habitats along with their surroundings are protected from any damage and disturbance.
Conservation of animals along with research study and educational purposes are the reasons why zoos are established.	The animals are provided a comfortable living with large rivers, large mountains and dense forests.
Animals are bred for conservation purposes.	It is one of the finest ways of protecting endangered species.
Example: Indira Gandhi Zoological Park, Andhra Pradesh, India	Example: Bharatpur Bird Sanctuary, Rajasthan, India

iv. Endangered species and extinct species

Endangered species	Extinct species
The species that are on the verge of extinction in the coming future either worldwide or in a particular region.	The death of the last ever individual of the species present in the world is known as extinction and the species is known as extinct.
Marked by red colour in Red Data Book.	Marked by black colour in Red Data Book

Endangered species	Extinct species
Captive breeding and habitat restoration are the target of extensive conservation efforts.	Pollution, deforestation, climate change and habitat destruction are the causes of extinct species.
Examples: Siberian tiger, Saimaa ringed seal	Examples: Tasmanian tiger, Dodo

3. The effects of deforestation are:
   - Loss of habitat and severe effect on survival due to no shelter and food leading to extinction
   - People of villages depend for food, shelter and many other things like cooking fuel (wood) on forests
   - Global warming and severe effect on ecosystem and loss of biodiversity, desertification and drought
   - Global warming, water scarcity, drought and loss of animal and plant habitat
   - Besides food, cities are dependent on plants for different materials. Apart from this, less rainfall and increase in global warming that leads to rise in sea level, may submerge most coastal cities in the near future.
   - No food, sufficient drinking water or favourable conditions will exist for the next generation to survive with excessive loss of flora and fauna with sea level rise
4. (i) If the habitat of animals are disturbed, directly affects the survival rate of the animals along with increase in competition between animals for different resources.

   (ii) If we keep destroying the trees and forests, there would be increase in global warming, soil erosion and extinction of flora and fauna.

   (iii) If the top layer of the soil is exposed then the fertility of the land is reduced and the land may eventually turn into a barren one.
5. (a) It is important for biodiversity to be conserved to avoid extinction of species that are vital for the balance of biodiversity.

   (b) Tribals live and depend on forests for all kind of resources from food to clothing to shelter.

   (c) A public document created to record the rare and endangered species of animals, plants and fungi along with local subspecies of a particular region is known as Red Data Book.

   (d) The protected forests maybe developed for the animals to be protected but the animals will still be exposed of all the natural threats like competition for food or the natural food chain.

   (e) The large-scale movement of species to a different environment is known as migration. It is a natural behaviour and a necessity for the life cycle of many species.
6. Trees, if cut for big projects, should be planted to balance the tree count. Sustainable development is the need of the hour.
7. Green wealth also known as forest and tree stretches could be achieved by afforestation and maintained by regular watering of plants.
8. Deforestation means cutting of trees. The rate of rainfall is directly affected by deforestation as plants increase the humidity of an area which leads to rainfall, but cutting of trees leads to less rainfall.
9. Paper is from the from barks of trees and millions of trees are cut every day for this. In order to save paper, the 3Rs: Reduce Reuse and Recycle, play an important role.

Competition Window

1. (d)	**2.** (b)	**3.** (c)	**4.** (c)	**5.** (a)
6. (d)	**7.** (d)	**8.** (c)	**9.** (c)	**10.** (a)
11. (c)	**12.** (b)	**13.** (b)	**14.** (b)	**15.** (d)
16. (d)	**17.** (d)	**18.** (b)	**19.** (c)	**20.** (a)